Transforming Language Education Through Mobile Technology Integration

Ehsan Namaziandost
https://orcid.org/0000-0002-8393-2537
Islamic Azad University, Iran

IGI Global
Scientific Publishing
Publishing Tomorrow's Research Today

Vice President of Editorial	Melissa Wagner
Director of Acquisitions	Mikaela Felty
Director of Book Development	Jocelynn Hessler
Production Manager	Mike Brehm
Cover Design	Jose Rosado

Published in the United States of America by
IGI Global Scientific Publishing
701 East Chocolate Avenue
Hershey, PA, 17033, USA
Tel: 717-533-8845 | Fax: 717-533-7115
Website: https://www.igi-global.com E-mail: cust@igi-global.com

Library of Congress Cataloging-in-Publication Data

Names: Namaziandost, Ehsan, 1985- editor
http://id.loc.gov/rwo/agents/n2025021618
Title: Transforming language education through mobile technology integration / edited by Ehsan Namaziandost.
Description: Hersey, PA : IGI Global Scientific Publishing, 2026. | Includes bibliographical references and index. | Summary: "This book aims to provide a comprehensive exploration of how mobile technology can be seamlessly integrated into language learning to improve outcomes for students and educators alike"-- Provided by publisher.
Identifiers: LCCN 2025019859 (print) | LCCN 2025019860 (ebook) | ISBN 9798337333663 hardcover | ISBN 9798337333670 paperback | ISBN 9798337333687 ebook
Subjects: LCSH: Language and languages--Computer-assisted instruction http://id.loc.gov/authorities/subjects/sh2008106644 | Mobile-assisted language learning http://id.loc.gov/authorities/subjects/sh2020008926 | LCGFT: Essays http://id.loc.gov/authorities/genreForms/gf2014026094
Classification: LCC P53.28 .T736 2026 (print) | LCC P53.28 (ebook) | DDC 418.0078/54167--dc23/eng/20250614
LC record available at https://lccn.loc.gov/2025019859
LC ebook record available at https://lccn.loc.gov/2025019860

British Cataloguing in Publication Data
A Cataloguing in Publication record for this book is available from the British Library.

Table of Contents

Preface xiii

Chapter 1
Integrating Mobile Technology Into Traditional Language Instruction: Challenges and Solutions 1
Mustafa Kayyali, Maaref University of Applied Sciences, Syria

Chapter 2
Incorporating Mobile Technology Into Pre-Service EFL Teacher Education: Opportunities and Challenges in the Moroccan Context 31
Rachid El Yazidi, Moulay Ismail University, Morocco

Chapter 3
Personalized Learning Pathways Through Mobile Devices: Enhancing Language Acquisition 83
Mustafa Kayyali, Maaref University of Applied Sciences, Syria

Chapter 4
An Analysis of Babbel Apps in Reading: Benefits and Drawbacks From the User Perspective 111
Sholihatul Hamidah Daulay, Universitas Islam Negeri Sumatera Utara, Indonesia
Hotma Berutu, Universitas Islam Negeri Sumatera Utara, Indonesia
Maslathif Dwi Purnomo, Universitas Islam Negeri Sumatera Utara, Indonesia
Muhammad Dalimunte, Universitas Islam Negeri Sumatera Utara, Indonesia
Eka Apriani, Institut Agama Islam Negeri Curup, Indonesia
Shaumiwaty Shaumiwaty, Institut Agama Islam Negeri Takengon, Indonesia
Julisah Izar, Universitas Jambi, Indonesia

Chapter 5
Mobile-Assisted Pronunciation Learning in an EFL Context: An Exploratory Study 137
Rizgar Qasim Mahmood, Wollongong University, Australia & Salahaddin University-Erbil, Iraq

Chapter 6
The Role of Gamification in Enhancing Mobile Language Learning Experiences 169
Mustafa Kayyali, Maaref University of Applied Sciences, Syria

Chapter 7
Digital Gamification in Times of Crisis: Perspectives of Pre-Service Elementary School Teachers on Language Teaching 189
Tetiana Holovatenko, V.N. Karazin Kharkiv National University, Ukraine

Chapter 8
Immersive Language Learning: Exploring the Role of AR and VR in Education 219
Laylo Rakhimova, Urgench State University, Uzbekistan
Zebo Saidjanova, Mamun University, Uzbekistan
Yunusjon Khakkulov, Termez University of Economics and Service, Uzbekistan
Munir Ahmad, Survey of Pakistan, Pakistan

Chapter 9
The Role of VR and AR in Immersive Language Education 245
Ferdi Çelik, Ondokuz Mayıs University, Turkey
Volkan Duran, Iğdır University, Turkey

Chapter 10
The Future of Language Learning: A Blend of Reality and Virtuality........ 285
Ali Rahmanipur, Bu-Ali Sina University, Iran
Moein Shokri, University of Salento, Italy

Chapter 11
Emotional Dynamics of EFL Students in Eportfolio-Based Speaking Assessment: An In-Depth Exploration of Positive and Negative Emotions...... 315
I. Putu Indra Kusuma, Universitas Pendidikan Ganesha, Indonesia
Luh Indrayani, Universitas Pendidikan Ganesha, Indonesia
Gres Desika Br Kaban, Universitas Pendidikan Ganesha, Indonesia
Belia Damayanti Br Sembiring, Universitas Pendidikan Ganesha, Indonesia

Chapter 12
An In-Depth Study on Enhancing the Effectiveness of ICT in CLT Through Mixed Method Research From Teachers' Perspectives 343
Nafis Mahmud Khan, TIIAME National Research University, Uzbekistan
Nargiza Arifovna Kadirova, TIIAME National Research University, Uzbekistan
Jasur Umirzokov, Tashkent State University of Economics, Uzbekistan

Chapter 13
The Potential of AI, Blockchain, and IoT in Language Education 379
S. Ida Evangeline, Government College of Engineering, Tirunelveli, India

Compilation of References 401

About the Contributors 463

Index 469

Detailed Table of Contents

Preface xiii

Chapter 1
Integrating Mobile Technology Into Traditional Language Instruction:
Challenges and Solutions 1
Mustafa Kayyali, Maaref University of Applied Sciences, Syria

As mobile technologies continue to reshape communication and learning landscapes, their integration into traditional language instruction remains both a promising frontier and a complex challenge. This chapter critically examines the intersection of mobile-assisted language learning (MALL) with conventional pedagogical frameworks, highlighting the friction points, institutional resistances, and cultural hesitations that educators often face. Drawing on real-world observations and emerging research, it explores the nuanced challenges of infrastructure limitations, teacher preparedness, and learner adaptability. At the same time, it proposes actionable solutions grounded in reflective practice, inclusive design, and context-aware innovation. Ultimately, it positions mobile integration not as a technological imperative, but as a pedagogical opportunity—one that demands sensitivity, patience, and a renewed understanding of what it means to teach and to learn in the 21st century.

Chapter 2
Incorporating Mobile Technology Into Pre-Service EFL Teacher Education:
Opportunities and Challenges in the Moroccan Context 31
Rachid El Yazidi, Moulay Ismail University, Morocco

The swift growth of mobile-assisted language learning (MALL) has transformed the horizons of language learning and teaching in various educational settings. In contrast, the incorporation of mobile technologies into teacher training, particularly in developing contexts, remains largely unstudied. This chapter looks at the prospects and obstacles in the use of mobile technologies for pre-service English as a Foreign Language (EFL) training in higher education institutions in Morocco. Grounded in a sociocultural and pedagogical framework, the chapter analyzes advanced mobile pedagogical practices and language teacher education to understand the impact of mobile technologies on the competencies and digital literacy frameworks of modern educators.

Chapter 3
Personalized Learning Pathways Through Mobile Devices: Enhancing Language Acquisition 83
Mustafa Kayyali, Maaref University of Applied Sciences, Syria

As mobile technologies continue to reshape the landscape of education, language learning has witnessed a profound transformation through personalized, adaptive experiences tailored to the needs of individual learners. This chapter explores how mobile devices can create meaningful and responsive learning pathways that enhance language acquisition. By leveraging personalization—ranging from app-based content to real-time feedback and data-driven customization—learners are increasingly empowered to direct their own progress, engaging with content that suits their pace, goals, and linguistic background. Drawing on both current research and real-world classroom applications, this chapter offers insights into the design and implementation of mobile-based personalized learning systems, while also examining their impact on learner motivation and autonomy. The discussion includes the practical and pedagogical implications of integrating mobile technologies in language education, highlighting both the promise and the complexities involved.

Chapter 4
An Analysis of Babbel Apps in Reading: Benefits and Drawbacks From the User Perspective 111
Sholihatul Hamidah Daulay, Universitas Islam Negeri Sumatera Utara, Indonesia
Hotma Berutu, Universitas Islam Negeri Sumatera Utara, Indonesia
Maslathif Dwi Purnomo, Universitas Islam Negeri Sumatera Utara, Indonesia
Muhammad Dalimunte, Universitas Islam Negeri Sumatera Utara, Indonesia
Eka Apriani, Institut Agama Islam Negeri Curup, Indonesia
Shaumiwaty Shaumiwaty, Institut Agama Islam Negeri Takengon, Indonesia
Julisah Izar, Universitas Jambi, Indonesia

This study aimed to investigate the benefits and drawbacks of the Babbel App based on users' perspectives on the Google Play Store. This study used interpretative qualitative as the research methodology. The data source used is screenshots of users' feedback on the Google Play Store for the Babbel App. The study came up with the finding that Babbel is a good tool for assisting users in learning a new language because it provides language learning classes in 14 different languages. Therefore, the users around the world benefit from the presence of Babbel. However, users also revealed some drawbacks of Babbel App such as (1) the tricky use of Babbel app (2) the limited language choices, especially for the 'native language' choice.

(3) users are required to understand English first before utilizing the app, and (4) it requires a high payment when the users want to jump to the next level.

Chapter 5
Mobile-Assisted Pronunciation Learning in an EFL Context: An Exploratory Study 137
Rizgar Qasim Mahmood, Wollongong University, Australia & Salahaddin University-Erbil, Iraq

Mobile-Assisted Pronunciation Learning (MAPL) provides digital tools with real-time feedback and exposure to native-like models. However, research remains limited in EFL contexts such as Iraqi Kurdistan. This chapter examines the engagement of Kurdish EFL learners with MAPL, focusing on their usage patterns, perceptions, motivations, and challenges, using surveys (n = 45) and interviews (n = 15). The data were analyzed descriptively and thematically. The study revealed that learners frequently utilized applications such as ChatGPT and Google Translate to practice vowel sounds, word stress, and connected speech. While learners valued MAPL's flexibility and effectiveness for segmental practice, they highlighted critical limitations, particularly in supporting suprasegmental features and providing complex feedback. Other limitations were also mentioned, such as technical shortcomings, distractions, and weak self-regulation. The study concludes with theoretical and pedagogical implications.

Chapter 6
The Role of Gamification in Enhancing Mobile Language Learning Experiences 169
Mustafa Kayyali, Maaref University of Applied Sciences, Syria

As mobile technology reshapes the landscape of language education, gamification has emerged as a powerful catalyst for transforming learner engagement. This chapter explores how game elements—such as points, levels, challenges, and rewards—can enhance motivation, increase persistence, and deepen cognitive engagement in mobile language learning environments. Drawing from both pedagogical theory and practical implementations, it examines how well-designed gamified experiences can personalize the learning journey, create emotional resonance, and foster a sense of agency and progress. Yet the chapter also offers a critical lens, considering the limitations of over-simplified or poorly contextualized gamification strategies. It underscores the importance of culturally responsive design and learner diversity in shaping meaningful experiences. Ultimately, this chapter argues for a balanced, intentional use of gamification—not as a superficial layer, but as an integrated framework that supports authentic, human-centered language learning through mobile technologies.

Chapter 7
Digital Gamification in Times of Crisis: Perspectives of Pre-Service Elementary School Teachers on Language Teaching 189
Tetiana Holovatenko, V.N. Karazin Kharkiv National University, Ukraine

This chapter investigates pre-service elementary school teachers' perspectives on digital gamification for English language teaching in the challenging context of wartime Ukraine. Quantitative data was gathered through surveying, and qualitative data was obtained through the analysis of artefacts (language teaching activities). Findings reveal strong positive attitudes toward gamification's effectiveness, particularly for lowering learners' Affective Filter and supporting trauma-informed pedagogy. The analysis showed that Ukrainian students accept the uncertainty of wartime teaching and gamify the activities using low-stress game mechanics and tend to use digital tools, which can be easily used in both online and offline settings. This study provides suggestions for teacher training programs, educational institutions and curriculum developers on the implementation of digital gamification in language teaching, with a critical focus on overcoming pedagogical barriers and developing trauma-informed education.

Chapter 8
Immersive Language Learning: Exploring the Role of AR and VR in Education 219
Laylo Rakhimova, Urgench State University, Uzbekistan
Zebo Saidjanova, Mamun University, Uzbekistan
Yunusjon Khakkulov, Termez University of Economics and Service, Uzbekistan
Munir Ahmad, Survey of Pakistan, Pakistan

This chapter examines the transformative potential of Augmented Reality (AR) and Virtual Reality (VR) in immersive language learning. By enabling experiential, context-rich, and culturally embedded interactions, AR/VR redefine traditional pedagogy and foster greater engagement, retention, and communicative competence. Grounded in experiential learning and situated cognition, these technologies support diverse learning styles and personalized pathways. Despite challenges in cost, accessibility, and teacher readiness, emerging innovations such as mobile-based VR, browser-native platforms, and AI-driven tutors are expanding access. The chapter envisions a future where immersive tools democratize language education, making it more inclusive, global, and learner-centered.

Chapter 9
The Role of VR and AR in Immersive Language Education 245
Ferdi Çelik, Ondokuz Mayıs University, Turkey
Volkan Duran, Iğdır University, Turkey

This chapter explores how virtual reality (VR) and augmented reality (AR) technologies transform language education through immersive, learner-centered experiences. Using a mixed-methods approach, bibliometric analyses of 634 publications (385 on VR, 249 on AR) and grounded theory, it finds that these technologies boost motivation, engagement, and contextual learning, though most studies prioritize technical development over pedagogical depth. VR simulates real-world and cultural scenarios, improving oral proficiency, vocabulary retention, and reducing anxiety. AR overlays digital content onto physical settings, enhancing interactivity and personalized feedback. However, challenges like hardware costs, teacher training, and scarce longitudinal data hinder adoption. Smaller research clusters (e.g., VR in healthcare language training or AR for special needs) highlight adaptability across contexts. Key next steps include integrating language acquisition theories, enhancing educator support, and pursuing large-scale comparative studies.

Chapter 10
The Future of Language Learning: A Blend of Reality and Virtuality.............. 285
Ali Rahmanipur, Bu-Ali Sina University, Iran
Moein Shokri, University of Salento, Italy

This chapter explores the transformative potential of augmented reality (AR) and virtual reality (VR) in mobile language education, arguing that these technologies address critical gaps in situational learning and cognitive load management. Drawing on theories of multimedia cognition and situated learning, it examines how AR/VR environments replicate authentic language contexts while optimizing mental engagement. The chapter analyzes current applications across formal education (e.g., classroom integration) and self-directed informal learning (e.g., gamified apps), highlighting their distinct advantages. Findings reveal that AR/VR not only enhances vocabulary retention and conversational fluency but also reduces learner anxiety through immersive, low-stakes practice. However, challenges like accessibility and teacher training persist. By synthesizing empirical evidence and practical implementations, this chapter underscores AR/VR's role in democratizing language acquisition and proposes frameworks for scalable adoption.

Chapter 11
Emotional Dynamics of EFL Students in Eportfolio-Based Speaking Assessment: An In-Depth Exploration of Positive and Negative Emotions...... 315
I. Putu Indra Kusuma, Universitas Pendidikan Ganesha, Indonesia
Luh Indrayani, Universitas Pendidikan Ganesha, Indonesia
Gres Desika Br Kaban, Universitas Pendidikan Ganesha, Indonesia
Belia Damayanti Br Sembiring, Universitas Pendidikan Ganesha, Indonesia

Previous studies of eportfolios in speaking courses suggest that this digital assessment technique significantly improves speaking skill improvement. However, what kind of positive and negative emotions experienced by the English as a foreign language (EFL) students remains unclear as previous studies did not profoundly discuss them. This study therefore purposed to explore the positive and negative emotions experienced by EFL students during the implementation of eportfolios in speaking courses. This study employed a phenomenology approach and recruited 20 Indonesian EFL students. The data were collected through semi-structured interviews and analyzed using inductive thematic analysis. The findings revealed that while eportfolios supported emotional growth through enjoyment, pride, motivation, and relief, they also triggered various negative emotions, including anxiety, embarrassment, frustration, and exhaustion. Four implications were drawn to effectively implement eportfolios in speaking courses.

Chapter 12
An In-Depth Study on Enhancing the Effectiveness of ICT in CLT Through Mixed Method Research From Teachers' Perspectives 343
Nafis Mahmud Khan, TIIAME National Research University, Uzbekistan
Nargiza Arifovna Kadirova, TIIAME National Research University, Uzbekistan
Jasur Umirzokov, Tashkent State University of Economics, Uzbekistan

The study aims to investigate English teachers' awareness of using different applications for personal and professional purposes at the secondary level, and explore the measures needed to adopt to enhance ICT's effectiveness in CLT. A mixed-method research design was employed to answer the research questions. In the first phase of the study, a random sampling technique was used to collect the data from the teacher participants of Bangladesh. In the second phase, a semi-structured interview was conducted to identify the dominant theme of this study. Results indicate that majority of the teacher participants are aware of using ICT materials for personal and professional purposes. The findings of the study reveal teacher participants' expectation from the school authorities for incorporating ICT in their classroom language teaching. Thus, the study provides meaningful insights for policymakers, stakeholders, and authorities to take timely necessary initiatives for the successful implementation of ICT in classroom language teaching.

Chapter 13
The Potential of AI, Blockchain, and IoT in Language Education 379
S. Ida Evangeline, Government College of Engineering, Tirunelveli, India

This chapter explores the transformative potential of artificial intelligence (AI), blockchain, and the Internet of Things (IoT) in reshaping language education through mobile technology integration. As digital learning environments continue to evolve, these emerging technologies offer new pathways for personalized instruction, secure credentialing, real-time assessment, and immersive learning experiences. The chapter begins by examining how AI enhances language learning through natural language processing, adaptive learning platforms, and speech recognition tools that personalize content and provide real-time feedback. It then investigates the role of blockchain in ensuring academic integrity, supporting decentralized learning networks, and enabling secure, tamper-proof credentialing for language proficiency. Finally, the discussion is extended to IoT, providing insight into how it is used in smart classrooms, wearable devices and context-aware environment for improving interactive and situated learning.

Compilation of References 401

About the Contributors 463

Index 469

Preface

INTRODUCTION

The rapid proliferation of mobile technologies has fundamentally transformed how languages are taught, learned, and experienced across global educational contexts. What was once confined to classrooms, textbooks, and fixed schedules has evolved into a dynamic, personalized, and increasingly immersive ecosystem—one shaped by mobile devices, intelligent applications, and emerging digital infrastructures. *Transforming Language Education Through Mobile Technology Integration* responds to this transformation by bringing together international scholarship that critically examines the pedagogical, technological, emotional, and ethical dimensions of mobile-enhanced language learning.

Edited by Ehsan Namaziandost, this volume positions mobile technology not merely as a tool, but as a catalyst for rethinking language education in the 21st century. The chapters collectively explore how mobile systems, applications, and emerging technologies intersect with pedagogy, inclusivity, teacher education, learner engagement, assessment practices, and future-oriented innovation. Contributors from diverse geographical, cultural, and institutional contexts provide empirical evidence, theoretical grounding, and reflective insights that speak to both global trends and localized realities.

The volume begins by establishing the technological foundations necessary to understand mobile learning environments, including mobile communication systems and network architectures. It then moves into pedagogical integration, examining how mobile technologies can be meaningfully incorporated into traditional language instruction and teacher education, particularly in under-researched and developing contexts. Subsequent chapters address learner-centered approaches such as inclusivity, personalization, pronunciation learning, and the use of commercial language-learning applications.

A significant portion of the book is devoted to engagement-driven innovations, including gamification and immersive technologies such as augmented reality (AR) and virtual reality (VR). These chapters highlight how experiential, game-based, and immersive learning environments can enhance motivation, reduce anxiety, and support authentic language use—even in times of crisis. The emotional dimensions of mobile-based assessment are also explored, offering nuanced insights into learners' affective experiences in digital environments.

The volume concludes by looking toward the future of language education through the lens of artificial intelligence, blockchain, and the Internet of Things, underscoring the growing convergence of mobile learning with intelligent, decentralized, and context-aware systems.

Together, the chapters in this book provide a comprehensive, interdisciplinary perspective on mobile technology integration in language education. This volume is intended for researchers, educators, teacher trainers, graduate students, curriculum designers, and policymakers who seek to understand, implement, and critically engage with mobile and emerging technologies in language learning contexts.

CHAPTER OVERVIEW

Chapter 1. Integrating Mobile Technology into Traditional Language Instruction: Challenges and Solutions

Examines institutional, cultural, and pedagogical challenges in integrating mobile technologies into conventional language classrooms, offering reflective and practical solutions.

Chapter 2. Incorporating Mobile Technology into Pre-Service EFL Teacher Education: Opportunities and Challenges in the Moroccan Context

Explores the role of mobile technologies in pre-service teacher education, highlighting impacts on digital literacy, pedagogical competence, and professional preparation.

Chapter 3. Personalized Learning Pathways Through Mobile Devices: Enhancing Language Acquisition

Focuses on adaptive and personalized mobile learning systems that empower learners through autonomy, motivation, and individualized learning trajectories.

Chapter 4. An Analysis of Babbel Apps in Reading: Benefits and Drawbacks from the Users' Perspective

Provides a user-centered evaluation of a popular commercial language-learning application, highlighting pedagogical benefits alongside practical limitations.

Chapter 5. Mobile-Assisted Pronunciation Learning in an EFL Context: An Exploratory Study

Investigates how mobile tools support pronunciation development, examining learner engagement, feedback mechanisms, and technological constraints.

Chapter 6. The Role of Gamification in Enhancing Mobile Language Learning Experiences

Explores how gamified design elements can foster engagement, motivation, and sustained participation in mobile language learning environments.

Chapter 7. Digital Gamification in Times of Crisis: Perspectives of Pre-service Elementary School Teachers on Language Teaching

Examines gamification as a trauma-informed and flexible pedagogical strategy in crisis contexts, with a focus on teacher perspectives during wartime education.

Chapter 8. Immersive Language Learning: Exploring the Role of AR and VR in Education

Analyzes how AR and VR technologies enable experiential, culturally embedded language learning and expand access through emerging mobile platforms.

Chapter 9. The Role of VR and AR in Immersive Language Education

Presents a comprehensive bibliometric and theoretical analysis of VR and AR research, identifying trends, gaps, and future research directions.

Chapter 10. Extended Learning: AR and VR as Game-Changers in Mobile Language Learning

Examines AR and VR as transformative tools for situational learning, cognitive engagement, and anxiety reduction in mobile language education.

Chapter 11. Emotional Dynamics of EFL Students in Eportfolio-Based Speaking Assessment

Explores learners' positive and negative emotional experiences in mobile-supported assessment, offering pedagogical implications for implementation.

Chapter 12. Enhancing the Effectiveness of ICT in Communicative Language Teaching: Teachers' Perspectives

Investigates teachers' awareness, expectations, and institutional needs regarding ICT integration in communicative language teaching.

Chapter 13. The Potential of AI, Blockchain, and IoT in Language Education

Concludes the volume by examining emerging technologies that promise to redefine personalization, assessment, credentialing, and smart learning environments.

In bringing together theoretical insights, empirical research, and reflective practice, this volume underscores the necessity of thoughtful, human-centered integration of mobile technologies in language education. While mobile and emerging technologies continue to evolve at an unprecedented pace, their educational value ultimately depends on pedagogical intentionality, inclusivity, and ethical implementation. By addressing technological foundations alongside emotional, cultural, and contextual dimensions of learning, this book invites readers to move beyond adoption toward meaningful transformation. It is our hope that this volume will inspire informed experimentation, critical dialogue, and collaborative innovation, contributing to more accessible, resilient, and future-ready language education systems worldwide.

Ehsan Namaziandost
Islamic Azad University, Iran

Chapter 1

Integrating Mobile Technology Into Traditional Language Instruction:
Challenges and Solutions

Mustafa Kayyali
https://orcid.org/0000-0003-3300-262X
Maaref University of Applied Sciences, Syria

ABSTRACT

As mobile technologies continue to reshape communication and learning landscapes, their integration into traditional language instruction remains both a promising frontier and a complex challenge. This chapter critically examines the intersection of mobile-assisted language learning (MALL) with conventional pedagogical frameworks, highlighting the friction points, institutional resistances, and cultural hesitations that educators often face. Drawing on real-world observations and emerging research, it explores the nuanced challenges of infrastructure limitations, teacher preparedness, and learner adaptability. At the same time, it proposes actionable solutions grounded in reflective practice, inclusive design, and context-aware innovation. Ultimately, it positions mobile integration not as a technological imperative, but as a pedagogical opportunity—one that demands sensitivity, patience, and a renewed understanding of what it means to teach and to learn in the 21st century.

DOI: 10.4018/979-8-3373-3366-3.ch001

INTRODUCTION

Language learning has always been deeply rooted in the rhythms of human interaction—spoken conversations, shared stories, gestures, expressions, and the subtle tones that carry culture across borders. For centuries, language instruction has followed relatively fixed models: classrooms where teachers guide learners through structured syllabi, textbooks, grammar drills, and face-to-face exchanges. The tradition of language education is, in many ways, built on continuity—a sense of stability that allows learners to build linguistic competence in stages, scaffolded by the presence of a knowledgeable teacher and a community of peers (Umarova, 2024). Yet the past two decades have brought about a seismic shift in how we communicate, access information, and relate to the world around us, largely driven by the rapid evolution of mobile technology. Smartphones, tablets, and mobile applications have become ubiquitous in daily life. From messaging platforms to language-learning apps like Duolingo, Memrise, or HelloTalk, mobile technology has unlocked new possibilities for learning beyond the walls of formal education (Vargas Florez, 2024). Learners can now access vocabulary lists, grammar tutorials, native speaker interactions, and pronunciation exercises at any time, from any place. The boundaries between classroom and everyday life have become increasingly porous. But with this expansion of access comes a profound pedagogical question: how can we thoughtfully and effectively integrate mobile technology into traditional language instruction without losing the human essence that defines it?

The answer is far from straightforward. While mobile tools promise personalization, flexibility, and autonomy, traditional classrooms offer structure, mentorship, and relational depth. Bringing the two into harmony is not a matter of simply digitizing worksheets or introducing an app during a lesson. It is a transformation that touches upon teaching philosophies, classroom dynamics, equity in access, and institutional readiness. Educators often find themselves in a liminal space—torn between the familiar practices they know and trust, and the disruptive potential of emerging mobile platforms (Barnard & Ferrier-Kerr, 2021). In many cases, teachers are expected to become tech-savvy innovators overnight, despite limited training or support. Meanwhile, learners face their own set of challenges, from navigating distractions on mobile devices to dealing with uneven digital literacy or internet access.

This chapter does not present technology as a cure-all, nor does it romanticize tradition. Instead, it begins from a place of humility: recognizing that both traditional methods and mobile tools have distinct affordances, and that meaningful integration requires more than enthusiasm—it requires strategy, sensitivity, and reflection. Drawing from research, classroom experiences, and practitioner insights, we will explore the core tensions that arise when mobile technology is introduced into conventional language instruction. We will examine the common barriers—

both technical and pedagogical—that educators encounter, and propose grounded solutions that do not undermine the foundational values of language education. Importantly, this chapter does not assume a one-size-fits-all approach. The challenges faced in urban schools with reliable infrastructure differ greatly from those in rural or under-resourced contexts. The expectations of adult learners learning English as a second language are different from those of teenagers learning French or Mandarin (Vicente González, 2021). The aim here is not to dictate a model, but to offer a thoughtful roadmap—one that invites educators to reflect, adapt, and co-create mobile integration strategies that are contextually relevant and ethically sound. As we navigate this evolving terrain, we must keep in mind that technology is not neutral. The ways we choose to adopt—or resist—it carry implications for pedagogy, equity, and the future of education itself. In language instruction, where identity, culture, and human connection are inseparable from grammar and vocabulary, the integration of mobile technology must be pursued with care. This chapter is an invitation to do just that: to pause, to question, and to imagine what is possible when tradition and innovation walk hand in hand.

RETHINKING LANGUAGE CLASSROOMS IN THE AGE OF MOBILITY

There was once a time when language learning unfolded in predictable ways. Rows of students sat with open textbooks, repeating phrases after their teacher in unison, copying grammar rules onto paper, and perhaps watching an occasional video to "spice up" a lesson. These were not necessarily ineffective approaches—they reflected the best available pedagogical practices of their time. But they were also static, uniform, and confined within the walls of the classroom. Language, however, has never been static. It breathes and evolves. It is carried through movement, through time, across borders and bodies. And now, in a world that is itself in constant motion—geographically, socially, and digitally—the space where language is taught is being reimagined. The age of mobility is not defined only by the devices we carry in our pockets. It is also marked by the pace of life, the fragmentation of attention, the disintegration of traditional boundaries between work and rest, home and school, learner and teacher. In this dynamic reality, where information is accessible at the swipe of a finger, language classrooms cannot remain spaces of passive reception (Tong et al, 2023). They must become fluid, responsive, and connected—anchored not in rigid curriculum delivery, but in the lived realities of learners who are increasingly shaped by mobile interactions.

To rethink the language classroom in this context is not to erase its past. On the contrary, it is to honor what has worked while remaining honest about what no longer serves. The teacher is still a central figure—not merely as a transmitter of knowledge but as a curator of learning experiences, a facilitator of dialogue, a guide in navigating both linguistic complexity and digital fluency (Kumar, 2023). In this new configuration, the classroom is not replaced by the phone, nor the teacher by the app. Rather, mobile tools become extensions of the learning environment—enablers of participation, personalization, and exploration. But this vision requires a radical departure from the notion that learning only happens in silence, in neat rows, or under the sole direction of a fixed syllabus. It requires classrooms to breathe—to stretch beyond scheduled hours, to reach into the messy spaces of learners' daily lives, and to recognize that language is acquired as much through practice and exposure as through instruction. A student on their commute, whispering vocabulary into their phone's speech-recognition app, is learning. A learner listening to a podcast while washing dishes is learning. A group of peers collaborating on a language challenge through a messaging app is learning. The question is no longer where learning happens—but how we embrace its ubiquity. This reimagining also calls into question our definitions of success. In traditional classrooms, assessment is often standardized, with discrete-point grammar tests and vocabulary quizzes. But mobile learning environments offer fluid, real-time demonstrations of skill—voice messages, video responses, spontaneous exchanges, location-based storytelling (Makhal, 2023). Shouldn't we value these demonstrations of language use as legitimate expressions of progress? Shouldn't we acknowledge that language is not only what one knows, but what one can do with what they know in authentic, often messy, real-world contexts?

Of course, this shift is not without tension. For one, it exposes the unequal footing on which many learners stand. Not all students have reliable access to smartphones or internet connections. Not all are comfortable navigating digital platforms. Rethinking the classroom in the age of mobility, then, cannot simply mean a rush toward the latest tools (Shew, 2023). It must be grounded in equity, in understanding the lived contexts of learners, and in ensuring that mobile integration expands opportunity rather than deepens divides. It also demands courage from educators—courage to unlearn, to experiment, to sit with discomfort. Many teachers were not trained to teach with mobile technology, nor were they prepared for the speed at which educational paradigms would shift (Atibuni et al, 2022). And yet, many are rising to the occasion, not by mastering every new app, but by reframing their role. They are learning to ask new questions: How can I use mobile tools to amplify student voice? How can I make grammar come alive outside the textbook? How can I help students transfer classroom learning into their everyday digital habits?

The transformation of language classrooms is also a transformation of time. No longer is the "class" limited to a one-hour slot on the schedule. With mobile integration, learning becomes asynchronous, fluid, and continuous. A learner might begin a vocabulary exercise in class, record a voice memo of new sentences on their phone during lunch, and receive peer feedback in the evening (Goodwin et al, 2022). This requires teachers and institutions to rethink their expectations—not only about where learning happens, but when. The rigidity of timetables must soften to allow for the natural ebb and flow of mobile-supported engagement. The reimagined classroom values *agency*. Traditional models often positioned students as passive recipients. In contrast, mobile tools allow learners to shape their learning pathways, to choose topics of interest, to engage at their own pace, and to revisit content on demand (McHaney, 2023). This autonomy can be both empowering and overwhelming. Not all learners are prepared to navigate it alone. That is why the role of the teacher becomes even more important—not as a controller, but as a mentor who scaffolds this autonomy with care, reflection, and trust.

Figure 1. Focus areas for mobile integration in language instruction

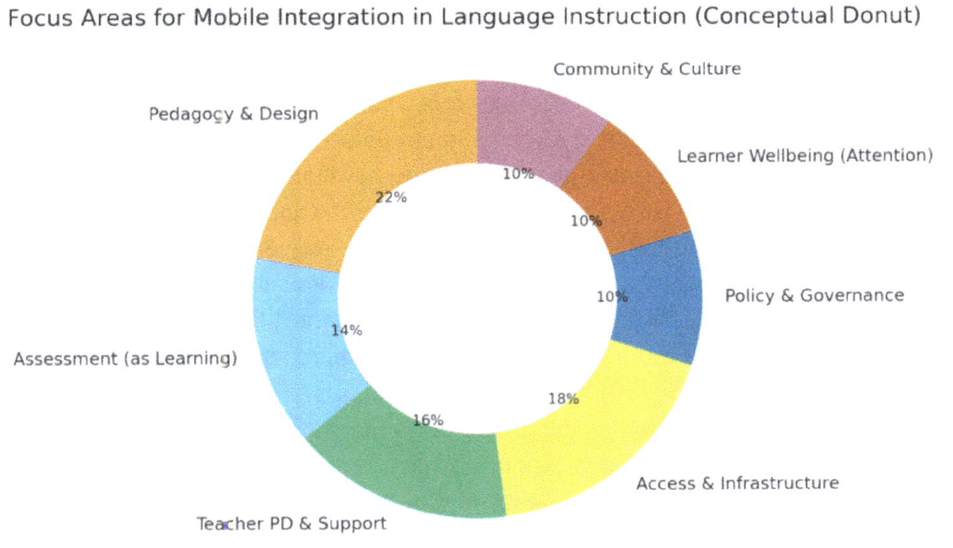

In thinking about the mobile classroom, we also begin to see a shift in the sources of authority. The textbook is no longer the only voice of truth. Learners can fact-check, translate, and explore multiple perspectives at the click of a button. They can interact with native speakers in real time, observe authentic language use in videos, and question prescribed norms. This democratization of knowledge can be unsettling for traditional education systems, but it can also lead to richer, more critical forms of learning (Sriprakash et al, 2024). It requires us to move beyond accuracy for its own sake, toward meaningful communication, intercultural understanding, and linguistic confidence. Culturally, too, the classroom must open itself. Mobile platforms expose learners to global discourses, voices from different regions, accents, expressions, and real-world language use that challenges textbook stereotypes (Evurulobi et al, 2024). Teachers must help students make sense of this diversity—to see it not as noise, but as the authentic texture of human communication. This is no easy task. It requires a pedagogy of openness, humility, and cultural responsiveness. And in all of this, the emotional dimension must not be forgotten. Mobile learning is not simply cognitive—it is deeply affective. It intersects with identity, with fear of judgment, with desire for belonging. A student may feel safer practicing pronunciation alone in their room before sharing in class (Baines et al, 2023). Another may find confidence in recording and re-recording messages before sending them to a group chat. These small moments matter. They reflect how mobile tools, when thoughtfully integrated, can create spaces for learners to take risks, to play, and to grow.

To rethink the language classroom, then, is not just to adopt technology—it is to reimagine pedagogy, space, time, authority, culture, and emotion (Barik, 2023). It is to move from control to collaboration, from content delivery to experience design, from uniformity to personalization. It is a shift that challenges us to be more humane, not less; more attentive to context, not detached by convenience. The age of mobility is not a temporary phase—it is the new fabric of life. And language classrooms that ignore this reality risk becoming irrelevant. But those that embrace it—not blindly, but critically and compassionately—have the chance to become places of incredible transformation (Kayyali, 2025). Places where learning is not confined to desks and bells, but flows into the lived world. Where students are not passive learners of a foreign tongue, but active participants in a global conversation. This transformation will not be without struggle. There will be false starts, missteps, resistance. But that is the nature of all meaningful change. What matters is not perfection, but persistence. Not uniformity, but intentionality. If we can keep the heart of language education—the human connection—at the center of this mobile evolution, then we have much to gain. Rethinking the language classroom is not about replacing tradition. It is about honoring it while acknowledging that the world has changed—and so must we. It is about creating spaces where learners are seen, heard, and empowered to use language not just to pass tests, but to shape their lives. And if mobile technology can help us do that—even imperfectly—then it is not just a tool. It is an invitation to rethink what learning truly means in a world always on the move.

PEDAGOGICAL TENSIONS: TRADITION MEETS TECHNOLOGY

There is something reassuring about a traditional classroom. Chalk in hand, the teacher stands before students whose eyes follow the rhythm of a familiar lesson structure: an introduction, a bit of controlled practice, a few exercises, and then perhaps a round of speaking drills. The pedagogical logic is sound, built on decades—if not centuries—of theory and practice. It's structured. It's predictable. It gives teachers a sense of control, and students a sense of direction. There's value in this kind of structure—especially in language education, where learners often need scaffolding, repetition, and clear models to build confidence and fluency. And yet, this traditional rhythm is being disrupted—not by pedagogical rebellion, but by a force that has seeped silently, almost invisibly, into every corner of modern life: technology (Ford, 2023). Mobile technology, in particular, doesn't arrive quietly. It beeps, buzzes, pings, and flashes. It tempts learners with distractions and invites them into a world of constant engagement. But it also holds immense promise: language apps that gamify vocabulary learning, real-time chat with native speakers halfway across the globe, instant translation, pronunciation tools, immersive augmented reality lessons, and personalized feedback systems. To deny this potential is to close the classroom door on a world that's already moving forward. And yet, to fully embrace it without caution risks turning pedagogy into novelty—replacing depth with dopamine, reflection with reaction.

The space where tradition meets technology is not a neutral one. It is filled with tension—philosophical, practical, emotional. These tensions are not inherently negative. On the contrary, they can become fertile ground for transformation if handled with care. But first, they must be named and understood. At the heart of this friction is a fundamental shift in the teacher's role. In a traditional setting, the teacher is often the primary source of knowledge and the orchestrator of learning (Song, 2021). The textbook is the spine of the curriculum, and lessons flow linearly. But in a digitally infused environment, the authority of the teacher is no longer absolute. Learners can search definitions faster than a teacher can explain them. They can access multiple perspectives on grammar, idioms, and pronunciation. They can find alternative resources—often ones more tailored to their interests or learning style. For some educators, this democratization of knowledge feels liberating. For others, it feels like erosion—a quiet unraveling of the expertise they spent years cultivating (Qutieshat, 2025). This can produce a sense of professional unease. When learners start turning to their phones for answers in the middle of a lesson, is that engagement or disrespect? When a student corrects a teacher based on something they read online, is that initiative or insubordination? These questions don't have simple answers. But they reflect a deeper anxiety: that technology might undermine the pedagogical authority upon which the traditional classroom is built.

Table 1. Mobile-integrated activity design matrix

Activity	Mobile Tool	Target Skill	Assessment Artifact
Voice-note Pronunciation Drills	Voice recorder / WhatsApp voice	Pronunciation & prosody	Audio samples over time
Microlearning Vocab (QR-linked)	QR codes + flashcard app	Lexis & spaced retrieval	Learner word bank + usage
Peer Story Circles (Messaging)	Class chat / groups	Fluency & interaction	Thread transcript + feedback
AR Label Hunt (Campus/ Neighborhood)	AR label app / camera + notes	Place-based pragmatics	Photo log + captions
Translation Comparison Task	Two translation apps + notes	Form meaning nuance	Annotated examples + rationale
Mobile E-Portfolio Check-in	Portfolio app / shared drive	Metacognition & reflection	Reflection entry + evidence links

Then there is the question of pacing. Traditional instruction often follows a fixed syllabus, progressing unit by unit, allowing time for reinforcement and assessment. Mobile learning, by contrast, is often nonlinear, fast-paced, and self-directed. Some learners may race ahead using apps and online materials, while others lag behind, overwhelmed by the sheer volume of content (McHaney, 2023). The classroom becomes fractured. The teacher must now balance the collective rhythm of the group with the individual trajectories of learners operating at different speeds, in different digital spaces. This can lead to feelings of pedagogical dissonance: a sense that the classroom is no longer cohesive, that learning is slipping beyond the teacher's grasp. Equally important is the issue of attention. Language learning requires focus. But focus is hard to sustain in a digital environment saturated with alerts, messages, and background noise. Some teachers, understandably, view mobile devices as intrusions. They ban phones in class, restrict app usage, and plead for eye contact. Others try to integrate mobile tools into the lesson, hoping that engagement will follow. But here lies another tension: can we truly foster deep language acquisition through tools designed for speed and stimulation? Can reflective language learning coexist with the rapid-scroll culture of mobile interfaces?

There are also pedagogical tensions around content itself. Traditional language teaching emphasizes grammar accuracy, structured progression, and often leans on print-based materials (Sattarova, 2024). Mobile platforms, on the other hand, emphasize immediacy, contextual learning, and microlearning—bite-sized lessons meant to be consumed quickly. This mismatch can lead to confusion. A learner might excel at gamified vocabulary challenges on an app but struggle with real-world conversations. Or they might rely heavily on translation features without developing their ability to think in the target language. In such cases, technology may give the illusion of progress without depth. Teachers must then reconcile these different modes of learning: ensuring that digital fluency complements rather than replaces foundational skills. Emotionally, there is another layer of complexity. Many teachers who have spent years refining their craft feel marginalized by the sudden imperative to "go digital." Professional development sessions often focus on tools rather than teaching, platforms rather than pedagogy (Hrastinski, 2021). Educators are expected to be constantly upskilling, experimenting, adapting. This can breed fatigue, even resentment. It's not that teachers are resistant to change—they are often deeply invested in their learners' success. But meaningful change must come with support, dialogue, and respect for the knowledge teachers already possess.

On the learner's side, technology can also produce a sense of overload. Students may feel pressure to use multiple apps, juggle different digital platforms, and engage with language 24/7 (Mustafoyeva, 2025). While autonomy is empowering, it can also be isolating. Learners might miss the structure of guided lessons, the reassurance of teacher feedback, the warmth of peer interaction. The assumption that digital natives automatically thrive in mobile learning environments is misguided. Digital fluency does not always translate into language learning efficacy. There are also deeper cultural tensions at play. Traditional classrooms are often shaped by local norms, values, and expectations. Mobile content, by contrast, is often globalized, designed by companies in Silicon Valley or European capitals (Jones & Sudlow, 2022). Learners in different regions may encounter materials that conflict with their cultural contexts or feel alien to their lived experiences. A gamified lesson about ordering wine in a Parisian restaurant may not resonate with learners in a conservative or non-Western context (Salimi, 2023). Teachers must then navigate these cultural mismatches—editing, adapting, or even rejecting digital content that does not align with their students' realities. And yet, despite all these tensions, it would be a mistake to retreat into nostalgia. Traditional methods, while comforting, are not always effective. Grammar-translation methods, rote memorization, and teacher-centered instruction can limit learners' communicative competence and motivation. Technology, for all its disruptions, opens new possibilities for interaction, authenticity, and personalization. It allows learners to encounter language as it is lived—not just studied. So what, then, is the way forward?

Perhaps it lies not in choosing between tradition and technology, but in reframing the question altogether. What if the goal is not to replace but to reconcile? To allow pedagogical integrity to guide technological innovation, rather than the other way around? This means asking hard questions at every step: What does this tool add to the learning experience? How does it support my learners' goals? Does it align with my teaching values? It also means being willing to experiment, to fail, to adjust. Integration is not a single event—it's an ongoing process, one that evolves with each class, each context, each learner. It also requires a new kind of professional development—one rooted in collaboration, reflection, and respect. Teachers need spaces to share strategies, express concerns, and explore tools without pressure (Vidergor, 2022). They need training that connects technology to pedagogy, not just features to functions. And perhaps most importantly, they need to feel trusted—not just as users of technology, but as critical thinkers who can discern what serves their students best. Ultimately, the intersection of tradition and technology is not a battleground—it's a meeting place. A space of possibility. Yes, the tensions are real. But so are the opportunities. Language classrooms can be places where printed textbooks live alongside mobile apps, where whiteboards coexist with voice recorders, where teachers still lead but learners also explore. In such classrooms, learning becomes a shared adventure—not into the digital unknown, but into a more connected, more responsive, and more human-centered future. The real challenge is not technological. It is pedagogical. And human. It is about navigating the delicate balance between structure and flexibility, between expertise and experimentation, between holding on and letting go. But that is what good teaching has always been about.

In this journey, it helps to remember: the best technology in the world cannot replace a teacher who knows how to listen. Nor can the most time-tested method survive if it refuses to grow. The future of language education lies not in choosing one side or the other, but in crafting a pedagogy that honors both—the wisdom of tradition and the promise of innovation.

NAVIGATING PRACTICAL BARRIERS TO MOBILE INTEGRATION

When conversations around mobile technology in education begin, they often start with the promise: improved access, personalized learning, dynamic engagement, and global connectivity. The optimism is understandable—after all, few other tools have transformed communication and information exchange as profoundly as mobile devices (Ndimbo et al, 2024). For language learning in particular, the mobile revolution offers striking potential: instant dictionaries, pronunciation apps, immersive translation tools, virtual pen pals, and access to authentic content in the target language. But while the possibilities are enticing, the realities of integrating mobile technology into traditional classrooms are far more complex. They are grounded not just in pedagogy, but in logistics, policy, infrastructure, and the often-overlooked emotional labor of adaptation. Navigating these practical barriers is not a simple matter of downloading apps or updating syllabi—it requires patience, flexibility, and a deep understanding of the nuanced ecosystems within which language learning takes place (Holozsai & József, 2024).

The first and perhaps most glaring obstacle is access. Although we live in what is often called a connected world, that connectivity is uneven. Students in urban, well-funded schools may have access to personal smartphones, stable Wi-Fi, and even institutional support for data plans or device loans (Weinberg, 2024). But for many others—particularly in rural areas, low-income communities, or developing regions—access to mobile technology is far from guaranteed. Even when students own mobile devices, those devices may be outdated, shared among family members, or reliant on intermittent internet access. This creates an immediate digital divide within classrooms, where some learners can fully engage with mobile-integrated lessons while others are left behind, through no fault of their own. Teachers, too, are not exempt from access limitations. Some may be working in schools where mobile use is officially restricted, where firewalls block essential apps, or where the infrastructure simply cannot support consistent connectivity (Azem Qashou et al, 2025). Others may be teaching in institutions that lack technical support staff or professional development programs that would help them use technology meaningfully. In such contexts, mobile integration risks becoming an additional burden—something else teachers are expected to manage without being given the tools or time to do so. It's easy to talk about innovation; it's much harder to implement it when basic conditions remain unmet.

Figure 2. Shift in learning time with mobile integration, 2021–2026

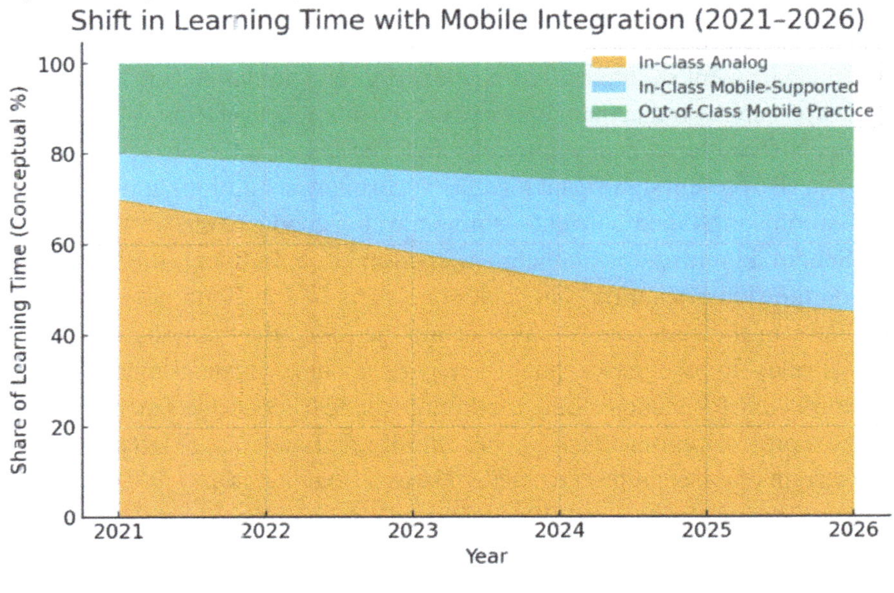

Beyond access, there are challenges of compatibility—both technological and curricular. Many mobile apps are designed for individual, self-paced learning. They work well for personal study, but less so in group settings. Trying to align them with the flow of a classroom lesson can feel awkward or disjointed. A vocabulary app may not match the unit a teacher is currently covering. A grammar game might introduce structures before students are ready. Furthermore, few apps are built with specific curricular standards in mind (Valentina, 2024). Teachers are then forced to either bend their curriculum to fit the tool or abandon the tool altogether. Neither option is ideal. In practice, this leads to a lot of wasted potential—tools that are rich in content but left unused because they don't "fit" the classroom structure. Language barriers in the most literal sense also appear. Many mobile apps are available only in major world languages, often English, and are designed with monolingual assumptions (Vargas Florez, 2024). This can make them difficult to use for learners who do not speak the app's primary interface language. It also places pressure on teachers to translate instructions, troubleshoot for students, or pre-teach how to use a platform before any actual language learning can take place. The result is an increase in cognitive load for both parties—and a decrease in the time available for meaningful pedagogical interaction. Then there is the problem of distraction. Mobile devices are not built solely for education; they are also portals to entertainment, social media, and a thousand other stimuli (McHaney, 2023). The line between focused learning and passive scrolling is thin. Teachers often express frustration with students who begin a pronunciation task on their phone, only to get sidetracked by notifications or drift into unrelated apps. Managing attention in a mobile-integrated classroom becomes a delicate dance—teachers must create structures that encourage on-task behavior without resorting to authoritarian control. This often means rethinking classroom management strategies, setting clear digital boundaries, and fostering a culture of responsibility rather than surveillance. But this, too, takes time—something in short supply for many educators already stretched thin.

Policy can also present significant roadblocks. In many institutions, mobile phone use in classrooms is banned outright—sometimes for good reason, sometimes out of habit. School administrators, concerned about misuse, bullying, or distraction, often err on the side of caution. This places teachers in a difficult position. Even if they believe in the pedagogical value of mobile tools, they may be limited by institutional rules. In other cases, policies may exist but be inconsistently applied, creating confusion for students and frustration for teachers. One teacher allows mobile-based projects, another bans phones entirely, and students are left to navigate a shifting terrain of expectations (Khalid & Owusu-Boateng, 2024). Without clear, supportive policy frameworks, mobile integration remains inconsistent, precarious, and dependent on individual initiative rather than systemic change. Professional development—or the lack thereof—is another recurring barrier. Many teachers were never trained to teach with mobile technology. Their teacher education programs may have emphasized face-to-face methods, textbook-based instruction, and analog materials. Even younger teachers, often assumed to be "digital natives," may struggle with translating personal tech fluency into pedagogical practice. Using a phone to message friends is not the same as designing a lesson that leverages mobile tools to deepen language skills. Teachers need support not just in how to use technology, but in *why* and *when* to use it. They need time to explore tools, reflect on their practice, and experiment without fear of failure. Unfortunately, professional development is often delivered as one-off workshops that focus on tools rather than pedagogy, platforms rather than learner outcomes. There is little follow-up, and little time for sustained exploration. As a result, many teachers remain hesitant, unsure, or skeptical—not out of resistance, but out of caution and fatigue.

Assessment adds another layer of complexity. Traditional assessments—multiple choice quizzes, grammar exercises, oral interviews—may not capture the breadth of skills students develop through mobile learning (Şi ianu & Pu ca u, 2024). How do we evaluate a student's progress if much of their learning happens through spontaneous conversations in a language exchange app? How do we assess engagement with authentic content like YouTube videos, podcasts, or mobile-based journaling? These are valid forms of language use, but they resist easy quantification. Schools still rely on grades, transcripts, and standard benchmarks. Teachers may struggle to align mobile-enhanced learning with these expectations (Ngwenya, 2022). Without flexible assessment frameworks, much of the value created through mobile tools remains invisible in official records—rendering it marginal or even irrelevant in the eyes of stakeholders. And we must not overlook the emotional and psychological dimension of these barriers. Change—even change aimed at improvement—can be overwhelming. Teachers who have spent years mastering their craft may feel disoriented when told to abandon tried-and-tested methods (Floor, 2023). They may fear losing control of the classroom, being judged by peers, or failing in front of their students. Learners, too, may feel exposed when using unfamiliar platforms, especially if they fear making mistakes in public digital spaces. There is vulnerability on all sides. Integrating mobile technology is not just a logistical shift—it is a human one. It calls for empathy, patience, and a culture of mutual respect.

Perhaps the most difficult barrier of all is time. Time to plan, to adapt, to troubleshoot. Time to develop new materials, to learn how a new tool works, to understand its quirks and potential. Time to rethink the lesson structure, the rhythm of activities, the flow of the week. Teachers already work under immense time pressure. Adding mobile integration without providing extra time for planning and reflection only exacerbates burnout and disillusionment (Wasie, 2024). And yet, despite all these barriers, educators continue to innovate. Not because they are asked to, but because they care—about their students, their craft, their communities. Many find ways to work around the obstacles. They develop simple, low-tech strategies that still tap into mobile potential: voice recordings instead of written homework, photo-based vocabulary challenges, QR codes for accessing reading materials. They use WhatsApp groups for peer feedback or Google Forms for quick, interactive assessments (Wulandari et al, 2021). They learn from each other, share ideas, and slowly build new habits. Integration doesn't always have to be flashy—it just has to be intentional. The road to meaningful mobile integration is neither straight nor smooth. It requires systemic support, institutional flexibility, and sustained investment—not just in tools, but in people. It requires policies that empower rather than restrict, training that inspires rather than overwhelms, and infrastructure that enables rather than limits. Above all, it requires a shift in mindset—from seeing mobile technology as an add-on or threat, to recognizing it as a partner in the evolving landscape of language education. No single solution will address every challenge. But small steps matter. A teacher finding one app that truly resonates with their learners. A school adjusting its policy to allow mobile use during specific learning activities. A student discovering the joy of practicing pronunciation on the bus ride home. These are not small victories (Wirentake & Arfani, 2024). They are seeds of transformation. Navigating the practical barriers to mobile integration is less about overcoming a checklist of problems and more about cultivating resilience, curiosity, and shared purpose. It's about asking: How can we create learning environments that reflect the way people live, communicate, and grow today? How can we honor the richness of human connection while embracing the tools that extend it?

The answers will differ by context, by classroom, by community. But the effort—to adapt, to include, to evolve—is what defines the heart of education. In that effort, mobile integration is not the final destination. It is part of a larger journey toward more responsive, more equitable, and more meaningful language learning.

STRATEGIES FOR SUSTAINABLE AND CONTEXT-SENSITIVE IMPLEMENTATION

Implementing mobile technology in language classrooms is not just a technical endeavor—it is a deeply human one. It requires more than infrastructure or devices. It calls for an attuned understanding of the educational culture, the socioeconomic realities of learners, the values held by teachers, and the rhythms of daily life within and beyond the classroom. Any strategy that overlooks these variables is likely to collapse under the weight of good intentions. The most successful integrations are those that are both sustainable—meaning they can endure over time without exhausting resources or morale—and sensitive to the unique context in which they unfold (Schweitzer et al, 2023). Sustainability in this domain does not mean flashy rollouts or one-time investments in digital tools. It means building something that teachers can maintain, learners can benefit from, and institutions can support without constant crisis management or overextension. It starts with clarity of purpose. Why introduce mobile technology into a specific classroom or program? What does it aim to solve, support, or transform? Without a clear pedagogical goal, mobile integration becomes another layer of noise—just another thing to manage. But when the purpose is rooted in the needs of learners, guided by the experience of teachers, and informed by the specific constraints of a local setting, it becomes more than innovation—it becomes transformation.

The first step is listening—truly listening—to the people involved. This includes students, who often have insight into how they already use mobile tools outside of class. Many learners come to the classroom with existing digital habits. Some already use apps for vocabulary, engage in social media with speakers of other languages, or participate in online communities that require them to code-switch or adapt linguistically (Slimane, 2024). Rather than assuming that technology is a foreign add-on, teachers can begin by asking: What are my students already doing with their devices? How can I build on those habits rather than introduce entirely new ones? Sustainable strategies often evolve from aligning learning goals with learners' real-world behaviors. Teachers themselves are at the heart of this process (Noy et al, 2021). They are not just implementers of strategies—they are co-designers. A sustainable approach respects their time, acknowledges their workload, and trusts their professional judgment (Kayyali, 2024). This means starting small. Instead of overhauling an entire curriculum overnight, effective implementation begins with a single lesson, a single activity, even a single class period. A short audio-based homework task using voice notes. A vocabulary game using an app that complements the textbook (Das, 2024). A class poll conducted via mobile form. These small interventions allow teachers to experiment without fear, reflect on what works, and adjust without burning out.

Institutional support plays a pivotal role here. Sustainability is difficult to achieve in isolation. Teachers need not only permission but encouragement to try new things (Parry & Metzger, 2023). They need time set aside for collaborative planning, access to training that is ongoing and practical, and platforms for sharing successes and failures. Schools and universities that create these conditions cultivate a culture of experimentation, one where mobile integration is not seen as a burden but as an evolving dialogue (Xiao, 2022). And that dialogue must be rooted in trust—not just in tools, but in people. Context-sensitivity demands that we resist one-size-fits-all solutions. What works in a private international school in Singapore may not translate to a public secondary school in rural Tunisia. What makes sense in a fully online adult language program may fall flat in a face-to-face elementary setting (Krengel, 2021). Strategies must be tailored not only to available resources but also to cultural norms, learner expectations, and community values. For example, in some settings, learners may feel uncomfortable recording their voice due to cultural norms around modesty or fear of embarrassment. In others, learners may lack privacy at home and prefer silent, text-based mobile activities. Implementation plans must be flexible enough to accommodate these lived realities.

Language learning is also embedded in identity, and mobile technology must support that identity rather than flatten it. For instance, using locally relevant examples, stories, and voices in mobile-based content can help students feel seen and respected. Encouraging learners to contribute their own materials—photos, audio messages, language memes, or digital journals—can make mobile learning personal, rather than just imported (Annamalai et al, 2024). The more learners see their worlds reflected in the tools they use, the more likely they are to engage with them meaningfully. Sustainability also depends on the kind of technology chosen. Not all mobile tools are created equal. Some demand constant updates, high data usage, or specific operating systems. Others are lightweight, accessible offline, and flexible across devices. In low-resource settings, favoring tools that work well with limited connectivity or minimal data consumption is not just practical—it's ethical. Choosing the right tool is not about using the newest or most popular app. It's about matching tools to the learning environment, and ensuring that no student is excluded because of financial or technical barriers.

Figure 3. Program readiness for mobile-supported language learning

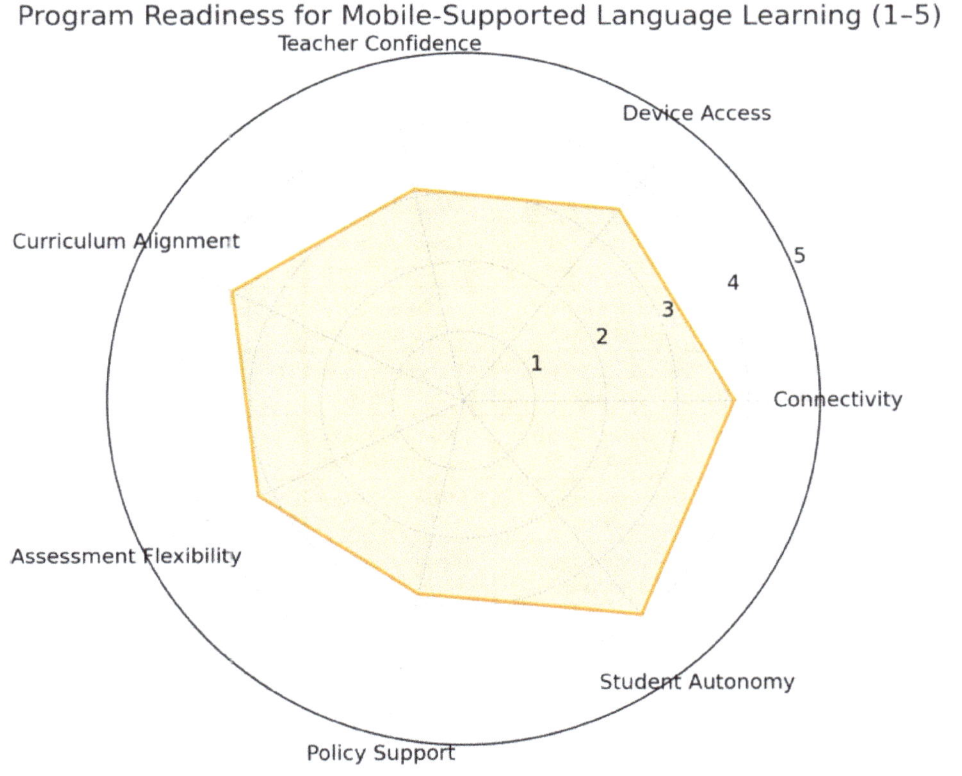

Furthermore, a sustainable strategy must integrate with pedagogy—not override it. Mobile tools should not dictate what is taught or how. They should support, enhance, and extend existing practices. This requires teachers to see technology not as an alternative to teaching, but as an assistant. A mobile quiz might reinforce vocabulary after a class discussion. A speech recording might offer students a chance to practice before an in-person dialogue. A collaborative writing app might allow for real-time feedback that is carried into class discussion (Bin Dahmash, 2021). When used this way, technology becomes part of a loop—reinforcing what is learned in class, and allowing for continuous practice outside of it. Feedback is central to sustainability. Mobile integration is not a static achievement but a living process. Teachers need to gather regular input from students: Are the tools helping? Are they confusing? Do learners feel more confident, more engaged, more in control? Listening to this feedback not only improves implementation—it deepens the learner-teacher relationship. It sends a clear message: This is your learning journey, and your voice matters. Similarly, peer feedback among teachers—what's working, what's not—should be encouraged and institutionalized. Communities of practice, whether formal or informal, help strategies evolve organically rather than becoming rigid or outdated.

Table 2. Barriers, solutions, supports, and placement

Barrier	Low-Tech Solution	High-Tech Option	Institutional Support Needed
Unequal Access to Devices/Data	Pair-share devices; offline packs	Loaner devices; zero-rated content	Procurement & maintenance plan
Distraction & Attention Drift	Set 'focus windows'; print tasks	Do-not-disturb automations	Digital wellbeing norms + training
Curriculum App Misalignment	Task wrappers around any app	LMS integration; LTI links	Curriculum mapping time
Teacher Confidence & Time	Start with 1 recurring routine	Micro-PD modules; template bank	Release time + coaching
Policy Restrictions on Phones	Scoped 'phones-on' windows	MDM per-class policy toggles	Clear, shared policy & signage
Assessment of Informal Learning	Audio/text logs; checklists	E-portfolios; rubrics in LMS	Assessment policy update
Privacy & Consent Concerns	No-login tools; local storage	Consent flows; data minimization	Privacy review + parent comms

Security and privacy are also important, though often under-discussed. Sustainable implementation must consider the protection of student data (Shi, 2024). Apps and platforms should be vetted not just for educational quality but for how they handle user information. Students should never be required to surrender their privacy or expose themselves to commercial exploitation in the name of learning. Transparent policies, consent-based practices, and awareness campaigns around digital safety must be part of any long-term plan. Assessment practices also need to shift toward sustainability. Traditional tests may not fully capture the diverse, real-time language use that mobile learning encourages. Teachers must be empowered to assess learning in ways that align with mobile experiences—audio reflections, text exchanges, digital portfolios, interactive tasks (Yadav, 2024). This not only makes assessment more holistic but also validates the learning that happens outside the formal classroom. It shows students that mobile-based efforts matter, that their work is visible, recognized, and connected to their overall growth. Crucially, sustainable and context-sensitive implementation also leaves room for imperfection. Not every activity will succeed. Not every tool will resonate. Not every student will engage the same way. That's not failure—it's part of the learning process, for both teachers and students. Strategies must include space for reevaluation, for iteration, for graceful exits from what doesn't work. The goal is not perfection, but progress:

a trajectory of learning, adapting, and refining that respects the lived conditions of those it seeks to serve.

In many ways, the most powerful strategy is humility. The humility to admit what we don't know, to ask for help, to listen to learners, to slow down when needed, to pause and reflect before moving forward (Welch, 2024). In a field so driven by innovation, humility may seem countercultural. But it is precisely what makes implementation sustainable—because it keeps us grounded in people, not just platforms. To implement mobile technology in language education sustainably is not to chase trends or adopt tools for their own sake. It is to ask: What helps my learners connect? What supports their growth, their confidence, their voice? What respects their realities? What supports my wellbeing as a teacher? These are not technological questions—they are pedagogical, ethical, and profoundly human. When we answer them with care, we craft strategies that last—not because they are flashy, but because they are rooted in the soil of real life. Sustainability is not measured in downloads or dashboards. It is felt in the moment a student, once silent, dares to record their voice. It is seen when a teacher, once skeptical, smiles at the engagement of her class during a mobile-based activity. It is lived in classrooms where technology does not overshadow the human, but instead helps it flourish—quietly, steadily, and with purpose.

REIMAGINING THE ROLE OF THE EDUCATOR IN MOBILE-SUPPORTED ENVIRONMENTS

To teach a language is to do more than convey vocabulary, grammar rules, or pronunciation guides. It is to open a door into another way of being, thinking, and relating to the world. Language teaching has always involved acts of interpretation, patience, encouragement, and negotiation. At its heart, it is an intensely human endeavor—one that relies not only on subject knowledge but on emotional intelligence, cultural awareness, and pedagogical agility (Arias, 2021). Yet in recent years, with the rapid rise of mobile technologies in education, the role of the educator has come under quiet—but significant—pressure to evolve. Not to disappear. Not to be replaced. But to be reimagined.

For decades, the teacher was the center of the classroom. The one who held the answers, decided the pace, designed the assessments, and guided the learning journey with a certain structured rhythm. Their presence provided stability and a sense of direction. But as mobile-supported environments become more common, the classroom has grown more porous. Learning doesn't begin and end with the school bell. It spills into the commute, into the home, into the pauses between daily routines (Shah, 2023). Students are no longer entirely dependent on their teachers for input or resources. They can search grammar explanations online, listen to native speakers on social platforms, or get instant corrections from an AI-based app. In this new landscape, the educator is no longer the sole gatekeeper of knowledge—and this is not a crisis. It's an invitation. An invitation to shift from authority to guide. From controller to collaborator. From content-deliverer to learning architect. This reimagining is not about de-skilling teachers or asking them to compete with algorithms. It is about reclaiming what machines cannot do: understand the subtle dynamics of a group of learners, foster trust, adapt with empathy, and create moments of joy, humor, and connection. Technology may provide tools, but it is still the educator who decides how, when, and why those tools are used (Kayyali, 2021). The value of the teacher doesn't diminish in a mobile-supported environment—it simply migrates into different spaces.

One of those spaces is that of facilitator. In mobile-enhanced settings, learners often work at different paces, engage with content in non-linear ways, and explore areas of personal interest. The teacher's task becomes less about delivering identical information to everyone, and more about helping each learner make sense of their own learning path (Doyle, 2023). It's about asking the right questions, providing timely feedback, offering choices, and curating a mix of activities that speak to diverse learning styles. It's also about noticing when a learner is stuck—not just technically, but emotionally. When they've lost motivation, when they're overwhelmed, when the tool is no longer serving the goal. These are moments that no app can catch. But a teacher can. The mobile-supported classroom also calls for a new kind of instructional design. Teachers must now think not only about what happens during class time, but what can happen outside of it—and how the two spaces connect. How can an activity begun in class continue seamlessly on a mobile platform? How can a task done individually at home feed into a group discussion the next day? This requires careful planning and a willingness to reimagine the temporal flow of learning. It also calls for flexibility: things don't always go as planned when devices are involved (Kalisz, 2024). Apps crash. Wi-Fi cuts out. Students forget passwords or lose interest. A teacher in this environment must be part educator, part improviser, and part problem-solver (Tromp & Baer, 2022). This evolving role also includes that of digital mediator. In an age where learners have access to vast amounts of information, not all of it reliable, the teacher becomes a crucial figure in helping them navigate the digital landscape critically. What makes a good language learning app? How do you evaluate whether an online explanation is accurate? When should you trust automatic translation, and when should you not? What do you do when the English subtitles on a video don't match what the speaker is actually saying? Teachers guide students not just in acquiring language, but in discerning the quality, bias, and appropriateness of the digital content they encounter. In a world overflowing with tools, the ability to choose wisely becomes a core part of learning—and teaching.

Perhaps one of the most overlooked aspects of the educator's role in mobile-supported environments is the work of humanizing the experience. Mobile devices are efficient but impersonal. They don't smile, hesitate, or adjust based on a learner's anxiety. They may track progress, but they don't celebrate small victories with genuine joy. Teachers do. In reimagining their role, educators must not surrender what makes them irreplaceable. They must double down on empathy, community-building, and relational presence. It is precisely because learners can now study alone that the need for connection becomes even more critical. The teacher creates a space—physical or virtual—where learners feel seen, supported, and safe enough to make mistakes. This redefined role also requires educators to embrace vulnerability. Not all teachers are comfortable with technology. Many were trained in methods that prioritized face-to-face instruction and see mobile learning as unfamiliar terrain. Others may feel that their authority is threatened by students who are more digitally fluent. The pressure to "keep up" can be exhausting, especially when professional development is inconsistent or irrelevant (Rinehart, 2023). Reimagining the teacher's role must include acknowledging these fears and providing real, sustained support. Teachers need time to explore new tools, to reflect, to fail without judgment. They need peer communities where experimentation is encouraged and where innovation is rooted in pedagogy, not performance.

Importantly, this shift is not a one-time change. It is ongoing. Teachers must learn to inhabit a role that is dynamic, elastic, and ever-evolving. The mobile-supported classroom is not a fixed destination—it is a living ecosystem. New tools will continue to emerge, student needs will change, and educational paradigms will shift (Chetry, 2024). What remains constant is the need for thoughtful, responsive educators who are willing to grow alongside their learners. The educator also becomes a learner. Not in the sense of mastering every technical detail, but in remaining curious—about how learners engage with new platforms, about which digital behaviors support or hinder language acquisition, and about how teaching can continue to adapt with integrity. This kind of learning is not linear. It's messy, recursive, and full of contradictions. But it's also deeply rewarding. It keeps teaching alive.

The educator in mobile-supported environments is a connector. Someone who bridges the gap between the formal and informal, between school and life, between standardized curricula and personalized learning (Jung, 2024). They help learners see the continuity between what they practice in an app and what they say to a friend, between the word list on a screen and the job interview they're preparing for. They draw attention to patterns, make space for cultural nuance, and help learners stitch together the fragmented threads of mobile interaction into a meaningful linguistic tapestry. Reimagining the teacher's role in this way is not about burdening them with more responsibilities (Goodwin et al, 2023). It's about recognizing the richness of what they already do, and giving them the tools, space, and trust to do it in a new way. It's about letting go of outdated binaries—online versus offline, tech-savvy versus traditional—and embracing a more holistic view of teaching: one that is grounded, flexible, and above all, human. Because at the end of the day, even in the most tech-enhanced classroom, it is the teacher's voice that can calm a nervous learner. It is the teacher's story that can bring a tense lesson to life. It is the teacher's eyes that notice when someone is drifting, when someone has a question they're too shy to ask. No notification will do that. No algorithm will replace that. Mobile technology may reshape the contours of the classroom, but it is the educator who animates it—who gives it purpose, warmth, and depth. As we move further into a world where learning is increasingly mobile, let us not forget that what matters most is not the tool in the hand, but the heart and mind of the one who guides the journey.

CONCLUSION

As we reach the end of this exploration, it becomes clear that integrating mobile technology into traditional language instruction is not a destination but a journey—one that must be walked thoughtfully, collaboratively, and iteratively. The promise of mobile-assisted language learning is not simply in its ability to digitize existing content or make learning more convenient. Its real value lies in how it can extend the classroom, deepen engagement, foster learner autonomy, and open doors to new forms of interaction that were previously unimaginable.

Yet, despite the enthusiasm surrounding mobile technology, the path forward is strewn with complexity. One of the most striking revelations throughout this chapter is how deeply rooted the challenges are—not only in infrastructure or access, but in mindsets, pedagogical habits, and institutional inertia. Traditional language instruction has long served as a stable foundation for learners. For many educators, especially those with decades of experience, the shift toward mobile integration can feel like a disruption of that foundation, rather than an evolution. This discomfort is not a flaw, but a reflection of the deep care and commitment that educators bring to their work. Change, especially in education, must respect that commitment. Technology integration is not equitable by default. In fact, if implemented carelessly, mobile solutions can amplify existing inequalities. Learners without stable internet connections, personal devices, or digital literacy skills may find themselves left behind. Teachers without training or institutional support may feel overwhelmed or disempowered. This is why solutions must be designed with empathy and inclusion at their core. Simply introducing mobile apps into a classroom is not enough. What's needed is a broader ecosystem of support—professional development, collaborative planning, thoughtful pedagogy, and policies that value both innovation and equity.

At the same time, this chapter has shown that integration does not require abandoning traditional methods. On the contrary, some of the most impactful approaches emerge from the blending of mobile tools with time-honored techniques: vocabulary games that build on word lists from class, mobile-based speaking tasks that complement in-person discussions, grammar exercises that reinforce concepts taught in textbooks. When used wisely, mobile technology does not replace the teacher—it extends their reach. It does not diminish face-to-face learning—it enhances it. What it does demand, however, is a shift in perspective: from control to facilitation, from information delivery to co-learning, from fixed outcomes to ongoing exploration. Looking ahead, the most sustainable forms of integration will be those that center the human experience. No matter how advanced the technology becomes, language learning will always be a deeply personal act—one that involves vulnerability, curiosity, and connection. As educators, our role is not to chase every new app or trend, but to remain grounded in pedagogy, while staying open to tools that serve the needs of our learners. Mobile technology is one such tool—but it must be guided by wisdom, not novelty. Ultimately, the question is not whether we should integrate mobile technology into traditional language instruction. That moment has already arrived. The real question is how we choose to do so—with care or convenience, with intention or impulse, with inclusion or oversight. This chapter urges us to choose care. To proceed not with haste, but with humility. To design not just for innovation, but for impact.

If we do, then perhaps we can shape language classrooms that are not only more connected, but more compassionate. Classrooms where technology amplifies—not replaces—the voices of learners and teachers alike. Classrooms that prepare students not just to speak a language, but to live it, in a world where human connection still matters more than ever.

REFERENCES

Annamalai, N., Mangaleswaran, S., Ab Rashid, R., Harb Alqaryouti, M., & Sadeq, A. E. (2024). Learner experiences of community of inquiry mode (CoI) in WhatsApp for English language learning. *SAGE Open*, *14*(2), 21582440241251944. DOI: 10.1177/21582440241251944

Arias, O. E. (2021). Learning through Transitions: Exploring the Relationships between Emotional Intelligence Competencies and Learning Agility (Doctoral dissertation, University of Pennsylvania).

Atibuni, D. Z., Manyiraho, D., & Nabitula, A. N. (2022). A Fourth Industrial Revolution Paradigm Shift in Teacher Education? *International Journal of African Higher Education*, *9*(2), 1–21. DOI: 10.6017/ijahe.v9i2.15365

Azem Qashou, A. M., Bahar, N., & Mohamed, H. (2025). Qualitative Exploration of Data Security Risks in Mobile Cloud Computing for Higher Education. *Security and Privacy*, *8*(2), e70001. DOI: 10.1002/spy2.70001

Baines, A., Medina, D., & Healy, C. (2023). *Amplify student voices: Equitable practices to build confidence in the classroom.* ASCD.

Barik, S. (2023). Embracing the transformative dynamics of English teaching and technology in higher education: A posthumanist perspective. Smart Moves Journal IJELLH, 17-31.

Barnard, M., & Ferrier-Kerr, J. (2021). Moving to an Innovative Learning Environment: Exploring Teachers' Liminal Space. Pedagogy and Partnerships in Innovative Learning Environments: Case Studies from New Zealand Contexts, 141-164.

Bin Dahmash, N. (2021). *Synchronous and asynchronous English writing classes in the EFL context: Students' practices and benefits. Arab World English Journal* (Vol. 12). AWEJ.

Chetry, K. K. (2024). Transforming education: Paradigm shifts in 21st century learning. *International Journal of Multidisciplinary Research*, *2*(5), 469–479.

Das, M. K. (2024). *Teacher education with ICT tools and techniques.* Academic Guru Publishing House.

Doyle, T. (2023). *Helping students learn in a learner-centered environment: A guide to facilitating learning in higher education.* Taylor & Francis. DOI: 10.4324/9781003445067

Evurulobi, C. I., Dagunduro, A. O., & Ajuwon, O. A.Chinasa Iroabughichi Evurulo-biAdebukola Olufunke DagunduroOlanike Abiola Ajuwon. (2024). Language learning technologies: A review of trends in the USA and globally. *World Journal of Advanced Research and Reviews*, *23*(03), 2697–2707. DOI: 10.30574/wjarr.2024.23.3.2851

Floor, N. (2023). *This is Learning Experience Design: What it is, how it works, and why it matters*. New Riders.

Ford, D. R. (2023). Arrhythmanalysis and revolutionary rupture of détournement: Lefebvre and the pedagogical rhythms of post-fordism. *Postdigital Science and Education*, *5*(2), 440–454. DOI: 10.1007/s42438-021-00285-x

Goodwin, A. L., Madalińska-Michalak, J., & Flores, M. A. (2023). Rethinking teacher education in/for challenging times: Reconciling enduring tensions, imagining new possibilities. *European Journal of Teacher Education*, *46*(5), 840–855. DOI: 10.1080/02619768.2023.2299926

Goodwin, B., Rouleau, K., Abla, C., Baptiste, K., Gibson, T., & Kimball, M. (2022). *The new classroom instruction that works: The best research-based strategies for increasing student achievement*. ASCD.

Holozsai, J., & József, H. (2024). Aspects of digital learning among foreign language learners in secondary education.

Hrastinski, S. (2021). Digital tools to support teacher professional development in lesson studies: A systematic literature review. *International Journal for Lesson & Learning Studies*, *10*(2), 138–149. DOI: 10.1108/IJLLS-09-2020-0062

Jones, H., & Sudlow, B. (2022). A contemporary history of Silicon Valley as global heterotopia: Silicon Valley metaphors in the French news media. *Globalizations*, *19*(7), 1122–1136. DOI: 10.1080/14747731.2022.2034308

Jung, I. (2024). Personalized education for all: The future of open universities. *Open Praxis*, *16*(1), 24–36. DOI: 10.55982/openpraxis.16.1.612

Kalisz, D. E. (2024). Beyond Education: Scenarios of the Re-Imagined Higher Education from the Future.

Kayyali, M. (2021). Positive impact of high technology on higher education.

Kayyali, M. (2024). Future directions towards sustainable educational equity: Policies and practices. In *Exploring Educational Equity at the Intersection of Policy and Practice* (pp. 81–100). IGI Global. DOI: 10.4018/979-8-3693-1614-6.ch005

Kayyali, M. (2025). Classroom Management: Techniques for Creating a Positive, Friendly Learning Environment. In Exploration of K-12 Teaching and Learning for Teacher Educators (pp. 53-78). IGI Global Scientific Publishing.

Khalid, A. M., & Owusu-Boateng, O. (2024). *Navigating through Technology in Modern Education*. Bentham Science Publishers. DOI: 10.2174/97898152387851240101

Krengel, M. E. (2021). *Instructor Perspectives of Digital Natives in Face-to-Face Classrooms: A Qualitative Descriptive Study*. Northcentral University.

Kumar, R. (2023). English language and digital literacy: Navigating the information age. Journal of International English Research Studies (JIERS), ISSN: 3048-5231, 1(2), 25-31.

Makhal, A. D. (2023). Role of Digital Social Stories in enhancing Campus Inclusivity.

McHaney, R. (2023). *The new digital shoreline: How Web 2.0 and millennials are revolutionizing higher education*. Taylor & Francis. DOI: 10.4324/9781003447979

Mustafoyeva, N. (2025). TEACHING FOREIGN LANGUAGES USING MOBILE APPS. *Modern Science and Research*, *4*(2), 181–187.

Ndimbo, G. K., Gu, J., Haulle, E., & Yu, L. (2024). Why mobile phone matters: The role of ICT in promoting farmers' access to agricultural information and extension services in a tea outgrowing scheme in Tanzania. *Information Development*, •••, 02666669241284235. DOI: 10.1177/02666669241284235

Ngwenya, S. C. (2022). Affordances of mobile learning technologies in addressing language issues in Grade 7 Natural Sciences Teaching and Learning (Doctoral dissertation, University of Johannesburg).

Noy, S., Capetola, T., & Patrick, R. (2021). The wheel of fortune as a novel support for constructive alignment and transformative sustainability learning in higher education. *International Journal of Sustainability in Higher Education*, *22*(4), 854–869. DOI: 10.1108/IJSHE-08-2020-0289

Parry, S., & Metzger, E. (2023). Barriers to learning for sustainability: A teacher perspective. *Sustainable Earth Reviews*, *6*(1), 2. DOI: 10.1186/s42055-022-00050-3

Qutieshat, A. (2025). Towards a Liberated Higher Education Landscape. In *Artificial Intelligence in Higher Education: A Contemporary Examination of Illich's Theories* (pp. 23–48). Springer Nature Singapore. DOI: 10.1007/978-981-97-9989-3_3

Rinehart, T. D. (2023). *Developing the Personal and the Professional Addressing Teacher Attrition Through Self-Care & Leisure Programming*. University of Dayton.

Salimi, H. (2023). Beyond Universal Gamification: Exploring Persian Gamification Strategies (Doctoral dissertation, Concordia University).

Sattarova, M. (2024). A COMPARATIVE ANALYSIS OF TRADITIONAL AND MODERN APPROACHES IN LANGUAGE TEACHING. *Mental Enlightenment Scientific-Methodological Journal*, *5*(09), 214–223.

Schweitzer, L., Lyons, S., & Smith, C. J. (2023). Career sustainability: Framing the past to adapt in the present for a sustainable future. *Sustainability (Basel)*, *15*(15), 11800. DOI: 10.3390/su151511800

Shah, K. B. (2023). Traversing the City: Examining the Impacts of Extended Commute on Student Wellbeing & High School Experience in Chicago.

Shew, A. (2023). *Against technoableism: rethinking who needs improvement*. WW Norton & Company.

Shi, W. (2024). Analysis of Local Governance and Legal Protection of Students' Information Security in the Context of Sustainable Development of Digital Education. Lex localis-Journal of Local Self-Government, 22(4), 435-458.

Şi ianu, A., & Pu ca u, A. (2024). Methods of assessing students' knowledge in english language lessons. *Journal of Social Sciences*, *7*(3), 234–242. DOI: 10.52326/jss.utm.2024.7(3).16

Slimane, H. K. (2024). Digital linguistic landscapes: Unveiling youth code-switching in computermediated communication. *Akofena Journal*, *12*(3), 297–308.

Song, Y. (2021). A review of how class orchestration with technology has been conducted for pedagogical practices. *Educational Technology Research and Development*, *69*(3), 1477–1503. DOI: 10.1007/s11423-021-10001-y

Sriprakash, A., Williamson, B., Facer, K., Pykett, J., & Valladares Celis, C. (2024). Sociodigital futures of education: Reparations, sovereignty, care, and democratisation. *Oxford Review of Education*, •••, 1–18.

Tong, G., Li, Y., Zhang, H., & Xiong, N. (2023). A fine-grained channel state information-based deep learning system for dynamic gesture recognition. *Information Sciences*, *636*, 118912. DOI: 10.1016/j.ins.2023.03.137

Tromp, C., & Baer, J. (2022). Creativity from constraints: Theory and applications to education. *Thinking Skills and Creativity*, *46*, 101184. DOI: 10.1016/j.tsc.2022.101184

Umarova, Z. I. (2024). THE IMPORTANCE OF THE CONTINUITY PRINCIPLE IN TEACHING FOREIGN LANGUAGES. *PEDAGOG*, *7*(11), 242–246.

Valentina, V. (2024). Teaching grammar to young learners.

Vargas Florez, J. L. (2024). Analysis of mobile applications for language learning: unpacking discourses about second language learning trends and marketing strategies.

Vicente González, M. E. (2021). Learning English as a Second Language: A Comparison between Spanish and Adopted Chinese Teenagers.

Vidergor, H. E. (2022). Coping with teaching in innovative learning spaces: Challenges, insights and practices. *Learning Environments Research*, *25*(3), 707–724. DOI: 10.1007/s10984-021-09396-5

Wasie, T. (2024). *Disconnect to reconnect: Finding balance to reclaim well-being*. Teshome Wasie.

Weinberg, L. (2024). *Smart University: Student Surveillance in the Digital Age*. JHU Press. DOI: 10.56021/9781421450018

Welch, B. F. (2024). In Conclusion, I Don't Know: Humility as the Beginnin g and End of Every Class. The Art of Teaching Philosophy, 173.

Wirentake, M. P., & Arfani, S. (2024). *Exploring English through tourist attractions: Motivating students to speak*. Jakad Media Publishing.

Wulandari, I., Purwati, O., Setiawan, S., & Anam, S. U. (2021). AN INQUIRY INTO EFL LEARNERS'PERCEPTIONS OF WHATSAPP FOR ORAL PEER ASSESSMENT. Academic Journal Perspective: Education. *Language and Literature*, *9*(1), 37–49.

Xiao, H. (2022). Innovation of digital multimedia VR technology in music education curriculum in colleges and universities. *Scientific Programming*, *2022*(1), 6566144. DOI: 10.1155/2022/6566144

Yadav, S. (2024). Designing Effective E-Portfolios for Authentic Assessment: Encouraging Reflection and Promoting Student Ownership. In Cases on Authentic Assessment in Higher Education (pp. 235-258). IGI Global.

KEY TERMS AND DEFINITIONS

Assessment as Learning: Capturing growth via mobile artifacts (audio, screenshots, portfolios) to guide reflection and next steps.

Blended Pedagogy: Integrating mobile tools with face-to-face instruction so apps enhance (not replace) teacher-led learning.

Digital Equity: Ensuring all learners have devices, connectivity, and skills so mobile integration narrows rather than widens gaps.

Digital Wellbeing: Structures that manage distraction and foster healthy, sustainable mobile use in language learning.

Microlearning: Short, focused activities (e.g., voice notes, QR tasks) that fit real-life moments and reinforce class content.

Mobile-Assisted Language Learning (MALL): Using smartphones/tablets to extend language practice beyond class with authentic, anytime activities.

Teacher Orchestration: The teacher's role in curating tasks, pacing, and norms across in-class and mobile spaces.

Chapter 2
Incorporating Mobile Technology Into Pre-Service EFL Teacher Education:
Opportunities and Challenges in the Moroccan Context

Rachid El Yazidi
https://orcid.org/0009-0004-0671-2428
Moulay Ismail University, Morocco

ABSTRACT

The swift growth of mobile-assisted language learning (MALL) has transformed the horizons of language learning and teaching in various educational settings. In contrast, the incorporation of mobile technologies into teacher training, particularly in developing contexts, remains largely unstudied. This chapter looks at the prospects and obstacles in the use of mobile technologies for pre-service English as a Foreign Language (EFL) training in higher education institutions in Morocco. Grounded in a sociocultural and pedagogical framework, the chapter analyzes advanced mobile pedagogical practices and language teacher education to understand the impact of mobile technologies on the competencies and digital literacy frameworks of modern educators.

DOI: 10.4018/979-8-3373-3366-3.ch002

1. INTRODUCTION

In the preceding twenty years, the development and popularity of mobile technology have dramatically changed educational practices around the world, with language education being one of the most impacted areas. The change has given rise to Mobile-Assisted Language Learning (MALL), which focuses on portability, accessibility, and interaction in real-time, a subsection of Computer-Assisted Language Learning (CALL) (Viberg & Grönlund, 2013; Xiangming & Song, 2017; Habib et al., 2022; Moya & Camacho, 2024). Although MALL endorses constructivist and collaborative pedagogical frameworks, it is seldom put into practice. This is often attributed to rigid teaching practices and enduring technological hurdles, including limited access to mobile devices, expensive internet access, and inconsistent hardware and software standards (Burston, 2014).

Mobile technology has become an integral component of modern language education, reshaping how teachers design instruction and how learners interact with linguistic content. As smartphones and tablets grow increasingly ubiquitous among students, educators worldwide are leveraging these devices to enhance multiple aspects of language learning, including content delivery, learner collaboration, and independent practice (Kukulska-Hulme & Viberg, 2018; Karakaya & Bozkurt, 2022). Beyond their functional affordances, mobile technologies also carry important sociocultural implications, as they mediate learners' identities, communication practices, and access to learning communities across digital spaces (Luo & Watts, 2022). In this sense, mobile-assisted language learning (MALL) represents not merely a technological innovation but a pedagogical and cultural shift that encourages personalized, context-aware, and socially connected learning experiences. This evolution underscores the need to examine how such tools are adopted and adapted within specific educational contexts—such as Moroccan pre-service EFL teacher education—where local realities shape both the opportunities and constraints of mobile integration.

From their perspective, the engagement of learners with smartphones for English language learning is almost fully established in personal and social contexts through the integration of personal choice, social participants, and culture. The authors describe a new phenomenon called Smartphone-Assisted English Language Learning (SAELL), which denotes language acquisition as an unending process blending formal and informal opportunities to learn. This model particularly highlights the ability of smartphones to enable interaction on the learner's terms, which is fundamentally different from what educators design, and thus poses serious challenges for educators and policymakers who wish to create dynamic and flexible language learning systems.

Expanding the research on MALL's evolution, Karakaya and Bozkurt (2022) address five key components from contemporary MALL research: fostering self-regulated learning, increasing learner motivation, personalization of learning through AI-driven mobile applications, facilitating spontaneous learning from daily life, and the incorporation of MALL into higher education. Although these points illustrate the growing adaptability of mobile technologies within education, the authors point out that most research is still dominated by linguistic productivity with little focus on extralinguistic factors of mobile tools and learner interaction. They advocate for a more comprehensive plan of action aimed at the synthesis of various themes in order to contribute more meaningfully to the area of focus.

Research has shown how mobile technologies can facilitate the implementation of a wide variety of educational technology strategies in language teaching and learning, such as gamification, differentiated instruction, and flipped classrooms (Godwin-Jones, 2011; Sung et al., 2016; Zou et al., 2020). Mobile applications, including Duolingo, Quizlet, Kahoot, Google Classroom, and WhatsApp, have been utilized for vocabulary acquisition, communicative competence, and learner motivation enhancement (Azabdaftari & Mozaheb, 2012; Burston, 2015; Loewen et al., 2019; Zou et al., 2020; Neureiter et al., 2020; Bueno-Alastuey & Nemeth, 2020; Tarteer et al., 2021; Phanwiriyarat et al., 2025). Although these developments are hopeful, there are still many persistent barriers that restrict equitable adoption of mobile learning technologies. These include gaps in digital literacy, insufficient infrastructure, and inequality of opportunity, and challenges are most acute in developing and post-colonial settings (Traxler, 2018).

In this regard, Traxler (2018) provides a critical review of paradigm-shifting mobile learning frameworks built from the perspective of the Global South, contending that they are dominated by Northern epistemologies which tend to ignore the diverse sociocultural contexts of the Global South. He argues that mobile learning models tend to perpetuate international inequalities by imposing educational and technological frameworks that are alien to the local context. Further, the crossroads of ethnography, policy, linguistic evolution, and international advancement makes the effective localization of mobile learning very difficult. Therefore, it is crucial to redefine mobile learning in terms of culture, context, and level of sustainability, while addressing mechanisms of power and advocating for fundamental change in educational access.

1.1 Rationale for Focusing on Moroccan Pre-Service EFL Teacher Education

Even though there is a substantial and growing literature on MALL, there is a gap in the literature regarding the integration of mobile technologies in the pre-service teacher education curriculum in the Global South, and more specifically, North Africa. Because of Morocco's ambitious national plans to modernize higher education, integrating information and communication technologies (ICTs) into the curriculum, it serves as a relevant and timely case for study (Higher Council of Education, Training and Scientific Research, 2015; National Telecommunications Regulatory Agency [ANRT], 2023)

The CSEFRS, in its strategic plans, has pointed out the necessity of the modernization of teaching and learning processes through the application of ICT tools. These reforms aim to foster inclusivity in a digitally rich educational setting by granting educators and learners equal access to digital tools (Higher Council of Education, Training and Scientific Research, 2015). The "Strategic Vision of the Reform 2015-2030" focuses on building a fully refined educational system that ensures equity, quality, and enhances human development. The reform agenda highlights the need for equitable access to education, improved teaching standards, and alignment with the overarching development efforts of the nation.

Recent educational and technological reforms in Morocco have positioned the country to benefit significantly from mobile-assisted approaches to learning. According to the National Agency for Telecommunications Regulation (ANRT, 2023), access to information and communication technologies (ICTs) has expanded dramatically, with the majority of households now owning smartphones and maintaining regular mobile internet connections. This rapid digital inclusion reflects not only socioeconomic modernization but also the growing feasibility of integrating mobile tools into teaching and learning processes. Within higher education, English as a Foreign Language (EFL) programs have experienced parallel expansion—driven by globalization, academic mobility, and the increasing value of English for employability and research participation. The convergence of these trends suggests that enhanced ICT access can serve as a catalyst for innovation in EFL pedagogy. Specifically, the widespread availability of smartphones and mobile internet provides fertile ground for integrating mobile-assisted language learning (MALL) into pre-service teacher education, thereby aligning Morocco's educational reforms with global movements toward digitalized, learner-centered pedagogy.

Consequently, English is now the preferred medium of instruction (EMI) in multiple academic fields, driven by students' desires to partake in English-language academic literature, seek international postgraduate programs, and improve their standing in international employment markets. The shift from the French language to English represents a significant change that illustrates Morocco's attempt to adapt its higher education system to global scientific and professional standards (Belhiah & Abdelatif, 2016).

Yet, one cannot disentangle the adoption of EMI in Morocco from its sociohistorical and political context. R'boul (2022) points out that the implementation of

EMI occurs within a postcolonial setting which considers English to be a less invasive language compared to French. This shift also raises the issue of dependency, inequality, and subjugated frameworks. The use of EMI may further marginalize local languages and cultures, inadvertently reinforcing social stratification and creating exclusionary barriers to knowledge. There is a gap in Moroccan higher education regarding multilingualism and linguistic justice that requires critical reflection through an EMI lens.

Within Morocco's broader educational reform agenda, pre-service EFL teacher education programs are increasingly expected to prepare teachers who can integrate English proficiency with digital pedagogical competence. This expectation aligns with the national emphasis on educational modernization and digital transformation outlined in policy frameworks such as *Vision Stratégique de la Réforme 2015–2030* (Conseil Supérieur de l'Éducation, de la Formation et de la Recherche Scientifique [CSEFRS], 2015) and the *Digital Morocco 2025* strategy. Despite these ambitions, persistent challenges—such as inadequate digital infrastructure in teacher-training institutions, limited professional development in technology-enhanced pedagogy, and socio-economic inequalities among pre-service teachers—continue to constrain the use of mobile technologies in teacher education (Maghnouj et al., 2018; Issa, 2020; Qasserras et al., 2023). These conditions underscore the urgency of examining how mobile technologies are currently being integrated into Moroccan pre-service EFL programs and of developing contextually grounded instructional frameworks that promote equitable and sustainable digital pedagogy.

The use of mobile technologies enhances communication as well as access to resources during the preparation phase of teachers. Their effective incorporation into the education of teachers may encourage innovation in teaching methods as well as foster professional development. However, some factors such as digital distractions, lack of institutional infrastructure, absence of adequate training, and others pose barriers to achieving optimal utilization. Addressing such gaps, as Qasserras et al. (2023) noted, requires the development of institutional frameworks and training programs specifically designed to eliminate mobile technology pedagogy gaps in teacher education programs.

Inspired by these concerns, the purpose of this chapter is to examine the lived experiences, challenges, and opportunities of Moroccan pre-service EFL teachers and their educators in the context of mobile technology integration into training. With an emphasis on educational reform and digital transformation, this chapter fills a gap in research from the Global South focused on language education while providing region-specific perspectives on the mobile technology-language education nexus.

1.2 Research Questions and Objectives of the Chapter

The chapter is guided by three overarching research questions:

1. What is the current landscape of mobile technology use in Moroccan pre-service EFL teacher education programs?
2. Which mobile tools and practices are perceived as most effective in supporting pedagogical development, lesson planning, and reflective teaching?
3. What challenges do pre-service teachers and their instructors face in integrating mobile technologies into teacher education?

Correspondingly, the study pursues the following three research objectives:

1. To map the extent and nature of mobile technology integration in Moroccan pre-service EFL teacher education programs.
2. To identify and analyze effective mobile applications and pedagogical practices that enhance pre-service teachers' instructional design, assessment, and reflective practice.
3. To examine the key structural, pedagogical, and socio-technical challenges constraining the use of mobile technologies in teacher training, and to propose informed strategies for their integration into the curriculum.

2. THEORETICAL AND CONCEPTUAL FRAMEWORK

The integration of mobile technology into pre-service EFL teacher education in resource-constrained and postcolonial contexts such as Morocco necessitates a robust theoretical framework that bridges pedagogical innovation with sociocultural realities. Drawing on sociocultural theory (Vygotsky, 1978), this study foregrounds the role of mediated learning and contextualized teacher development. The Mobile-Assisted Language Learning (MALL) paradigm (Kukulska-Hulme & Shield, 2008) provides a lens through which mobile technologies are not merely tools but dynamic agents of learner autonomy and contextual engagement. Complementing this, the Technological Pedagogical Content Knowledge (TPACK) framework (Mishra & Koehler, 2006) offers a structured approach to understanding how digital competence intersects with pedagogical content knowledge in teacher preparation. These frameworks are further situated within the systemic challenges of educational reform in the Global South, where issues of equity, infrastructure, and postcolonial legacies shape the implementation of digital initiatives (Rocco & Plakhotnik, 2009). This integrated framework thus enables a nuanced exploration of how mobile technologies can be meaningfully embedded in teacher education programs that are both context-sensitive and future-oriented.

2.1 Sociocultural Theory and Digital Pedagogy

The sociocultural theory (Vygotsky, 1978) places emphasis on social interaction and culture as paramount drivers of the acquisition of advanced cognitive skills. According to Vygotsky (1978), cognitive processes do not happen in isolation, but rather in concert with the social and historical-cultural context within which a person is situated. At the center of this sociocultural paradigm is the Zone of Proximal Development ZPD, which refers to the gap between what a learner can do unaided and what they can do with the assistance of more capable peers. This framework describes socially grounded instruction and its significance, paying particular attention to the function of interaction and communication as methods of learning and marks of understanding (Vygotsky, 1978).

Within this perspective, learning a language goes beyond just a purely mental, intellectual activity. It also incorporates the culture, history, and relationships that are active and surrounding. Viewing mobile technologies as mediational tools, these new technologies possess the potential to increase the learner's ZPD through the ability to participate, provide feedback, and use language meaningfully within context (De Costa, 2007; Lantolf & Thorne, 2006; Warschauer & Liaw, 2011). Such tools act as conduits for scaffolding, which enable learners to progress from their actual competences to a higher level of capabilities through collaboration with peers and instructional support.

The ZPD, as defined by Vygotsky (1978), represents "the distance between the actual developmental level as determined by independent problem solving and the level of potential development as determined through problem solving under adult guidance or in collaboration with more capable peers" (p. 86). This notion can be beneficial in teaching languages, and in particular for English Language Learners (ELLs) and Multilingual Learners (MLLs) because it calls for instruction that is tailored and centered on the learners and their unique cognitive and linguistic assets. The ZPD is useful in identifying the ranges within which learners can act independently across the different activities in communication that they have been taught.

In this regard, mobile technologies have the greatest potential for helping learners to move along the ZPD. Their affordances, like communication in real time, scaffolding opportunities, and access to real-life linguistic environments, make learning active and participatory. The MALL framework studies show that these technologies increase collaboration, bringing in the possibility of individualized instruction as feedback, which leads to the development of learners within socially-mediated contexts (Li et al., 2022; Kukulska-Hulme & Shield, 2008). Kukulska-Hulme and

Shield (2008) highlighted the potential of mobile technologies for promoting learning-related activities because they enable students to participate in situated activities away from the classroom, which bridges sociocultural orientations to education. Xiangming and Song (2017) note this shift in MALL from providing static content to include interactivity and collaboration, using mobile devices for real-time and asynchronous oral language instruction.

While mobile and emerging technologies offer unprecedented opportunities for innovation in language education, their pedagogical value depends largely on how purposefully and critically they are integrated into instructional practice. As Kessler (2018) notes, tools such as social media platforms, artificial intelligence applications, big data analytics, and augmented reality open new avenues for interaction, personalization, and engagement. However, these affordances can easily lead to superficial implementation when technology is adopted for its novelty rather than for its pedagogical purpose. The main challenge lies in aligning technological use with clear instructional goals, ensuring that mobile tools serve as vehicles for meaningful, culturally responsive, and learner-centered education rather than ends in themselves. Thus, effective mobile integration in teacher education requires moving beyond enthusiasm for innovation toward pedagogical intentionality, critical reflection, and contextual adaptation.

From 2000 to 2012, Duman et al. (2015) analyzed several studies and illustrated the expansion and diversification of MALL during this time frame. The acquisition of vocabulary seemed to be the most researched area, with less attention given to grammar and writing. Moreover, many of these studies were poorly developed conceptually, although some studies applied learning theories like constructivism and sociocultural theory. The approaches more commonly used were applied and design-based, with some descriptive qualitative and experimental research. Most studies relied on mobile phones and Personal Digital Assistants (PDAs) as tools, concentrating on their pedagogical usability. However, the field of study continues to struggle with methodological rigor, lack of overarching theory, and inadequate empirical outcomes critical to pedagogical value.

Mobile technologies, in particular, support learners by providing them with authentic language experiences. It allows learners to interact with language resources in rich, contextual settings, including place-based activities and augmented reality applications, as cited by Yudintseva (2023). Such contextualization enhances cognitive processing and supports the internalization of linguistic structures. Unlike traditional classroom learning which is often abstract, mobile technologies offer learners greater relevance and meaning for language acquisition.

Virtual Reality (VR) technologies offer immersive and interactive environments that stand to greatly benefit oral communication in ESL classrooms. The real-time feedback, content co-creation, cultural exposure, and other aspects of VR make it easier to motivate and boost the confidence of learners, making them more willing to communicate. Still, empirical data regarding its impact on oral fluency is inconclusive. Other factors such as cognitive overload, virtue of limited hardware, motion sickness, and lack of alignment with pedagogical tasks may restrain effectiveness. Integrating VR in instruction poses challenges that advancing the instructional design will solve by aligning technological affordances and communicative learning goals (Yudintseva, 2023).

In addition, educator feedback, language learning applications, and even AI-driven writing aids transform into mobile-assisted language learning tools that sustain students' active participation, foster metalinguistic consciousness, and facilitate learner autonomy. Godwin-Jones (2011) draws attention to the advancement of self-directed learning that mobile technologies afford, as learners can assess and modify their language use on the spot. Stockwell and Hubbard (2013) endorse the use of mobile technology for language instruction, as it promotes level-appropriate practice, adaptive learning, and individualized feedback.

The sociocultural theory provides a context-specific and comprehensive framework for the use of mobile technology in language education. It highlights the importance of social interaction as well as cultural mediation and responsive scaffolding that aid in learner development. This theory has particular relevance in education, like in the case of Moroccan teacher education programs, where lectures remain the primary mode of instruction. Using mobile technologies within a sociocultural approach can change the use of these technologies in language teaching, making it more collaborative, learner-centered, and integrated with the context.

2.2 Mobile Assisted Language Learning (MALL) Frameworks

The evolution of MALL marks a new shift in mobile-enabled language learning, integrating mobile technologies like smartphones, tablets, and handheld computers into both formal and informal learning environments (Kukulska-Hulme & Shield, 2008; Burston, 2014; Karakaya & Bozkurt, 2022). MALL is particularly noted for the shifts it brings towards situated, personalized, and ubiquitous learning, principles that are in tandem with learner-centered pedagogies, allied towards fostering lifelong learning. Burchard and Burston (2015) remark that, unlike traditional classroom settings where boundaries are encountered, mobile technologies nurture learning engagement across time and space through what is termed as "ubiquitous learning." MALL promotes situated learning by having learners engage with authentic linguistic environments which increases the applicability and relevance of English language use (Stockwell & Hubbard, 2013; Reinders & White, 2010). Moreover, the ease of mobile-based systems promises the greatest flexibility for personalized learning by adapting tasks to the learner, thus fostering self-directed learning (Kinshuk,

2016). These insights highlight the impact that MALL framework has on mobility's versatility in modern language education as well as the pedagogical shift towards flexible, learner-centered paradigms.

Stockwell and Hubbard (2013) provide some of the most important theoretical contributions toward MALL by creating a conceptual framework with three interrelated dimensions: learner mobility, device portability, and personalization of learning. Mobility of the learner, which includes physical movement between sociocognitive spaces (environments) and task thinking transitions, is quite prominent in settings like Moroccan higher education, where pre-service teachers seem to shuttle around university courses, teaching practice, and self-study. There is evidence suggesting that mobile technologies enhance the ease with which this type of learning can be accomplished by providing uninterrupted access to instructional materials throughout different environments, which in turn supports retention and application of knowledge (Traxler & Kukulska-Hulme, 2015).

Portability of devices, the second dimension, increases accessibility and spontaneity of learning as users can access educational materials and engage with them irrespective of location (Moya & Camacho, 2024). This heightened mobility permits greater immediacy, ubiquity, and interactivity, which in turn support essential educational affordances such as feedback, collaboration, and learning situated in context. These functionalities enable active engagement and interaction with peers and are aligned with socially constructivist educational frameworks and pedagogical approaches (Moya & Camacho, 2024).

In the third dimension of the personalization of learning, the personalization of learning as a custom fit supports differentiated instruction. It helps educators address learners' varied needs, along with preferences and pacing. As Kinshuk (2016) and Katona and Gyonyoru (2025) argue, adaptive learning technologies can customize instructional content to optimize engagement and understanding. This level of flexibility is especially important for pre-service teachers who are required to fulfill academic work as well as teach during practicum placements. All these aspects demonstrate how MALL frameworks are highly attentive to address their attention to higher education, particularly in cases where flexibility and customization are prioritized.

An extra layer of sophistication within MALL frameworks is related to two concepts, namely context awareness and seamless learning, which focus on transcending learning experiences across different devices, platforms, and environments (Wong & Looi, 2011). Context awareness refers to how mobile learning environments can automatically adjust to a learner's physical, temporal, social, and technological context (Hwang & Tsai, 2011). This means mobile platforms that modify content because of constraints like location, time, activities of the learner, and personal choices. Such responsiveness promotes situated learning through fostering learners to engage with linguistic content in real-time and in authentic contexts. For example, location-based tasks, like identifying objects in one's environment, asking or giving directions, or describing landmarks (Ogata & Yano, 2004; Ogata et al., 2005; Bachmair & Pachler, 2015).

These adaptive skills are reinforced by the ubiquity of mobile devices and gadgets which enable learners to retrieve information at their convenience (Bachmair & Pachler, 2015). This level of flexibility is advantageous for developing core language skills such as vocabulary and listening comprehension and just-in-time learning, particularly in multilingual or immersive learning environments (Burston, 2014). Moreover, context-aware systems provide additional possibilities for real-time feedback and interaction, personalized scaffolds, offering responsiveness aligned with constructivist and sociocultural perspectives of second language learning (Lantolf & Thorne, 2006).

The concept of seamless learning developed by Wong and Looi (2011) can further augment context-awareness, capturing the integration of learning activities across the physical, digital, and social realms as continuous. In a MALL environment, learners could start a task in class, continue in transit using a smartphone, and complete it at home on a tablet. This scenario demonstrates how learners can maintain cognitive flow across contexts, which enhances synthesis and engagement. This model of learning across contexts and devices increases learner autonomy and enhances opportunities for reflection and sustained practice (Sharples et al., 2010).

Empirical studies highlight that motivation, engagement, and retention are improved when learning is done seamlessly, particularly when the tasks are collaborative and assisted by cloud-synchronized tools (Wong & Looi, 2011). For learners of languages, this means being able to access rich multimedia content such as videos and podcasts, receive and produce outputs such as voice recordings and written responses, in addition to providing and receiving formative feedback, all in a continuous and unrestricted cycle (Godwin-Jones, 2011). Such experiences help learners to adopt a greater responsibility towards their educational paths whilst encouraging them to shape a more personalized and engaging learning experience.

As noted, the combination of context-aware and seamless learning principles into MALL marks an epistemological change towards ecological views of language learning. Such views conceptualize learning as distributed, mobile, and emergent through the interaction with people, tools, and environments (Rinda, 2019; Novawan et al., 2019) and therefore challenging language acquisition models confined to classrooms. Instead, learning is described as being socially situated, technologically mediated, and fluid (Kramsch & Vork Steffensen, 2008).

Given the context of Morocco, where differences in the availability of digital resources and institutional support are prevalent, the use of a context-aware MALL approach offers a more effective and adaptable solution for improving language teacher education.

Even with persisting infrastructure and digital literacy issues, the Moroccan Strategic Vision 2015–2030 promotes the use of digital resources to eliminate disparities in education and enhance teaching training (Moumen, 2024). In this context, digital inclusion programs specifically sought to increase technological access and foster a culture of IT competence among educators (Bakkali et al., 2023). In addition, the overall digitalization of higher education in Morocco has been presented as an auxiliary system designed to foster accessibility, facilitate the tailoring of educational pathways, and enable ongoing career development for aspiring teachers (Maghni & El Khannouss, 2024). With these changes in mind, MALL creates new opportunities for developing adaptive and self-directed systems that are responsive to the chronic educational needs of different learners and educators, thereby ensuring ongoing professional development.

2.3 Teacher Digital Competence and the TPACK Model

As with every educational level, mobile technologies need to be integrated into a teacher's education seamlessly. Mishra and Koehler (2006) proposed the Technological Pedagogical Content Knowledge (TPACK) model, which seeks to explain the interrelationship among technology, pedagogy, and content expertise. According to this framework, effective teaching with technology requires an integrated understanding of three interdependent domains of knowledge. The first is Content Knowledge (CK), which refers to a teacher's mastery of the subject matter to be taught. The second is Pedagogical Knowledge (PK), encompassing the methods, strategies, and principles of effective teaching and learning. The third is Technological Knowledge (TK), which involves familiarity with digital tools and platforms, including their affordances and constraints. The TPACK model emphasizes that meaningful technology integration in education depends on teachers' ability to synthesize these domains rather than treat them as separate or isolated areas of expertise.

The model also places emphasis on TPACK as an amalgamated form of knowledge teachers possess, which is necessary for developing and executing meaningful lessons that integrate technology (Koehler et al., 2017).

In the TPACK framework, as with most concepts regarding technology instruction in pre-service teacher education, enhancing TPACK goes beyond mere technical training; it entails the skillful analysis and application of technology in relation to specific pedagogical goals and the dynamic needs of learners. Similar studies have shown that when pre-service teachers are trained in TPACK, their self-perception and actual level of competence in employing mobile technologies in language classrooms increase (Lehtinen et al., 2016; Valtonen et al., 2017; Ghomi & Redecker, 2019). In the Moroccan context, the inclusion of the TPACK model into the teacher training curriculum enhances the framework for developing digital competence in teachers from a pedagogical perspective while ensuring contextual relevance (Bouzidi, 2024).

The TPACK framework, which integrates technological, pedagogical, and content knowledge, has been regarded as a useful model for training teachers to use technologies in their practices (Mishra & Koehler, 2006). Studies of teacher education in Morocco note the lack of digitally focused training exercises as a deficiency and consider the lack of infrastructure and institutional support as barriers to technology use (Benali et al., 2018). By enhancing adaptive teaching methods and digital literacy, a context-aware implementation of TPACK guarantees that pre-service teachers receive focused instruction that aligns with regional educational issues (Rodríguez Moreno et al., 2019). Also, some studies suggest that teachers who participate in TPACK-based training programs display greater self-efficacy regarding the use of educational technologies and apply them more actively in teaching lessons (Oved & Alt, 2025). The incorporation of TPACK into teacher training programs can help Moroccan institutions address the disparity between theoretical education and practical digital skills, thus improving the quality of education in developing technological contexts.

2.4 Relevance to Teacher Education in Developing and Postcolonial Contexts

With respect to mobile technologies, the history, economy, and sociocultural factors of the region still make the reality of education possible in postcolonial and developing mobile contexts. In such under-resourced environments where infrastructure, digital access, and languages are hierarchically dominated, a universal, overlooking approach to technology is unsustainable (Selwyn, 2011; Unwin, 2009). This has prompted the more socially responsive model of using technology in education, which is anchored on specific contexts.

As for communications and information technologies for development, Unwin (2009) contends that equipping economically backward areas with digital resources in education, healthcare, governance, and enterprises has to be context-specific if it is to be sustainable. While the promise of addressing educational access and social empowerment inequalities through ICTs exists, the lack of adequate design, community participation, and addressing fundamental access barriers often stifles change in some African, Asian, and Latin American countries. The provision of technological tools, as Unwin (2009) argues, should be coupled with socio-political frameworks aimed at equitable governance, socio-political commitment, and long-term viability if the projects are to succeed.

In attention to Selwyn's work (2011), aside from discussing the impact of technology integration in teaching and learning, there is an important attention given towards the transformative features that claim educational technology offers. Selwyn (2011) elaborates on the relationship between the use of the various forms of digital tools and the institutional, pedagogical, and socio-economic structures that mediate their use in practice. Some of the major arguments include the impact of technology on teaching in terms of reinforcing or displacing operational roles of the institution, and whether it tends to alleviate or increase inequality in education. A critical approach is propounded by Selwyn (2011), who underscores local agency by proposing a critical sociotechnical framework within which actors are situated alongside the socio-material relations of technologies.

From a postcolonial perspective, Canagarajah (1999) warns against the indiscriminate application of Western teaching methods and information technologies in the Global South. He advocates for the indigenization of educational documents pertaining to the epistemic communities, linguistic environments, and sociocultural developmental frameworks of the relevant communities. These issues are particularly pronounced in Morocco, where the associative colonial impact and globalization influence pluralistic linguistic elements. The dominance of French and Arabic at institutional levels, coupled with the more recent expansion of English in higher education and global business, embodies both historical struggles and hopes for economic development (Marley, 2004; Zouhir, 2013; Kabel, 2021). The rich, yet problematic, phenomena of multilingualism give rise to issues of linguistic disparity and inequality of access.

In this context, MALL has emerged as a possibly inclusive approach due to its flexible and context-sensitive tools that aid language learning in various sociolinguistic contexts (Karakaya & Bozkurt, 2022). Nonetheless, MALL tools must be designed and developed in a socioculturally appropriate way so as not to exacerbate differences. Chakrani (2020) highlights the need for an effective and inclusive multicultural approach to be embedded in mobile language applications alongside multiple languages. Therefore, within the context of Morocco's changing educational system, mobile technologies offer possibilities not only as teaching aids, but also as tools for language empowerment—if informed by the local context.

Resta and Laferrière (2015) have noted the lack of equitable access to education in marginalized and rural regions, deepening the divide that mobile technologies can fill, especially when inclusivity frameworks are applied. Digital equity and intercultural education, as they point out, are essential elements of knowledge societies but are persistently undermined by deep-rooted barriers such as inadequate digital infrastructure, limited access to trained teachers, and insufficient nurturing of digital skills. There has been some progress made towards global mobile connectivity; however, large portions of the population remain excluded from the digital world, widening knowledge divides and obstructing intercultural understanding. The authors advocate for active inclusion and relevance in curriculum content, urging educational technology policies to promote culturally responsive and equitable participation within digital learning ecosystems.

Regarding the Moroccan context of MALL in pre-service teacher training, mobile technologies can enable learner-centered pedagogies, professional participation, and mitigate institutional barriers. Such benefits, however, depend on efficient resource allocation, planning, appropriate infrastructure, ample materials, and pedagogical training. As Qasserras et al. (2023) noted, Moroccan pre-service teachers face serious challenges in implementing MALL due to insufficient digital infrastructure and inadequate institutional support. However, mobile tools allow for multi-faceted teaching activities that encourage autonomy, student participation, and differentiated learning (Kuru Gönen & Zeybek, 2022).

Additionally, mobile integration fosters professional growth by linking Moroccan teachers to international communities, providing new instructional methods and global insights (Mountasser, 2023). Nonetheless, Lablidi et al. (2013) argue that the advantages of MALL for teacher education are an assumption that cannot be taken for granted. Without a defined policy framework that focuses on accessibility, culturally situated instructional design, and specific MALL teacher preparation programs, MALL may unintentionally worsen disparities instead of serving as a catalyst for revolutionary educational change.

Equity, contextualization and culturally responsive teaching must underpin the incorporation of mobile technologies into the teacher education curriculum in developing and postcolonial countries. For Morocco, this means an interplay of multilingualism with historical and infrastructural constraints. Mobile technologies, when implemented thoughtfully, can significantly enhance language teaching and learning, promote professional development, foster educational inclusivity, and support local educational initiatives as well as global educational objectives.

3. LITERATURE REVIEW

The present section provides an overview of Mobile-Assisted Language Learning (MALL) as the central construct of this study and reviews relevant empirical works examining its integration into English as a Foreign Language (EFL) education and pre-service teacher training. The review aims to outline the theoretical development of MALL, highlight its pedagogical affordances and limitations, and synthesize empirical findings to identify the specific research gap addressed by the present study.

3.1 The Concept and Evolution of Mobile-Assisted Language Learning (MALL)

Mobile-Assisted Language Learning (MALL) emerged in the early 2000s as an extension of Computer-Assisted Language Learning (CALL), reflecting a paradigm shift toward flexibility, learner mobility, and ubiquitous access to learning environments (Kukulska-Hulme & Shield, 2008; Burston, 2014). The rapid proliferation of smartphones, tablets, and handheld devices transformed traditional language education by extending learning beyond the classroom into authentic and contextualized environments (Stockwell & Hubbard, 2013; Reinders & White, 2010). The foundational theories underlying MALL, including sociocultural and constructivist perspectives, emphasize that mobile technologies can act as mediational tools that promote interaction, collaboration, and experiential learning (Kukulska-Hulme, 2013; Lantolf & Thorne, 2006).

MALL has been characterized by several key features that distinguish it from earlier forms of digital language learning. It allows for mobility, both physical and cognitive, enabling learners to access instructional materials anytime and anywhere (Traxler & Kukulska-Hulme, 2015). It also emphasizes portability, ensuring that learning resources are available across devices and contexts (Moya & Camacho, 2024). Furthermore, MALL supports personalization, offering adaptive content that caters to individual learners' needs, pace, and preferences (Kinshuk, 2016). Through these characteristics, MALL fosters autonomous, context-aware, and seamless learning experiences, where learners move fluidly between classroom and real-world interactions (Wong & Looi, 2011; Bachmair & Pachler, 2015).

Despite these strengths, MALL is not without limitations. Challenges such as insufficient theoretical grounding, infrastructural constraints, and pedagogical misalignment remain prevalent, particularly in developing contexts (Duman et al., 2015; Resta & Laferrière, 2015). While MALL can enhance learner engagement and motivation, its success depends heavily on educators' ability to integrate mobile tools purposefully into instruction rather than using them as ends in themselves (Kessler, 2018). Furthermore, Traxler (2018) and Selwyn (2011) warn that uncritical adoption of Northern-centric technological paradigms may reproduce inequalities rather than mitigate them, particularly in postcolonial settings with uneven access to digital infrastructure.

Overall, the literature on MALL highlights a tension between its pedagogical potential and the structural barriers that shape its implementation. The evolution of MALL from isolated, content-delivery models to more interactive, collaborative, and context-aware frameworks signals an important shift toward sustainable, learner-centered pedagogies aligned with sociocultural realities. This conceptual understanding provides the foundation for examining its application within Moroccan pre-service EFL teacher education.

3.2 Empirical Studies on MALL in EFL and Teacher Education Contexts

Empirical research on MALL has largely focused on its impact on language skill development, learner motivation, and teacher professional growth. In EFL settings, numerous studies have demonstrated the effectiveness of mobile applications in enhancing vocabulary acquisition, listening comprehension, and communicative competence (Lu, 2008; Mahdi, 2017; Soleimani et al., 2022). For instance, interactive applications offering repeated exposure to new vocabulary across diverse contexts significantly improve retention (Basal et al., 2016), while mobile-based audio tools have shown measurable gains in listening proficiency and learner autonomy (Basaran & Cabaroglu, 2014; Andujar & Hussein, 2019). Similarly, gamified mobile learning platforms that integrate feedback, rankings, and rewards have been found to sustain learners' motivation and engagement (Yang & Gottlieb, 2023; Ding & Zhu, 2025).

Beyond learner outcomes, a growing body of research examines the role of mobile technologies in teacher education. Kearney and Maher (2019) highlight that mobile tools facilitate contextualized, reflective learning and help pre-service teachers develop professional competencies through digital storytelling, microteaching, and online collaboration. Mobile platforms also enable pre-service teachers to document and reflect on their instructional practices, thereby fostering metacognitive awareness and continuous professional growth (Leinonen et al., 2014). These findings affirm that MALL not only enhances language proficiency but also provides a practical avenue for developing the digital and pedagogical skills required for 21st-century teaching.

However, empirical evidence also reveals persistent contextual and infrastructural challenges. Limited access to devices, weak Internet connectivity, and insufficient institutional support continue to constrain MALL integration in developing and low-resource contexts (Burston, 2014; Unwin, 2009; Dolezal et al., 2025). In sub-Saharan Africa and North Africa, such issues are compounded by socioeconomic disparities and uneven digital literacy among educators (Cappelli & Akkari, 2025). These conditions often result in fragmented or superficial use of mobile tools rather than systematic pedagogical integration.

Specifically regarding Morocco and North Africa, the existing literature remains sparse. While recent studies have examined the use of ICTs and digital technologies in Moroccan higher education (Belhiah & Abdelatif, 2016; Chakrani, 2020; Rhaya & El Ouazzani, 2025; Maghni & El Khannouss, 2024; Fadli & El Mediouni, 2024), few have directly investigated MALL within pre-service teacher education (Qasserras et al., 2023; Amaaz et al., 2024). This gap is striking given Morocco's ongoing digital transformation and national efforts to integrate mobile technologies into education through frameworks such as the Strategic Vision 2015–2030 and the ESRI 2030 Pact (Rhaya and El Ouazzani, 2025; Moumen, 2024). Despite these initiatives, studies continue to report deficiencies in digital competence, infrastructural readiness, and the pedagogical use of mobile tools in teacher training programs (Benali et al., 2018; Mountasser, 2023).

3.3 Research Gap and Rationale for the Present Study

In sum, while global research provides substantial evidence on the educational affordances of MALL, the contextual realities of developing and postcolonial nations, particularly Morocco, remain underexplored. The reviewed studies demonstrate that although MALL fosters learner autonomy, engagement, and professional growth, its application in pre-service EFL teacher education has received limited empirical attention in North African settings. The scarcity of research examining the intersection between digital competence, pedagogical design, and mobile technology use in Moroccan teacher education justifies the need for a context-specific inquiry. Accordingly, this study aims to fill this gap by investigating how mobile technologies are integrated, experienced, and conceptualized within Moroccan pre-service EFL teacher training programs.

4. METHODOLOGY

This section details the methodological framework adopted to explore the integration of mobile technologies into pre-service EFL teacher education programs in Morocco. It explains what the study investigated, why a qualitative design was selected, and how the research was conducted. The overarching aim was to generate an in-depth understanding of how pre-service teachers and teacher educators perceive, adopt, and experience mobile-assisted language learning (MALL) within Moroccan higher education institutions.

A qualitative case study approach was employed because it enables the researcher to explore contemporary educational phenomena within their real-life contexts (Yin, 2018; Merriam & Tisdell, 2016). This design aligns with the exploratory nature of the study's objectives, which seek to capture complex interactions between technology, pedagogy, and institutional practices. The study focused on three public Moroccan universities offering pre-service EFL teacher education programs, using semi-structured interviews, classroom observations, and reflective field notes to collect data. These tools were selected to allow for methodological triangulation and to ensure a rich, multi-perspective account of the lived experiences of participants.

In this section, the methodological process is organized into four parts:

(1) the research design, which justifies the qualitative case study approach;
(2) the context and participants, describing the institutional and demographic characteristics;
(3) the data collection methods, which include interviews and classroom observations; and
(4) the ethical considerations and data analysis approach, detailing how data were handled, coded, and interpreted.

4.1 Research Design: Qualitative Case Study

This study adopts a qualitative case study design to investigate how mobile-assisted language learning (MALL) is integrated into pre-service EFL teacher education in Moroccan higher education institutions. The "case" in this research is defined as the bounded system (Yin, 2018; Stake, 1995) comprising three Moroccan public universities that offer pre-service EFL teacher education programs between 2023 and 2024. The case is delimited by three core parameters:

(1) the institutional context, including infrastructure, curriculum, and technological resources;
(2) the participants, who are pre-service teachers and their instructors engaged in MALL-based activities; and
(3) the temporal frame, which captures practices and perceptions within one academic year.

The choice of a case study design is particularly suitable for this research because it allows for a holistic, in-depth examination of a contemporary phenomenon namely the integration of mobile technologies in language teacher education within its authentic educational environment (Baxter & Jack, 2015; Merriam & Tisdell, 2016; Denzin & Lincoln, 2017; Yin, 2018. Creswell & Poth, 2018). This design enables the exploration of "how" and "why" questions central to this study: how mobile tools are adopted and used pedagogically, and why certain institutional and socio-technical factors facilitate or hinder their implementation.

Unlike ethnography, which primarily seeks to describe cultural groups through prolonged immersion, this study does not aim to construct a cultural portrait but to analyze a bounded educational system where MALL practices intersect with pedagogical innovation. Similarly, it differs from action research, which involves researcher-led intervention and cyclical change. This research, by contrast, is observational and exploratory, seeking to interpret existing practices rather than to implement or evaluate interventions.

Therefore, the qualitative case study provides a flexible yet rigorous structure for examining the dynamic relationship between technological affordances, pedagogical practices, and institutional realities within Moroccan teacher education. It enables thick description, interpretive analysis, and contextual understanding—all of which are essential for capturing the complexity of mobile learning in evolving educational systems.

4.2 Context and Participants

This study was conducted across three Moroccan public higher education institutions that house faculties of education and language studies offering pre-service EFL teacher training programs. These institutions were selected through purposive sampling to ensure representation of Morocco's major geographical regions (northern, central, and southern) and to capture variation in technological infrastructure and institutional culture (Creswell & Poth, 2018; Ahmad & Wilkins, 2024). Public institutions were prioritized because they serve the majority of Morocco's pre-service teacher population and are directly affected by national policies on educational reform and digitalization (MENPS, 2023). Private institutions were excluded since their training frameworks, funding models, and access to technological resources differ significantly, which could compromise the comparability of data.

The case study involved a total of 24 participants, comprising 21 pre-service EFL teachers in their final year of training and three teacher educators who specialized in English language pedagogy and had experience integrating ICT into instruction. Participants were selected based on their active engagement with mobile technologies for pedagogical or learning purposes within their coursework or practicum experience. This inclusion criterion ensured that all participants had relevant, first-hand exposure to MALL-based activities.

Additional selection criteria included:

- Enrollment in an accredited pre-service EFL teacher training program;
- Availability and willingness to participate in interviews and classroom observations;
- Basic digital literacy enabling the use of mobile learning platforms.

Diversity within the sample was also considered to capture variations in perspectives across age, gender, regional background, and ICT familiarity. Table 1 below provides a demographic summary of the participants.

Table 1. Summary of participant demographics

Participant Group	N	Gender (M/F)	Age Range	ICT Familiarity	Institutional Region
Pre-service EFL Teachers	21	8M / 13F	21–27	Moderate to High	North, Center, South
Teacher Educators	3	2M / 1F	35–54	High	North, Center, South
Total	**24**	**10M / 14F**	—	—	—

All participants voluntarily consented to take part in the study after being informed of its aims, procedures, and confidentiality safeguards. Participation was entirely voluntary, and participants were assured that they could withdraw at any point without repercussions.

This diverse sample enabled the study to capture a nuanced understanding of mobile technology integration in Moroccan teacher education, reflecting not only institutional diversity but also the interplay of demographic and technological factors shaping pre-service teachers' learning experiences.

4.3 Data Collection Methods

To ensure methodological triangulation and depth of understanding (Sciberras & Dingli, 2023), this study employed two primary qualitative data collection methods: semi-structured interviews and classroom observations. These methods were selected to capture both participants' articulated experiences and observable classroom practices, allowing for a nuanced exploration of how mobile technologies were integrated into pre-service EFL teacher education.

4.3.1 Semi-Structured Interviews

Semi-structured interviews were conducted with both pre-service teachers (n = 21) and teacher educators (n = 3) to explore their perceptions, experiences, and practices related to Mobile-Assisted Language Learning (MALL). This method was chosen for its capacity to elicit rich, reflective, and contextually grounded data that reveal how participants make sense of technological integration in their pedagogical development (Adams, 2015).

Each interview lasted between 25 and 45 minutes and was conducted face-to-face in the respective institutions. However, three interviews were conducted online via Google Meet due to participants' scheduling constraints. All interviews were audio-recorded with participants' consent and subsequently transcribed verbatim.

The interview protocol, developed based on existing MALL literature (Burston, 2014; Kukulska-Hulme & Viberg, 2018; Kukulska-Hulme & Shield, 2008), focused on five thematic areas: participants' background and digital literacy, current use and perceptions of mobile technology, pedagogical development and reflective practice, challenges and barriers, and future recommendations. The full interview protocol is provided in the Appendix.

This instrument's semi-structured nature enabled flexibility in following participants' narratives while ensuring coverage of key areas aligned with the study's objectives. Probing questions were used to clarify or deepen responses, and iterative adjustments were made to the interview sequence based on emerging insights from early interviews.

4.3.2 Classroom Observations

Classroom observations were conducted across the three participating institutions over a period of two months, spanning March to May 2024. Each pre-service teacher was observed at least twice, resulting in a total of 42 classroom observation sessions and approximately 84 hours of recorded observation data. These sessions included teaching methodology classes, practicum activities, and collaborative projects involving mobile applications such as Kahoot, Padlet, Quizlet, WhatsApp, and Google Classroom.

The observation sampling was guided by the principle of maximum variation, ensuring that diverse instructional settings, technological tools, and teaching stages were represented. An observation protocol was used to document patterns of mobile device utilization, teacher-student interaction, peer collaboration, and student engagement.

To supplement observation data, field notes were taken following Spradley's (1980) ethnographic approach, focusing on real-time and contextual dynamics, including non-verbal cues, classroom atmosphere, and the sociomaterial context of mobile use. These notes provided critical contextual information for triangulating interview data and capturing discrepancies between self-reported and enacted practices.

4.4 Ethical Considerations and Data Analysis Approach

After describing the study to the participants, they were assured of confidentiality, anonymity, and the right to withdraw at any time without any repercussions. Participant codes were assigned to all data sets, and all materials were anonymized and securely stored on password-protected devices accessible only to the researcher.

Although formal Institutional Review Board (IRB) procedures are not mandated in Moroccan universities, the researcher obtained oral ethical approval from the heads of the participating departments and secured informed consent from all participants prior to data collection. These procedures align with international ethical standards for qualitative educational research.

Data were analyzed thematically following Braun and Clarke's (2019) six-phase framework: (1) familiarization with the data, (2) generating initial codes, (3) searching for themes, (4) reviewing themes, (5) defining and naming themes, and (6) producing the final report. Thematic analysis was chosen for its flexibility and capacity to capture nuanced patterns of meaning across interviews and classroom observations.

An inductive–deductive hybrid coding approach was used. Inductive codes emerged directly from participants' narratives (e.g., *"using mobile apps for reflective learning," "peer collaboration through WhatsApp," "limited institutional Wi-Fi"*), while deductive codes were derived from the literature and research objectives (e.g., *"digital literacy," "curriculum alignment," "technological scaffolding"*). These codes were then grouped into broader themes corresponding to the study's research questions:

- Theme 1: Pedagogical Opportunities and Best Practices (RQ1 and RQ2)
- Theme 2: Structural and Pedagogical Barriers to Mobile Integration (RQ3)
- Theme 3: Transformative Learning and Teacher Identity Development (RQ1 and RQ2).

Analytic verification was conducted by cross-checking interview transcripts and observation notes to identify recurring patterns and contradictions. This process ensured that findings remained empirically grounded while theoretically informed by sociocultural and MALL frameworks.

To enhance the trustworthiness of the analysis (Ahmed, 2024), the study incorporated member checking, peer debriefing, and data triangulation. Member checking allowed participants to review and validate their interview transcripts and emerging interpretations (Kullman & Chudyk, 2025), while peer debriefing with external researchers helped refine coding schemes and thematic coherence. Using multiple data sources strengthened the credibility and confirmability of the findings (Lincoln & Guba, 1985).

When it comes to the researcher's reflexivity and positionality, as the principal investigator, the researcher occupied an "informed insider" position (being a Moroccan university educator with extensive experience in EFL teacher education and digital pedagogy). This dual role provided a nuanced understanding of institutional realities but also required conscious reflexivity to avoid interpretive bias. To mitigate potential subjectivity, the researcher maintained a reflexive journal throughout data collection and analysis, noting assumptions, emotional responses, and interpretive choices. Regular discussions with peer reviewers further ensured analytic transparency and critical distance from the data.

5. FINDINGS AND DISCUSSION

This section presents the findings of the study, derived from semi-structured interviews and classroom observations with 21 pre-service EFL teachers and three teacher educators across three Moroccan higher education institutions. Following Braun and Clarke's (2019) six-step thematic analysis, both inductive and deductive approaches were employed to develop themes that emerged from the data and theoretical frameworks discussed earlier.

Three interrelated thematic clusters were identified:

1. **Opportunities and Best Practices**,
2. **Challenges of Integration**, and
3. **Case Narratives and Voices of the Participants.**

These themes collectively address the three guiding research questions:

- **RQ1:** What is the current landscape of mobile technology use in Moroccan pre-service EFL teacher education programs?
- **RQ2:** Which mobile tools and practices are perceived as most effective in supporting pedagogical development, lesson planning, and reflective teaching?
- **RQ3:** What challenges do pre-service teachers and their instructors face in integrating mobile technologies into teacher education?

The thematic map figure illustrates the alignment between these clusters and the research questions.

Figure 1. Thematic map linking research questions and emergent themes

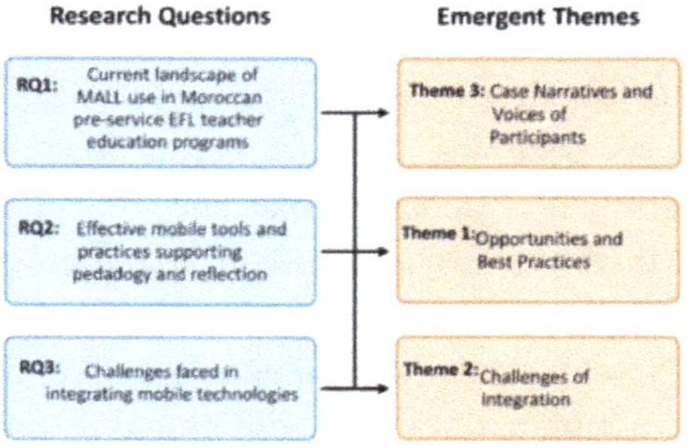

Together, these themes illustrate how opportunity and constraint interact dynamically in the development of mobile-assisted EFL teacher education in Morocco.

5.1 Opportunities and Best Practices

Effective Mobile Applications

From the participants' perspective, several mobile applications were identified as instrumental in enhancing interactivity and promoting a more student-centered approach to language teaching and learning. These tools were consistently praised for their availability, user-friendliness, and pedagogical versatility, particularly in fostering engagement, collaboration, and creativity (Tarteer et al., 2021; Bueno-Alastuey & Nemeth, 2020; Gorbunova et al., 2025; Filiz et al., 2025; Neureiter et al., 2020; Albarqi, 2023). Building on this overall perception, the following examples illustrate how specific mobile applications were applied in practice and the distinct pedagogical affordances each offered in the Moroccan pre-service EFL teacher education context:

1. **Google Classroom** was widely used as a communication and management tool, facilitating writing activities, task assignment, feedback exchange, and blended learning that extended beyond the physical classroom (Tarteer et al., 2021; Pham & Nguyen, 2024).
2. **Quizlet** was reported to sustain learners' attention and motivation through gamified vocabulary and grammar activities that turned remediation into an engaging experience (Bueno-Alastuey & Nemeth, 2020).
3. **Canva for Education** contributed to developing pre-service teachers' digital literacy and creativity as they designed infographics and lesson plans (Gorbunova et al., 2025; Durgungoz & Durgungoz, 2025).
4. **Padlet** facilitated brainstorming, collaborative writing, lecture recording, and other interactive tasks that align with 21st-century pedagogical objectives (Durgungoz & Durgungoz, 2025).

Lesson Planning, Assessment, and Learner Engagement

Mobile tools went beyond being simple delivery devices to serve as scaffolds for planning lessons, formative assessment, and learner engagement. According to the reports of pre-service teachers, mobile application software was used to create differentiated lesson plans, interactive quizzes, and real-time student monitoring systems. Observations conducted in the classroom noted greater student participation and attention during MALL-based activities compared to traditional methods.

This supports what Hampel (2009) and Stockwell and Hubbard (2013) claim about appropriately incorporated mobile technologies into instruction: integrated mobile technologies increase participation with the content and encourage more flexible ways of teaching.

Fostering Collaboration and Autonomy

Particularly noticeable was the way MALL supported collaborative learning and autonomy. Peer collaboration on lesson planning and microteaching was possible through WhatsApp groups, Edmodo threads, and Google Docs. Participants reported greater ownership and initiative regarding their learning processes, which resonates with Burston (2014), Kukulska-Hulme & Viberg (2018), and Kukulska-Hulme and Shield (2008)—mobile learning fosters independent and collaborative agency.

5.2 Challenges of Integration

Digital Literacy Gaps Among Students and Faculty

As smartphones and internet access continue to expand in urban regions in Morocco, the level of digital literacy within the educated population remains startlingly low. Some pre-service teachers, notably those hailing from rural regions, had difficulty with certain applications. Likewise, teacher educators acknowledged that their limited training in digital pedagogy severely restricted their ability to MALL model instruction. This supports the claims made by Hockly (2015), Tomeschek et al. (2024), Bülbül & Özelçi (2025) and Redecker (2017) in their calls for enhancing teacher education programs by incorporating advanced digital literacy competencies, thus addressing the pedagogical-technology gap.

Uneven Access to Devices and Internet

Disparities in the availability of mobile devices and reliable internet connections were addressed by participants, especially among students from low socioeconomic backgrounds. These factors affected the equitable application of MALL resources and, at times, interrupted group activities (Unwin, 2009). Institutions did not have comprehensive policies or forms of institutional aid designed to address the digital divide, a concern described in other low-resourced educational settings (Mirazchiyski, 2024).

Resistance to Change and Reliance on Traditional Pedagogies

Some participants in the study expressed resistance from certain faculty perspectives towards mobile technology as an academic distraction. This attitude is further compounded by stubborn adherence to the lecturing and testing teaching model. Some pre-service teachers embraced such conservatism, showing hesitation to utilize mobile applications during their practicum. This reflects the adoption of a broader postcolonial outlook in which innovation is readily used but suspiciously scrutinized, albeit deeply entrenched in a colonial legacy of pedagogical practices (White, 1987; Pennycook, 2017).

5.3 Case Narratives and Participant Voices

This subsection foregrounds the voices of the pre-service EFL teachers who participated in the study. Their experience, gained through interviews and classroom observations, provides critical perspectives on the pedagogical uses, benefits, and transformations associated with MALL. These narratives help to illustrate mobile technology's role as a professional development tool within the Moroccan setting.

Illustrative Examples from Pre-Service Teachers

During practicum placements, different participants were able to showcase creative and contextualized applications of mobile technologies. For example, *Participant 1*, a pre-service teacher from a regional training college, discussed how Google Classroom helped her with weekly lesson planning. She elaborated on how she was able to 'assign' grammar exercises asynchronously, and provide feedback in real-time: "At first, the students were shy to submit their writing online, but once they saw I was commenting and correcting their drafts directly, they became more motivated to revise. It made feedback more meaningful." *Participant 2*, another participant teaching in a suburban school, highlighted his use of Quizlet Live for vocabulary drills: "My students were laughing while competing in teams. They debated the meanings of words and even devised their own sets later on. Vocabulary was a chore, but now it seemed like a game." As for *Participant 3*, he elaborated on how he used Padlet in a low-resourced class with limited Wi-Fi to enable collaborative reading response activities: "We connected using the mobile data from my phone. I uploaded the story, and students used one device to give me the responses. It was slower, but they felt great knowing their ideas were displayed like a digital wall."

These examples highlight the adaptive capabilities of pre-service teachers, who managed to employ MALL in innovative ways despite having limited resources.

Classroom Moments of Mobile-Enhanced Instruction

Observation of classroom activities showed that mobile technology integrated into the instruction aids in shifting the reliance from the teacher to the learners. For example, in the university practicum seminar, students created digital posters using Canva that explained the difference between simple past and present perfect tenses. The posters were presented in small groups and students self-assessed using rubrics that had been provided via Google Forms.

Another pre-service teacher, *participant 4*, created WhatsApp groups not only to pass class announcements but to motivate students to send their voice notes, practicing pronunciation. She reflected in her own words: "It was easier for some shy students to practice their English when they were by themselves, so relaxed. They would listen to each other's recordings, and imitate accents, intonation…".

In another classroom, a group activity was to create a prompt-based role-play, which they were to perform using scripts submitted through Edmodo, and record their dialogues using voice memos for submission to their instructor. The instructor provided this feedback in the form of formative assessments to monitor progress over time.

Reflections on Training and Personal Growth

In the course of the interviews, an observable shift in instructional strategies alongside self-reported professional efficacy stood out. Initially, some participants regarded mobile devices as an ancillary tool or even as distracting. Eventually, they began to view mobile technology as an exploration of learning and an expression of one's self.

For instance, *Participant 5*, who initially claimed that mobile devices "didn't belong in class," later shared: "When I watched my learners use Padlet to compose poems and provide commentary on each other's lines, I saw that this device can foster creativity and not just entertainment."

In the same way, *Participant 6*, who did his teaching practice at a poorly equipped school, remarked: "I assumed mobile learning was the sole purview of private schools. But using my phone to show short grammar videos and collecting answers through Kahoot made the class more interactive than ever."

Another striking comment was from *Participant 7*, who said that he designed a mobile portfolio of his teaching practicum, including lesson plans, work samples, and reflective voice notes. He explained, "This portfolio was helpful during employment interviews. It demonstrated my technological capabilities alongside my teaching philosophy. I understand now that I was not learning how to teach, but becoming a teacher in a digital world."

Others also suggested that collaboration with peers through online platforms enhanced participants' sense of community identity as a community of practitioners. This echoes the model of legitimate peripheral participation (Lave and Wenger, 1991). They offered each other support, teaching resources, and provided critiques on lesson plans through WhatsApp group chats and Google Drive folders.

These accounts illustrate the expansive impact of mobile technologies, which extends beyond the classroom. They facilitated the process of identity construction, pedagogical experimentation, and self-empowerment, especially among those participants who had considered themselves outsiders to discourse on digital innovation and technological advancement.

Synthesis and Interpretation

The case narratives corroborate the game-changing impact MALL has on reflective, practice-based teacher education within a teaching case context. Even though there are contextual challenges like infrastructure and policy constraints, the narratives demonstrate that pre-service teachers are not simply passive users of technology. Rather, they adapt and innovate. Their experimentation with new tools, responding to learner needs, and reflections on their teaching identities provide hope for mobile integration in Moroccan EFL teacher education.

These narratives also illustrate a further gap in the literature, highlighting the relevance of the TPACK model and sociocultural theory with regard to digital pedagogical development, especially in developing and post-colonial countries where most innovations regionally arise out of necessity rather than institutional frameworks.

6. PROPOSED PRACTICE-ORIENTED INTEGRATION MODEL

Moroccan pre-service teacher education requires a MALL integration strategy that is systemic, context-based, and responsive to pedagogical and infrastructural needs simultaneously. Based on the findings of this study and other literature (Kukulska-Hulme & Shield, 2008; Hampel, 2009; Stockwell & Hubbard, 2013; Burston, 2014; Kukulska-Hulme & Viberg, 2018; Traxler, 2018), this section proposes a model for the implementation of mobile technology in teacher education programs consisting of three sequential stages, which includes distinct approaches for educators and institutional stakeholders.

6.1 Stages for Integrating Mobile Technology into Pre-Service Teacher Education

The Proposed Integration Model unfolds through three interconnected stages: Awareness and Orientation, Pedagogical Implementation and Sustainable Integration and Innovation.

Stage 1: Awareness and Orientation

This stage is fundamental as it concentrates on the sociocultural aspects, such as promoting digital readiness, explaining mobile learning, and fostering favorable perceptions towards technology use in language instruction and learning. The concept of Mobile-Assisted Language Learning (MALL) is being discussed, along with some ethical aspects of mobile tools and their pedagogical usefulness.

Key Activities:

- Technical workshops on mobile literacy and professional digital identity.
- Guided use of Google Classroom, Quizlet, Padlet, Canva and other mobile learning tools.
- Sociocultural reflective discussions on the role of mobile technology in education.

Learning Outcomes:

- Improved confidence in the employment of mobile tools for teaching.
- Awareness of the application of mobile tools in educational settings and addressing learner participation.

Stage 2: Pedagogical Implementation

Full mobile integration commences at this stage: pre-service teachers start to use mobile devices during micro-teaching, practicum, and collaborative lesson planning. The focus is on encouraging application through active learning and critical self-evaluation.

Key Activities:

- Developing and implementing mobile pedagogy lesson plans.
- Evaluation of the lesson results using reflective journal analysis.
- Peer-assisted evaluation through digital platforms, shared files, and WhatsApp Peer-to-Peer Approach.

Learning Outcomes:

- Mastery of practical strategies for teaching and assessing via mobile devices.
- TPACK competency development (Mishra & Koehler, 2006).

Stage 3: Sustainable Integration and Innovation

In the Integration Phase, a new level of contextual innovation and critical thought about mobile technology is integrated. This also means teaching mobile pre-service teachers to act as digital change leaders in their future schools.

Key Activities:

- Mobile technology capstone projects focusing on teaching inclusively.
- Developing mobile-supported action research or case study projects.
- Documenting and disseminating results in local conferences, webinars, or institutional symposiums.

Learning Outcomes:

- Ability to advocate for and implement mobile learning in education at all levels.
- Advocate for the ownership of mobile pedagogy linked to reflective professional identity.

6.2 Suggested Strategies for Educators and Program Designers

The adoption of mobile integration requires pre-service teachers to be equipped with the necessary tools. As their mobile technology literacy requires enhancement, these strategies are recommended along with the observed practices undertaken in Morocco:

1. Contextualized App Selection: Select apps and tools that achieve curricular relevancy within local languages and that operate in low connectivity settings (e.g., offline compatible through modes such as Kahoot offline modules or Google Docs with offline editing).
2. Scaffolded Digital Assignments: Propose basic level competencies that build up incrementally, commencing with exploration of applications, transitioning to modification of lessons, and finally implementation in teaching sessions.
3. Group-based Collaborative Learning models: Use group projects to foster knowledge creation and peer teaching from curated resources and apps like Padlet, Canva, and WhatsApp.
4. Critical media literacy: Use great sociopolitical considerations of the digital world to make future teachers aware of the phenomena that impact education technology, such as the digital divide, mechanisms of monitoring, and the ethics of communication over the internet.
5. Integrated Assessment practices: Create innovative and integrated formative assessments that reflect process and outcome data of mobile teaching, e.g., using e-portfolios, screencasts, and digital reflective logs.

These strategies are flexible to ensure they shift alongside technological advancements, local, and contextual classroom realities without becoming prescriptive checklists.

6.3 Institutional Recommendations: Infrastructure, Training, and Policy Support

Support from educational institutions and policymakers is crucial for enabling successful incorporation, which may be aided by strategic infrastructure augmentations at Moroccan universities. The remarks listed here are based on regional reports (UNESCO, 2022; Issa, 2020).

1. Infrastructure and Accessibility

- Wi-Fi Enhancement: Universities need to upgrade the Wi-Fi infrastructure of their practicum schools, particularly those in remote regions.
- Device Provision Policies: Facilitate equal access to mobile devices through shared digital labs or subsidized programs that promote device access.

2. Faculty Professional Development
    - Ongoing Training: Provide periodic professional development activities on the pedagogical application of mobile apps in teaching within the TPACK framework and 21st-century teaching methodologies.
    - Communities of Practice: Form interdisciplinary digital teaching circles for peer-to-peer learning across universities for the purposes of sharing best practices, case studies, and providing technical assistance.
3. Curriculum and Policy Alignment
    - Curriculum Development: MALL-related competencies should be included within the prescribed outcomes of teacher education programs.
    - Accreditation and QA: Compliance of digital teaching methodologies to national policies on teacher accreditation and quality assurance (e.g., CNEC benchmarks for Morocco).
4. Research and Innovation Incentives
    - Investment in Pedagogical Research: Grant mini-funds and guidance for research on mobile teaching methodologies conducted by students.
    - Collaboration with Educational Technology Organizations: Stimulate partnership between Moroccan universities and commercial technology developers to tailor and adapt materials and local content.

The model designed above serves as a guide for systematically incorporating mobile technologies into pre-service EFL teacher training in Morocco. It is based on sociocultural theory and takes into account the infrastructural, pedagogical, and sociopolitical aspects of Morocco. As discussed earlier, technology as such cannot remedy all educational challenges, however, its purposeful application, when managed by well-trained and sensitized educators, can enhance language teaching classrooms into vibrant, holistic, and contextualized spaces of learning.

7. IMPLICATIONS FOR STAKEHOLDERS

The sociocultural and digital pedagogical theories grounded in empirical evidence suggest that Mobile-Assisted Language Learning (MALL) has significant implications for pre-service teacher education. It is, however, important to note that with the current structure integrated within the education system, multiple stakeholders need to come together within their contexts to make tangible change. This section highlights key implications for curriculum developers and teacher trainers, educational policymakers and university administrators, and international collaborators, including mobile technology providers.

7.1 For Curriculum Developers and Teacher Trainers

Particular attention should be paid to teacher training in order to develop digital skills for digitally competent educators. It is therefore vital to integrate mobile technology into pedagogy and instructional practices at every level within the curriculum.

Curriculum Design Considerations

- MALL objectives Should Be Embedded: Provide learning outcomes as part of the program with emphasis placed on digital teaching capabilities, aligning with the TPACK model of Mishra and Koehler (2006) as well as the UNESCO Framework for Teachers Competencies (2017).
- Iterative Integration: Starting from the basic level of literacy on digital tools, MALL activities and tools should be introduced gradually throughout the coursework, culminating in application during the practicum phase.
- Assessment Strategies: Candidates should have their level of reflective and practical thinking evaluated through mobile-enhanced lesson plans and technology-mediated reflections; however, including the use of digital portfolios should also be considered.

Trainer Competence and Pedagogical Models

- Faculty Development: As noted by van Meerkerk (2017), teacher trainers require professional development in mobile technologies to incorporate these tools into their teaching in innovative ways.
- Pedagogical Modeling: Trainers need to provide mobile-enhanced instruction and a model for pre-service teachers so that effective scaffolding aids are available for the learners.

7.2 For Educational Policymakers and University Leadership

For policymakers and institutional leaders, policies that govern the use of mobile technologies in teacher education are crucial, especially in low-resourced settings such as Morocco and other countries in the Global South.

Policy-Level Interventions

- Digital Policy Frameworks: National and institutional policies should have well-defined goals targeting the comprehensive digital transformation of teaching and education, encompassing device provision, content creation, and pedagogical innovation.
- Alignment with National Priorities: For sustainability and government support, MALL initiatives should be aligned with Morocco's vision of educational transformation (ESRI 2030 PACT) as well as the 2020–2027 Moroccan Digital Strategy.

Infrastructure and Equity

- Resource Mobilization: Collaboration among universities and the government and private sectors is necessary for effective internet provision, especially in rural areas where students undertake teaching placements during their training.
- Digital Inclusion: There is a need to provide access to students from marginalized groups, which may include financial assistance to obtain the necessary devices or digital resource materials (Maghnouj et al., 2018).

7.3 For International Collaboration and Mobile Tech Providers

Equity-responsive mobile learning solutions for teacher education in Morocco and other countries can be supported by mobile learning educational technology developers. International organizations and developmental agencies have an important role to play.

Collaborative Program Development

- University Sponsorships: International research and development universities and centers can participate in the implementation of training and professionals' co-constructed mobility programs referred to as MALL.
- Funding and Technical Assistance: Special agencies, for example, the British Council and UNESCO, can offer some funding for training and pilot projects from their preferred region, especially in marginalized areas.

Culturally Responsive Technology Design

- Content Customization: Educational mobile app developers need to collaborate with teachers in Morocco to develop instructional materials that are culturally and linguistically appropriate.

- Offline Functionality: Given intermittent connectivity in many regions, mobile schools in the region should build tools that are available without the need for online connectivity.

Capacity Building

- Transformative Teaching Course: Some of the models may be sustained by grant resources from international partners that may build out MALL resources within Morocco.
- Research Collaborations: International partnerships can promote cross-Southern Global knowledge exchange by conducting longitudinal research on the effect of mobile learning on education.

Effective use of mobile technologies in pre-service teaching of EFL in Morocco needs the joint and coordinated efforts of a number of people and groups.

As educators and trainers develop digital pedagogical skills, policymakers should also work on facilitating policies and supporting infrastructure development. Alongside these efforts, global stakeholders and educational technology vendors also have a moral obligation to engage fairly. This means ensuring that the tools and interventions are designed with the actual experiences of learners and educators in postcolonial and financially diverse contexts. Together, these actions can foster educational systems that embrace diversity, innovation, and proactive measures, preparing for the future.

8. CONCLUSION

Summary of Key Findings

In this chapter, we examined a qualitative case study exploring the adoption of mobile technology in pre-service EFL teacher training in Morocco from a sociocultural perspective employing the TPACK model. The study revealed both encouraging gaps and persistent challenges in the educational utilization of MALL tools such as Google Classroom, Quizlet, Canva, Padlet, and WhatsApp.

Key findings include:

- Pedagogical Utility: Mobile applications enabled enhanced lesson planning, cooperative activities, timely feedback, and learner independence. Pre-service teachers noted increased participation, the ability to reflect digitally, and greater self-efficacy towards practice innovation through technology.

- Challenges: Participants reported a lack of adequate digital literacy, sufficient reliable internet and access to mobile devices, curricular alignment, and willingness to accept change as common barriers. The absence of coherent organizational policy and continuous instructional staff development was cited as exacerbating these challenges.
- Contextual Influences: The study highlighted the impact of the postcolonial educational environment of Morocco, characterized by gaps in digital resources and pedagogical paradigms requiring responsive and context-specific approaches to educational integration.
- Participant Narratives: The provided narratives illustrated the adoption and adaptation processes of mobile technologies, alongside the resistance encountered by pre-service teachers in real classroom situations.

Reflection on the Transformative Potential of Mobile Integration in Moroccan Teacher Education

Considering these findings, the research underscores the need for metamorphic changes in integrating technology into education in Morocco to transform teacher education. The research concludes that carefully planned and integrated mobile technologies have the potential to improve access and the overall quality of teacher education in Morocco. Particularly, MALL tools facilitate the transition from teacher-focused instruction to more interactive and participatory, learner-centered paradigms. It integrates digital pedagogical creativity for pre-service teachers who can critique their teaching and interact with learners in more complex ways (Kukulska-Hulme et al, 2015; Walsh, 2010).

Additionally, the integration of mobile technologies has profound potential to solve systemic inequities in higher education in Morocco. Mobile learning is one of the most adaptable, affordable, and scalable means of closing the urban-rural gap, improving access to quality instructional materials, and advancing inclusive teaching and learning strategies. Still, the success of such changes is dependent on system-wide policy, including curriculum changes, policy support, investment from the institution, and international collaboration (UNESCO, 2018).

As highlighted in this study, there is also a call for a decolonial digital pedagogy that does not simply transplant technologies from the Global North, but rather fosters mobile learning that is caring, culturally and linguistically relevant to the lived experiences of Moroccan educators and learners (Phirangee and Foster, 2024; Traxler & Vosloo, 2014).

Future Research Directions

While this study helps to understand the context of MALL in the education of EFL student teachers in Morocco, it also leaves several issues to be addressed in further research:

1. **Longitudinal Studies:** Particularly where pre-service teachers are concerned, there is a gap in longitudinal studies researching how their digital practices change over time into in-service teaching roles within Moroccan schools.
2. **Comparative Studies:** MALL research conducted in a North African or Francophone African country could shed more light on digital pedagogical practices in other regions that pedagogues would find useful.
3. **Policy and Infrastructure Studies:** Studies that deal with the relationship between policies related to education in the digital era, funding policies, and the institutional policy of digitization of education tend to craft better implementation models.

Closing Remarks

This chapter illustrates that mobile technology has the potential to transform pedagogical practices within teacher education in Morocco, but only if integrated purposefully, institutionally supported, and enduring frameworks are put in place. Such innovation cannot afford to ignore the socio-political context of the educators involved or the frameworks that shaped it, which need to focus on equity and inclusion. If mobile learning is placed at the apex of pre-service teacher training, Morocco and other developing nations will enable their educators to face the linguistic, technological, and intercultural challenges of the 21st-century classroom.

REFERENCES

Adams, W. C. (2015). Conducting semi-structured interviews. In Newcomer, K. E., Hatry, H. P., & Wholey, J. S. (Eds.), *Handbook of practical program evaluation* (4th ed., pp. 492–505). Jossey-Bass. DOI: 10.1002/9781119171386.ch19

Ahmad, M., & Wilkins, S. (2024). Purposive sampling in qualitative research: A framework for the entire journey. *Quality & Quantity*. Advance online publication. DOI: 10.1007/s11135-024-02022-5

Ahmed, S. K. (2024). The pillars of trustworthiness in qualitative research. *Journal of Medicine, Surgery, and Public Health*, *2*, 100051. DOI: 10.1016/j.glmedi.2024.100051

Albarqi, G. (2023). Padlet as a Formative Assessment Tool in the Online Language Classroom: Action Research. In Chong, S. W., & Reinders, H. (Eds.), *Innovation in Learning-Oriented Language Assessment*. New Language Learning and Teaching Environments. Palgrave Macmillan., DOI: 10.1007/978-3-031-18950-0_11

Amaaz, A., Mouradi, A., & Erradi, M. (2024). Pre-service Elementary Teachers' Attitudes Toward the Use of Mobile Learning in Elementary Science Education. In Motahhir, S., & Bossoufi, B. (Eds.), *Digital Technologies and Applications. ICDTA 2024. Lecture Notes in Networks and Systems* (Vol. 1099). Springer., DOI: 10.1007/978-3-031-68653-5_2

Andujar, A., & Hussein, S. A. (2019). Mobile-mediated communication and students' listening skills: A case study. *International Journal of Mobile Learning and Organisation*, *13*(3), 309. DOI: 10.1504/IJMLO.2019.100432

Azabdaftari, B., & Mozaheb, M. A. (2012). Comparing vocabulary learning of EFL learners by using two different strategies: Mobile learning vs. flashcards. *The EuroCALL Review*, *20*(2), 47. DOI: 10.4995/eurocall.2012.11377

Bachmair, B., & Pachler, N. (2015). Framing Ubiquitous Mobility Educationally: Mobile Devices and Context-Aware Learning. In Wong, L. H., Milrad, M., & Specht, M. (Eds.), *Seamless Learning in the Age of Mobile Connectivity*. Springer., DOI: 10.1007/978-981-287-113-8_3

Bakkali, I., Pérez-Torregrosa, A. B., Cebrián-Robles, V., & Cebrián-de-la-Serna, M. (2023). Digital Inclusion in the Moroccan Context: A Review of Research and Educational Initiatives in the Twenty-First Century. In Tomczyk, Ł., Guillén-Gámez, F. D., Ruiz-Palmero, J., & Habibi, A. (Eds.), *From Digital Divide to Digital Inclusion. Lecture Notes in Educational Technology*. Springer., DOI: 10.1007/978-981-99-7645-4_17

Basal, A., Yilmaz, S., Tanriverdi, A., & Sari, L. (2016). Effectiveness of Mobile Applications in Vocabulary Teaching. *Contemporary Educational Technology*, *7*(1), 47–59. DOI: 10.30935/cedtech/6162

Basaran, S., & Cabaroglu, N. (2014).Language Learning Podcasts and Learners' Belief Change.TESL-EJ, 17(4), n4 retrieved from http://eric.ed.gov/?id=EJ1024978 on 30.11.2015

Baxter, P., & Jack, S. (2015). Qualitative case study Methodology: Study design and implementation for novice researchers. *The Qualitative Report*. Advance online publication. DOI: 10.46743/2160-3715/2008.1573

Belhiah, H., & Abdelatif, A. (2016). English as a medium of instruction in Moroccan higher education. SSRN *Electronic Journal*. https://doi.org/DOI: 10.2139/ssrn.2895569

Benali, M., Kaddouri, M., & Azzimani, T. (2018). Digital competence of Moroccan teachers of English. *International Journal of Education and Development Using Information and Communication Technology*, *14*(2), 99–120. https://files.eric.ed.gov/fulltext/EJ1190022.pdf

Bouzidi, H. E. I. (2024). Advancing an educational framework to enhance students' learning experiences. *E-Learning, 2*(11). ISCAE. DOI: 10.48423/IMIST.PRSM/rmere-v2i11.46258

Braun, V., & Clarke, V. (2019). Reflecting on reflexive thematic analysis. *Qualitative Research in Sport, Exercise and Health*, *11*(4), 589–597. DOI: 10.1080/2159676X.2019.1628806

Bueno-Alastuey, M. C., & Nemeth, K. (2020). Quizlet and podcasts: Effects on vocabulary acquisition. *Computer Assisted Language Learning*, *35*(7), 1407–1436. DOI: 10.1080/09588221.2020.1802601

Bülbül, A. E., & Özelçi, S. Y. (2025). The mediating role of mobile learning motivation in the relationship between pre-service teachers' digital literacy and pre-service teacher identity. *Education and Information Technologies*, *30*(2), 1729–1743. DOI: 10.1007/s10639-024-12868-9

Burston, J. (2014). MALL: The pedagogical challenges. *Computer Assisted Language Learning*, *27*(4), 344–357. DOI: 10.1080/09588221.2014.914539

Burston, J. (2015). Twenty years of MALL project implementation: A meta-analysis of learning outcomes. *ReCALL*, *27*(1), 4–20. DOI: 10.1017/S0958344014000159

Canagarajah, A. (1999). *Resisting linguistic imperialism in English lan guage teaching*. Oxford University Press.

Cappelli, M. A., & Akkari, A. (2025). Academic ICTs training in South Africa, Cameroon, and Nigeria. Strategies for ICT training course design. *Education and Information Technologies, 30*(8), 10463–10510. DOI: 10.1007/s10639-024-13200-1

Chakrani, B. (2020). Languages in Conflict: EXAMINING THE STATUS OF STANDARD ARABIC AND FRENCH IN MOROCCO. *Al-'Arabiyya, 53*, 37–56. https://www.jstor.org/stable/48659135

Creswell, J. W., & Poth, C. N. (2018). *Qualitative Inquiry and Research Design Choosing among Five Approaches* (4th ed.). SAGE Publications, Inc.

De Costa, P. I. (2007). J. P. Lantolf and S. L. Thorne: Sociocultural Theory and the Genesis of Second Language Development. Oxford University Press, 2006. *Applied Linguistics, 28*(3), 477–480. DOI: 10.1093/applin/amm027

Denzin, N. K., & Lincoln, Y. S. (Eds.). (2017). *The SAGE handbook of qualitative research* (5th ed.). SAGE Publications.

Ding, Y., & Zhu, J. (2025). Mobile based peer feedback in EFL speaking: Learners' motivation, behavioral engagement in feedback provision, and achievement. *Asian. J. Second. Foreign. Language and Education, 10*(1), 10. DOI: 10.1186/s40862-024-00314-9

Dolezal, D., Motschnig, R., & Ambros, R. (2025). Pre-Service Teachers' Digital Competence: A Call for Action. *Education Sciences, 15*(2), 160. DOI: 10.3390/educsci15020160

Duman, G., Orhon, G., & Gedik, N. (2015). Research trends in mobile assisted language learning from 2000 to 2012. *ReCALL, 27*(2), 197–216. DOI: 10.1017/S0958344014000287

Durgungoz, F. C., & Durgungoz, A. (2025). "Interactive lessons are great, but too much is too much": Hearing out neurodivergent students, Universal Design for Learning and the case for integrating more anonymous technology in higher education. *Higher Education*. Advance online publication. DOI: 10.1007/s10734-024-01389-6

Fadli, S., & El Mediouni, A. (2024). Digital Transformation in Higher Education: Moroccan University Students' Perceptions and Expectations on Learning Analytics Implementation for Learning Optimization. In Adoui, A. (Ed.), *Internationalization of Higher Education and Digital Transformation*. Palgrave Macmillan., DOI: 10.1007/978-3-031-76444-8_14

Filiz, O., Kaya, M. H., & Adiguzel, T. (2025). Teachers and AI: Understanding the factors influencing AI integration in K-12 education. *Education and Information Technologies*, *30*(13), 1–37. DOI: 10.1007/s10639-025-13463-2

Ghomi, M., & Redecker, C. (2019). Digital Competence of Educators (DigCompEdu): Development and Evaluation of a Self-assessment Instrument for Teachers' Digital Competence. *In Proceedings of the 11th International Conference on Computer Supported Education (CSEDU 2019), Pages* 541-548. DOI: 10.5220/0007679005410548

Godwin-Jones, R. (2011). Mobile apps for language learning. *Language Learning & Technology*, *15*(2), 2–11. https://scholarspace.manoa.hawaii.edu/bitstream/10125/44244/1/15_02_emerging.pdf. DOI: 10.64152/10125/44244

Gorbunova, N., Boginskaia, Y., & Khrulyova, A. (2025). Analysis of the Benefits of Digital Services in Inclusive Higher Education. In: Lapina, M., Prakasha, G.S., Grigoriev, S. (eds) International Conference on Innovative Approaches to the Application of Digital Technologies in Education and Research. SLET 2022. Lecture Notes in Networks and Systems, vol 1222. Springer, Cham. https://doi.org/DOI: 10.1007/978-3-031-78776-8_48

Habib, S., Haider, A., Suleman, S. S. M., Akmal, S., & Khan, M. A. (2022). Mobile Assisted Language Learning: Evaluation of Accessibility, Adoption, and Perceived Outcome among Students of Higher Education. *Electronics (Basel)*, *11*(7), 1113. DOI: 10.3390/electronics11071113

Hampel, R. (2009). Training teachers for the multimedia age: Developing teacher expertise to enhance online learner interaction and collaboration. *Innovation in Language Learning and Teaching*, *3*(1), 35–50. DOI: 10.1080/17501220802655425

Higher Council of Education, Training and Scientific Research. (2015). *For a school of equity, quality and promotion: A strategic vision of reform 2015–2030*. https://www.csefrs.ma/wp-content/uploads/2015/05/Re%CC%81sume%CC%81-vision-Anglais.pdf

Hockly, N. (2015). Developments in online language learning. *ELT Journal*, *69*(3), 308–313. DOI: 10.1093/elt/ccv020

Hwang, G. J., & Tsai, C. C. (2011). Research Trends in Mobile and Ubiquitous Learning: A Review of Publications in Selected Journals from 2001 to 2010. *British Journal of Educational Technology*, *42*(4), E65–E70. DOI: 10.1111/j.1467-8535.2011.01183.x

Issa, D. A. (2020). *Morocco - Financial and Digital Inclusion Development Policy Financing Project: Appraisal Program Information Document (PID) - Financial and Digital Inclusion Development Policy Financing - P171587 (English).* World Bank Group. Retrieved from https://documents.worldbank.org/curated/en/690571587680057013

Kabel, A. (2021). 'The tide is coming in fast': Ideologies of English, global linguistic coloniality and decolonial pluriversalingualism. In Tupas, R., & Rubdy, R. (Eds.), *World Englishes: Ideologies* (pp. 43–63). Bloomsbury.

Karakaya, K., & Bozkurt, A. (2022). Mobile-assisted language learning (MALL) research trends and patterns through bibliometric analysis: Empowering language learners through ubiquitous educational technologies. *System, 110*, 102925. DOI: 10.1016/j.system.2022.102925

Katona, J., & Gyonyoru, K. I. K. (2025). Integrating AI-Based Adaptive Learning into the Flipped Classroom Model to Enhance Engagement and Learning Outcomes. *Computers and Education Artificial Intelligence, 100392*, 100392. Advance online publication. DOI: 10.1016/j.caeai.2025.100392

Kearney, M., & Maher, D. (2019). Mobile learning in pre-service teacher education: Examining the use of professional learning networks. *Australasian Journal of Educational Technology, 35*(1). Advance online publication. DOI: 10.14742/ajet.4073

Kessler, G. (2018). Technology and the future of language teaching. *Foreign Language Annals, 51*(1), 205–218. DOI: 10.1111/flan.12318

Kinshuk, (2016). Designing Adaptive and Personalized Learning Environments (1st ed.). Routledge. DOI: 10.4324/9781315795492

Koehler, M. J., Mishra, P., & Cain, W. (2017). What is Technological Pedagogical Content Knowledge (TPACK)? Journal of Education, 193(3), 13-19. (Original work published 2013)DOI: 10.1177/002205741319300303

Kramsch, C., & Vork Steffensen, S. (2008). Ecological Perspectives on Second Language Acquisition and Socialization. In Hornberger, N. H. (Ed.), *Encyclopedia of Language and Education*. Springer., DOI: 10.1007/978-0-387-30424-3_194

Kukulska-Hulme, A. (2013). Mobile-assisted language learning. In Chapelle, C. (Ed.), *The encyclopedia of applied linguistics* (pp. 3701–3709). Wiley.

Kukulska-Hulme, A., Norris, L., & Donohue, J. (2015). Mobile pedagogy for English language teaching: a guide for teachers. *British Council*. https://www.teachingenglish.org.uk/publications/case-studies-insights-and-research/mobile-pedagogy-english-language-teaching-guide

Kukulska-Hulme, A., & Shield, L. (2008). An overview of mobile assisted language learning: From content delivery to supported collaboration and interaction. *ReCALL, 20*(3), 271–289. DOI: 10.1017/S0958344008000335

Kukulska-Hulme, A., & Viberg, O. (2018). Mobile collaborative language learning: State of the art. *British Journal of Educational Technology*, *49*(2), 207–218. DOI: 10.1111/bjet.12580

Kullman, S. M., & Chudyk, A. M. (2025). Participatory Member Checking: A Novel Approach for Engaging Participants in Co-Creating Qualitative Findings. *International Journal of Qualitative Methods*, *24*, 16094069251321211. Advance online publication. DOI: 10.1177/16094069251321211

Kuru Gönen, S. İ., & Zeybek, G. (2022). Training on multimodal mobile-assisted language learning: A suggested model for pre-service EFL teachers*. *Computer Assisted Language Learning*, *37*(7), 2202–2223. DOI: 10.1080/09588221.2022.2157016

Lablidi, A., Nachit, B., Abourriche, A., Namir, A., & Talbi, M. (2013). ICT Practice in Morocco's innovative teachers. *American Journal of Engineering Research (AJER)*. https://www.ajer.org/papers/v2(5)/T0251450150.pdf

Lantolf, J., & Thorne, S. L. (2006). *Sociocultural Theory and the genesis of second language development*. Oxford University Press.

Lave, J., & Wenger, E. (1991). *Situated Learning: Legitimate Peripheral Participation*. Cambridge University Press., DOI: 10.1017/CBO9780511815355

Lehtinen, A., Nieminen, P., & Viiri, J. (2016). Preservice teachers' TPACK beliefs and attitudes toward simulations. *Contemporary Issues in Technology & Teacher Education*, *16*(2). https://citejournal.org/volume-16/issue-2-16/science/preservice-teachers-tpack-beliefs-and-attitudes-toward-simulations

Leinonen, T., Keune, A., Veermans, M., & Toikkanen, T. (2014). Mobile apps for reflection in learning: A design research in K-12 education. *British Journal of Educational Technology*, *47*(1), 184–202. DOI: 10.1111/bjet.12224

Li, F., Fan, S., & Wang, Y. (2022). Mobile-assisted language learning in Chinese higher education context: A systematic review from the perspective of the situated learning theory. *Education and Information Technologies*, *27*(7), 9665–9688. DOI: 10.1007/s10639-022-11025-4

Lincoln, Y. S., & Guba, E. G. (1985). *Naturalistic inquiry*. Sage. Crossref. DOI: 10.1016/0147-1767(85)90062-8

Loewen, S., Crowther, D., Isbell, D. R., Kim, K. M., Maloney, J., Miller, Z. F., & Rawal, H. (2019). Mobile-assisted language learning: A Duolingo case study. *ReCALL, 31*(3), 293–311. DOI: 10.1017/S0958344019000065

Lu, M. (2008). Effectiveness of vocabulary learning via mobile phone. *Journal of Computer Assisted Learning, 24*(6), 515–525. DOI: 10.1111/j.1365-2729.2008.00289.x

Luo, Y., & Watts, M. (2022). Exploration of university students' lived experiences of using smartphones for English language learning. *Computer Assisted Language Learning, 37*(4), 608–633. DOI: 10.1080/09588221.2022.2052904

Maghni, A., & El Khannouss, A. (2024). *Digitalisation de l'enseignement supérieur au Maroc: Approche théorique et conceptualisation*. International Journal of Accounting, Finance, Auditing, Management and Economics., DOI: 10.5281/zenodo.14253421

Maghnouj, S., Bélanger, J., Clarke, M., Fordham, E., Kitchen, H., & McGregor, I. (2018). *Examens de l'OCDE du cadre d'évaluation de l'éducation*. OCDE., DOI: 10.1787/9789264301832-fr

Mahdi, S. H. (2017). Effectiveness of mobile devices on vocabulary learning: A meta-analysis. *Journal of Educational Computing Research, 56*(1), 134–154. DOI: 10.1177/0735633117698826

Marley, D. (2004). Language attitudes in Morocco following recent changes in language policy. *Language Policy, 3*(1), 25–46. DOI: 10.1023/B:LPOL.0000017724.16833.66

Merriam, S. B., & Tisdell, E. J. (2016). *Qualitative Research: A Guide to Design and Implementation* (4th ed.). Jossey Bass.

Mirazchiyski, P. V. (2024). *Contemporary gaps in research on digital divide in education: a literature review*. Univ Access Inf Soc., DOI: 10.1007/s10209-024-01166-3

Mishra, P., & Koehler, M. J. (2006). Technological Pedagogical Content Knowledge: A Framework for Teacher Knowledge. Teachers College Record, 108(6), 1017-1054. (Original work published 2006)DOI: 10.1111/j.1467-9620.2006.00684.x

Moumen, A. (2024). TICE and Their Role in Achieving the Objectives of Morocco's Strategic Vision 2015-2030. *International Journal of Scientific Research and Innovative Studies, 3*(4), 82–85. DOI: 10.5281/zenodo.14178955

Mountasser, T. (2023). Smartphones Usefulness in Mobile Learning: The Moroccan University Case - Students of High School of Education and Training, Agadir. In: Ben Ahmed, M., Boudhir, A.A., Santos, D., Dionisio, R., Benaya, N. (eds) Innovations in Smart Cities Applications Volume 6. SCA 2022. Lecture Notes in Networks and Systems, vol 629. Springer, Cham. https://doi.org/DOI: 10.1007/978-3-031-26852-6_39

Moya, S., & Camacho, M. (2024). Leveraging AI-powered mobile learning: A pedagogically informed framework. *Computers and Education Artificial Intelligence*, *7*, 100276. DOI: 10.1016/j.caeai.2024.100276

National Telecommunications Regulatory Agency (ANRT). (2023). ICT *Equipment and uses for households and individuals: Survey*. Morocco. www.anrt.ma/sites/default/files/2024-06/enquete_tic_2023_-_synthese_ang.pdf

Neureiter, D., Klieser, E., Neumayer, B., Winkelmann, P., Urbas, R., & Kiesslich, T. (2020). Feasibility of Kahoot! as a Real-Time Assessment Tool in (Histo-)pathology Classroom Teaching. *Advances in Medical Education and Practice*, *11*, 695–705. DOI: 10.2147/AMEP.S264821 PMID: 33117039

Novawan, A., Aisyiyah, S., Miqawati, A. H., Wijayanti, F., & Rinda, R. K. (2019). Implementing an Ecological Teaching through Mobile-Assisted Language Learning in EFL Education. *Advances in Social Science, Education and Humanities Research*, *254*. Advance online publication. DOI: 10.2991/conaplin-18.2019.247

Ogata, H., Akamatsu, R., & Yano, Y. (2005). Computer Supported Ubiquitous Learning Environment for Vocabulary Learning Using RFID Tags. In Courtiat, J. P., Davarakis, C., & Villemur, T. (Eds.), *Technology Enhanced Learning. IFIP WCC TC3 2004. IFIP International Federation for Information Processing* (Vol. 171). Springer., DOI: 10.1007/0-387-24047-0_10

Ogata, H., & Yano, Y. (2004). Context-aware support for computer-supported ubiquitous learning. In *Proceedings of the 2nd IEEE International Workshop on Wireless and Mobile Technologies in Education* (pp. 27–34). IEEE. DOI: 10.1109/WMTE.2004.1281330

Oved, O., & Alt, D. (2025). Teachers' technological pedagogical content knowledge (TPACK) as a precursor to their perceived adopting of educational AI tools for teaching purposes. *Education and Information Technologies*, *30*(10), 14095–14121. Advance online publication. DOI: 10.1007/s10639-025-13371-5

Pennycook, A. (2017). The cultural politics of English as an international language. In *Routledge eBooks*. https://doi.org/DOI: 10.4324/9781315225593

Pham, A. T., & Nguyen, T. B. (2024). English as a foreign language students' acceptance of Google Classroom in writing classes: A case study in Vietnam. *Heliyon*, *10*(8), e29832. DOI: 10.1016/j.heliyon.2024.e29832 PMID: 38699708

Phanwiriyarat, K., Anggoro, K. J., & Chaowanakritsanakul, T. (2025). Exploring AI-powered gamified flipped classroom in an English-speaking course: A case of Duolingo. *Cogent Education*, *12*(1), 2488545. Advance online publication. DOI: 10.1080/2331186X.2025.2488545

Phirangee, K., & Foster, L. (2024). Decolonizing digital learning: Equity through intentional course design. *Distance Education*, *45*(3), 357–364. DOI: 10.1080/01587919.2024.2362413

Qasserras, L., Qasserras, M., Anasse, K., & Slimani, N. (2023). Exploring Pre-Service Teachers' Experience With The Use Of Smartphones During The Internship Period - CASE STUDY: CRMEF OF RABAT, MOROCCO. *European Journal of Open Education and E-learning Studies*, *8*(2). Advance online publication. DOI: 10.46827/ejoe.v8i2.4765

Redecker, C. (2017). Digital Competence of Educators. https://doi.org/DOI: 10.2760/159770

Reinders, H., & White, C. (2010). The theory and practice of technology in materials development and task design. In Harwood, N. (Ed.), *Materials in ELT: Theory and practice* (pp. 58–80). Cambridge University Press.

Resta, P., & Laferrière, T. (2015). Digital equity and intercultural education. *Education and Information Technologies*, *20*(4), 743–756. DOI: 10.1007/s10639-015-9419-z

Rhaya, F., & El Ouazzani, Y. (2025). ICT and the Digital Divide: Analyzing ICT Indicators in Morocco Using PCA. In Bach Tobji, M. A., Jallouli, R., Sadok, H., Lajfari, K., Mafamane, D., & Mahboub, H. (Eds.), *Digital Economy. Emerging Technologies and Business Innovation. ICDEc 2024. Lecture Notes in Business Information Processing* (Vol. 531). Springer., DOI: 10.1007/978-3-031-76368-7_11

Rinda, R. (2019). Implementing an Ecological Teaching through Mobile-Assisted Language Learning in EFL Education. *Proceedings of the Eleventh Conference on Applied Linguistics (CONAPLIN 2018).*

Rocco, T. S., & Plakhotnik, M. S. (2009). Literature Reviews, Conceptual Frameworks, and Theoretical Frameworks: Terms, Functions, and Distinctions. *Human Resource Development Review*, *8*(1), 120–130. DOI: 10.1177/1534484309332617

Rodríguez Moreno, J., Agreda Montoro, M., & Ortiz Colón, A. M. (2019). Changes in Teacher Training within the TPACK Model Framework: A Systematic Review. *Sustainability (Basel)*, *11*(7), 1870. DOI: 10.3390/su11071870

Sciberras, M., & Dingli, A. (2023). Research Analysis—Triangulation Approach. In *Investigating AI Readiness in the Maltese Public Administration. Lecture Notes in Networks and Systems* (Vol. 568). Springer., DOI: 10.1007/978-3-031-19900-4_9

Selwyn, N. (2011). *Education and Technology: Key Issues and Debates*. Continuum International Publishing Group.

Sharples, M., Taylor, J., & Vavoula, G. (2010). A Theory of Learning for the Mobile Age. In Bachmair, B. (Ed.), *Medienbildung in neuen Kulturräumen*. VS Verlag für Sozialwissenschaften., DOI: 10.1007/978-3-531-92133-4_6

Spradley, J. P. (1980). *Participant observation*. Holt, Rinehart and Winston.

Stake, R. E. (1995). *The art of case study research*. SAGE Publications.

Stockwell, G., & Hubbard, P. (2013). *Some Emerging Principles for Mobile-Assisted Language Learning*. The International Research Foundation for English Language Education., https://www.tirfonline.org/wp-content/uploads/2013/11/TIRF_MALL_Papers_StockwellHubbard.pdf

Sung, Y.-T., Chang, K.-E., & Liu, T.-C. (2016). The effects of integrating mobile devices with teaching and learning on students' learning performance: A meta-analysis and research synthesis. *Computers & Education*, *94*, 252–275. DOI: 10.1016/j.compedu.2015.11.008

Tarteer, S., Badah, A., & Khlaif, Z. N. (2021). Employing Google Classroom to Teach Female Students during the COVID-19 Pandemic. *Computers in the Schools*, *38*(4), 300–321. DOI: 10.1080/07380569.2021.1988318

Tomeschek, S., Schaberreiter, T., Kandlhofer, M., & Lieckfeld, A. S. (2024). Assessing data literacy and AI education in digital education policies—Insights from the TrainDL Erasmus+ project. *Informatik Spektrum*, *47*(3-4), 113–119. DOI: 10.1007/s00287-024-01576-w

Traxler, J. (2018). Learning with mobiles: The Global South. *Research in Comparative and International Education*, *13*(1), 152–175. DOI: 10.1177/1745499918761509

Traxler, J., & Kukulska-Hulme, A. (Eds.). (2015). *Mobile Learning: The Next Generation* (1st ed.). Routledge., DOI: 10.4324/9780203076095

Traxler, J., & Vosloo, S. (2014). Introduction: The prospects for mobile learning. *Prospects*, *44*(1), 13–28. DOI: 10.1007/s11125-014-9296-z

UNESCO. (2018). *ICT competency framework for teachers*. https://unesdoc.unesco.org/ark:/48223/pf0000265721

UNESCO. (2022). *Reimagining our futures together: A new social contract for education*. https://unesdoc.unesco.org/ark:/48223/pf0000379707

Unwin, T. (Ed.). (2009). *ICT4D: Information and Communication Technology for Development*. Cambridge University Press.

Valtonen, T., Sointu, E., Kukkonen, J., Kontkanen, S., Lambert, M. C., & Mäkitalo-Siegl, K. (2017). TPACK updated to measure pre-service teachers' twenty-first century skills. *Australasian Journal of Educational Technology*, *33*(3). Advance online publication. DOI: 10.14742/ajet.3518

van Meerkerk, E. (2017). Teacher Logbooks and Professional Development: A Tool for Assessing Transformative Learning Processes. *International Journal of Qualitative Methods*, *16*(1), 1609406917735255. Advance online publication. DOI: 10.1177/1609406917735255

Viberg, O., & Grönlund, Å. (2013). Cross-cultural analysis of users' attitudes toward the use of mobile devices in second and foreign language learning in higher education: A case from Sweden and China. *Computers & Education*, *69*, 169–180. DOI: 10.1016/j.compedu.2013.07.014

Vygotsky, L. S. (1978). *Mind in Society: Development of Higher Psychological Processes* (Cole, M., Jolm-Steiner, V., Scribner, S., & Souberman, E., Eds.). Harvard University Press., DOI: 10.2307/j.ctvjf9vz4

Walsh, M. (2010). Multimodal Literacy: What does it mean for classroom practice? *Australian Journal of Language and Literacy*, *33*(3), 211–239. DOI: 10.1007/BF03651836

Warschauer, M., & Liaw, M.-L. (2011). Emerging Technologies for Autonomous Language Learning. *Studies in Self-Access Learning Journal*, 107–118. DOI: 10.37237/020302

White, R. (1987). Managing innovation. *ELT Journal*, *41*(3), 211–218. DOI: 10.1093/elt/41.3.211

Wong, L.-H., & Looi, C.-K. (2011). What seams do we remove in mobile-assisted seamless learning? A critical review of the literature. *Computers & Education*, *57*(4), 2364–2381. DOI: 10.1016/j.compedu.2011.06.007

Xiangming, L., & Song, S. (2017). Mobile technology affordance and its social implications: A case of "Rain Classroom.". *British Journal of Educational Technology*, *49*(2), 276–291. DOI: 10.1111/bjet.12586

Yang, L., & Gottlieb, M. (2023). Gamification Mobile Applications: A Literature Review of Empirical Studies. In Auer, M. E., Pachatz, W., & Rüütmann, T. (Eds.), *Learning in the Age of Digital and Green Transition. ICL 2022. Lecture Notes in Networks and Systems* (Vol. 633). Springer., DOI: 10.1007/978-3-031-26876-2_88

Yin, R. K. (2018). *Case Study Research and Applications: Design and Methods* (6th ed.). Sage.

Yudintseva, A. (2023). Virtual reality affordances for oral communication in English as a second language classroom: A literature review. *Computers & Education: X Reality, 2*, 100012. https://doi.org/DOI: 10.1016/j.cexr.2023.100012

Zou, D., Luo, S., Xie, H., & Hwang, G. J. (2020). A systematic review of research on flipped language classrooms: Theoretical foundations, learning activities, tools, research topics and findings. *Computer Assisted Language Learning*, *35*(8), 1811–1837. DOI: 10.1080/09588221.2020.1839502

Zouhir, A. (2013). Language situation and conflict in Morocco. In Orie, O. O., & Sanders, K. W. (Eds.), *Selected proceedings of the 43rd annual conference on African linguistics* (pp. 271–277). Cascadilla Proceedings Project., http://www.lingref.com/cpp/acal/43/paper2975.pdf

APPENDIX

Interview Protocol: Semi-Structured Interviews with Pre-Service EFL Teachers and Teacher Educators

Section 1: Background and Demographics

1. Can you briefly describe your role and experience in EFL teacher education (as a pre-service teacher or a teacher educator)?
2. What type of mobile devices do you typically use in your academic or teaching practice (e.g., smartphones, tablets, laptops)?
3. Have you received any formal training or instruction in using mobile technologies for teaching purposes?

Section 2: Current Use and Perceptions of Mobile Technology

4. In your experience, how are mobile technologies currently being used in your teacher education program (e.g., for lesson planning, classroom activities, assessment, etc.)?
5. Which mobile applications or platforms do you find most useful in your learning or teaching practice? Why?
6. Can you describe a specific instance when you used mobile technology effectively during a practicum or in a methodology course?

Section 3: Pedagogical Development and Reflective Practice

7. In what ways has mobile technology supported your understanding of lesson design, instructional strategies, or learner engagement?
8. Have you used mobile tools to reflect on your teaching practices (e.g., digital journals, audio/video feedback, peer discussions)? If so, how did this affect your development as a teacher?
9. How do mobile technologies contribute to collaborative learning with peers and instructors in your program?

Section 4: Challenges and Barriers

10. What are the main challenges you have faced in using mobile technologies during your training (e.g., access, skills, support, curriculum)?

11. How would you describe the level of digital literacy among your peers and instructors in your institution?
12. Are there institutional policies or support systems in place that either facilitate or hinder the use of mobile technologies in your program?

Section 5: Recommendations and Future Perspectives

13. What kind of training, tools, or resources do you believe would help better integrate mobile technologies into your teacher education experience?
14. From your perspective, what role should mobile technology play in the future of EFL teacher education in Morocco?
15. Do you have any additional thoughts or suggestions for improving the use of mobile technologies in pre-service teacher education?

Chapter 3
Personalized Learning Pathways Through Mobile Devices:
Enhancing Language Acquisition

Mustafa Kayyali
https://orcid.org/0000-0003-3300-262X
Maaref University of Applied Sciences, Syria

ABSTRACT

As mobile technologies continue to reshape the landscape of education, language learning has witnessed a profound transformation through personalized, adaptive experiences tailored to the needs of individual learners. This chapter explores how mobile devices can create meaningful and responsive learning pathways that enhance language acquisition. By leveraging personalization—ranging from app-based content to real-time feedback and data-driven customization—learners are increasingly empowered to direct their own progress, engaging with content that suits their pace, goals, and linguistic background. Drawing on both current research and real-world classroom applications, this chapter offers insights into the design and implementation of mobile-based personalized learning systems, while also examining their impact on learner motivation and autonomy. The discussion includes the practical and pedagogical implications of integrating mobile technologies in language education, highlighting both the promise and the complexities involved.

DOI: 10.4018/979-8-3373-3366-3.ch003

INTRODUCTION

In a world increasingly defined by rapid connectivity and ubiquitous digital access, the traditional paradigms of education are being questioned, challenged, and, in many ways, reimagined. Nowhere is this more evident than in the domain of language learning. Language, as both a social practice and cognitive skill, requires engagement, interaction, and consistent exposure—elements that can now be delivered in personalized and dynamic ways thanks to mobile technology (Procel et al., 2024). For decades, language education operated within rigid classroom structures, often adopting one-size-fits-all approaches that left many learners disengaged or underserved. However, the emergence of mobile-assisted language learning (MALL) has introduced not just a new set of tools, but an entirely different philosophy: one that centers the learner, adapts to their unique profile, and evolves in real-time based on interaction, interest, and need (Iftikhar, 2025). This chapter begins with a simple but powerful question: What does it mean to truly personalize language learning in a digital age? The answer is not merely about customizing exercises or translating content into multiple languages. It is about rethinking the very foundations of language education. Personalized learning through mobile devices is not just a method—it is a shift in mindset that respects the individuality of each learner, recognizes diverse learning trajectories, and leverages the power of technology to adapt, support, and enhance the journey. It also speaks to an evolving educational ethic: one that prioritizes learner agency, autonomy, and voice.

The promise of personalization lies in its ability to address the challenges that conventional language instruction often fails to resolve. Learners struggle with varying paces of acquisition, differing levels of prior exposure, contrasting motivations, and distinct cultural or linguistic backgrounds (Alexandra, 2024). Mobile platforms, through intelligent algorithms, data tracking, gamified interfaces, and on-demand resources, have the potential to accommodate this diversity, creating a learning environment that is responsive, flexible, and deeply human-centered. These platforms do not merely deliver lessons; they create pathways—learning ecosystems where content adapts to performance, feedback is immediate, and motivation is sustained through relevance and engagement. But the journey toward meaningful mobile integration is not without its complications. Issues of digital equity, screen fatigue, over-reliance on algorithms, and the potential for shallow engagement pose real challenges (George, 2024). Personalization, if poorly designed, can devolve into mere content recommendation systems rather than pedagogically sound learning architectures. Therefore, the goal of this chapter is not to celebrate technology uncritically, but to explore how mobile devices—when thoughtfully integrated—can facilitate authentic language acquisition experiences tailored to the learner.

This inquiry is grounded in current research, pedagogical theory, and emerging practices from classrooms and informal learning spaces around the world. It also reflects the lived realities of learners and educators navigating the tensions between innovation and implementation. From language apps that offer AI-powered feedback to mobile platforms that track learner growth over time, we will explore how mobile technology can act not as a replacement for traditional instruction, but as a partner in a more holistic, humanized learning process. In the pages that follow, we unpack what it truly means to build *personalized learning pathways*. We examine not just the technical infrastructure that enables such pathways, but the pedagogical, psychological, and ethical considerations that must accompany them. At its core, this chapter is a reflection on potential—on the possibility that, with the right design and intention, mobile technologies can bring language learning closer to the learner than ever before: into their hands, into their lives, and ultimately, into their sense of identity and belonging as language users.

RETHINKING LANGUAGE LEARNING IN THE MOBILE AGE

To rethink language learning in the mobile age is not merely to integrate new devices into old methods; it is to confront a quiet revolution—one that reshapes how, where, and why people engage with language. At its core, language is about connection, communication, and the navigation of identity in an ever-evolving social world. Traditionally, language learning occurred in structured environments: classrooms filled with textbooks, blackboards, and timed drills (Ihnatova et al, 2021). It was a system often designed more for control and standardization than for curiosity and growth. Today, however, mobile technology has shifted the axis of learning, dispersing it into everyday life. It has invited language learning into bus stops, kitchens, bedrooms, and lunch breaks. And with that shift comes the need to rethink not only our pedagogies but our philosophies. Mobile devices, once seen as distractions, are now portals to immersive, personalized language experiences (Lăpădat, 2023). They are not simply delivery tools for static content—they are dynamic companions in the learning journey. With the swipe of a screen, learners can converse with native speakers thousands of miles away, receive tailored feedback based on their unique errors, or revisit grammar explanations in the quiet solitude of late-night study. But beyond convenience, mobile learning challenges us to reconsider the nature of instruction itself. What does it mean to be a teacher in a world where learners can access an ocean of content with no more than a voice command? And what does it mean to be a learner when every interaction, search, or spoken phrase becomes part of one's linguistic data profile? In many ways, mobile technology has blurred the boundaries between formal and informal learning. The walls of the classroom are no longer the sole container of linguistic development. Language is being acquired during a walk with a podcast, over coffee with a language exchange partner on a video call, or during a gamified vocabulary challenge on an app. This decentralization of learning offers unprecedented freedom, but also requires a recalibration of responsibility. Learners must become self-directed, discerning not just what they learn, but how and why they do so (Hamlin, 2022). At the same time, educators are no longer just transmitters of knowledge, but facilitators, curators, and guides in a far more complex and fluid learning environment.

Figure 1. Personalized pathway components

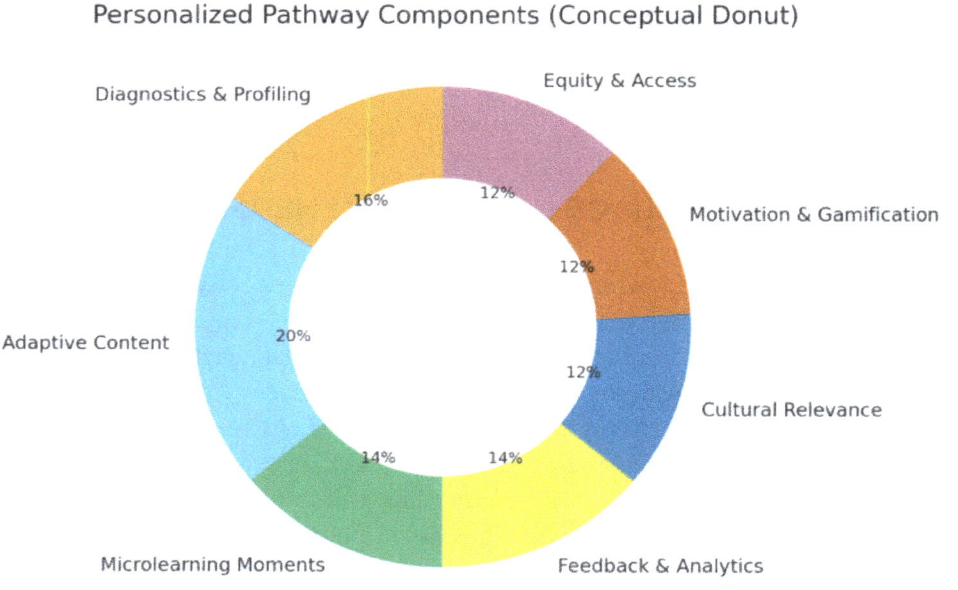

One of the most compelling aspects of mobile-based language learning is its capacity for personalization. Where traditional curricula often forced learners into rigid sequences regardless of ability or interest, mobile technologies can now adapt to the individual—offering lessons tailored to one's pace, strengths, and even mood. An app may detect that a learner is struggling with verb conjugation and adjust its content accordingly. A platform may notice that a learner retains vocabulary better when it is embedded in stories rather than lists, and thus prioritize narrative-based exercises (EFTHYMIA-AGAPI, 2024). In this sense, the mobile age is not just a change in medium—it is a transformation in methodology. It allows education to listen to the learner, rather than the other way around. This rethinking also requires us to interrogate our assumptions about what success in language learning looks like. The mobile age encourages progress that is nonlinear and multidimensional. Fluency is no longer the sole benchmark. Competency, comfort, cultural understanding, and communicative confidence take center stage (Permatasari & Wulandari, 2025). Learners may not aim to become grammarians or professional translators. Instead, they may seek the ability to connect meaningfully with family, to travel with ease, to engage with foreign media, or to simply expand their worldview. These motivations—deeply personal and often dynamic—deserve recognition and validation. And mobile technology, with its capacity to respond to these varied aims, has the potential to honor the multiplicity of learner goals in ways that traditional approaches could not (Chen, 2022). Yet with all these possibilities come important tensions. The mobile age has also ushered in a wave of instant gratification. Learners can receive points, badges, and rewards for completing lessons—but are they truly internalizing what they learn? There is a risk that language acquisition becomes reduced to streaks, scores, and superficial engagement. The challenge lies in designing mobile experiences that are both engaging and pedagogically rich. It requires developers and educators to collaborate, ensuring that learning remains meaningful, contextual, and grounded in real communication rather than mechanical repetition.

Another tension arises from the question of access. While mobile technology promises democratization, the digital divide persists. Not all learners have equal access to high-speed internet, updated devices, or the time and space to engage with mobile learning tools. Rethinking language education in this context must also involve a reckoning with inequality—ensuring that the shift to mobile does not deepen existing educational gaps, but rather bridges them (Yılmaz & Söğüt, 2022). Equity must be designed into the system from the ground up, with considerations for linguistic diversity, socioeconomic status, and accessibility needs. We must also consider the human aspects that technology cannot replace. Empathy, encouragement, humor, and spontaneous feedback are difficult to simulate. A mobile app may correct your pronunciation, but it cannot laugh with you when you mispronounce something in a funny way. It cannot see the doubt in your eyes, or the pride in your expression when you finally get it right. In rethinking language learning, we must be careful not to lose the relational essence of education—the sense of community, shared struggle, and mutual growth that comes from learning together. Mobile learning should supplement, not supplant, these experiences. It should enhance human connection, not replace it. The future of language learning in the mobile age, then, is not about choosing between technology and tradition. It is about integration. It is about building ecosystems where mobile devices, human mentors, cultural experiences, and personal motivations all work in harmony. It is about designing learning journeys that honor the learner's voice, leverage the strengths of mobile technology, and remain deeply rooted in the human need for meaning and connection.

We also must be open to rethinking assessment. Standardized testing, for all its familiarity, often fails to capture the subtlety of language competence in mobile contexts (Park et al, 2023). How do we measure growth when learning happens in fragmented, spontaneous moments? How do we recognize progress when it is spread across different platforms, languages, and modes? The mobile age calls for new ways of validating learning—through portfolios, self-reflection, peer feedback, and dynamic performance-based assessments that reflect the real-world use of language. Rethinking language learning in the mobile age is an invitation. It is a call to step outside the comfort of traditional paradigms and embrace a more holistic, learner-centered approach (Kumar et al, 2023). It is a chance to reimagine what it means to learn a language—not as a race to fluency, but as a lifelong relationship with communication, culture, and self-expression. And it is a reminder that at the heart of every technological advancement should lie a commitment to the learner—not just as a user, but as a whole person, seeking connection, growth, and belonging. As we move forward into this mobile future, our role is not to dictate a singular path, but to offer many paths—and the freedom to choose among them. It is to design not just apps, but experiences. Not just curricula, but communities. Not just systems, but stories. In doing so, we do more than modernize language education—we humanize it. And in a world that is increasingly digital, that human touch may be the most transformative innovation of all.

THE ROLE OF PERSONALIZATION IN ENHANCING LANGUAGE ACQUISITION

To speak about language learning today without speaking about personalization is to overlook one of the most significant pedagogical transformations of our time. The journey of acquiring a new language has always been personal in nature—shaped by each learner's background, motivation, and context (Caldwell-Harris & MacWhinney, 2023). Yet, for much of modern educational history, language instruction has been standardized, curriculum-driven, and uniform. This mismatch between the deeply individual experience of learning and the mass-produced models of instruction has often led to frustration, disengagement, and uneven outcomes. Personalization, particularly when paired with mobile technologies, represents a return to honoring the learner as an individual, not a demographic (Kayyali, 2024 a). It offers a path forward that is both more humane and more effective. To begin, it is important to recognize that language acquisition is not a linear process. It is messy, emotional, and often nonlinear. Learners plateau, regress, surge ahead, and struggle—sometimes all within the same week. Traditional educational models tend to treat learners as cohorts: progressing at the same pace, mastering skills in the same sequence, and being assessed against the same benchmarks. Personalization disrupts this model by asking a different question: not “Where should the class be?” but “Where is the learner right now, and what do they need next?” This shift is not just technical—it is philosophical. It centers the learning process on the learner’s current abilities, preferences, and goals, rather than a predetermined syllabus.

Personalized learning in language acquisition involves more than simply adapting content difficulty. It encompasses a broader spectrum of factors—learning style, preferred modalities (audio, visual, textual, interactive), cultural background, motivation, and even the learner’s emotional state. A personalized approach seeks to weave all of these elements into a learning pathway that feels both relevant and empowering. In practice, this could mean that a learner with a strong auditory memory is provided with more podcast-based exercises, while a visually inclined learner might thrive through infographics and video content (Kayyali, 2024 a). Similarly, a learner interested in travel might receive vocabulary and contexts related to airports, navigation, and cultural interaction, while someone aiming for academic study may receive exposure to formal structures and academic discourse. The significance of such personalization becomes even more profound in the realm of mobile learning. Mobile platforms have the unique capacity to collect data passively, respond in real time, and adapt continuously (Kayyali, 2025 b). Unlike a textbook, a mobile app can notice that a learner consistently struggles with irregular verbs and provide targeted practice until improvement is observed. It can track the time of day the learner is most active and send reminders accordingly. It can gauge which types of activities maintain engagement and subtly offer more of those, avoiding fatigue and monotony. In this way, the mobile device does not just deliver content—it becomes a responsive companion in the learning journey.

But perhaps the most transformative aspect of personalization is the sense of ownership it instills in learners. When learners feel that the learning path is theirs—not imposed, but designed with them in mind—they are more likely to stay engaged, take risks, and persist through challenges. Language acquisition, especially beyond childhood, requires perseverance. It involves moments of vulnerability, embarrassment, and self-doubt. Personalized learning can ease this process by offering a supportive environment tailored to the learner’s pace and comfort level. For instance, an introverted learner might benefit from asynchronous speaking practice via voice recordings before engaging in live conversations. A learner anxious about grammar accuracy may feel more confident working through error-feedback loops in private before showcasing their skills publicly (LIU, 2024). This sense of psychological safety—knowing that the learning experience is forgiving, adaptive, and non-judgmental—can have a profound impact on outcomes. Learners are more likely to persevere when they know the system is built to respond to their difficulties rather than penalize them. In this sense, personalization is not simply an efficiency mechanism—it is an act of empathy, a way of recognizing that behind every learner is a complex human being with fears, hopes, and aspirations.

Equally important is the cultural dimension of personalization. Language is not just a code to be deciphered—it is a cultural artifact, embedded with values, histories, and worldviews. Personalized learning allows for culturally relevant instruction, which is especially crucial in linguistically and culturally diverse contexts (Yangambi, 2024). For example, idioms, humor, social norms, and politeness strategies differ widely across cultures. A one-size-fits-all model risks flattening these nuances or, worse, alienating learners by reinforcing content that feels distant or irrelevant. With personalization, learners can engage with content that resonates with their identity, making the process not only more effective but more meaningful. A mobile learning app, when thoughtfully designed, can offer region-specific variations, dialectal differences, or even integrate cultural knowledge as part of the language learning process (El Zahraa, 2025). Personalization also plays a key role in sustaining motivation—a notoriously fragile element in long-term language learning. While the initial excitement of picking up a new language can be strong, it often fades in the face of irregular verbs, unfamiliar sounds, or plateaus in progress. A personalized system, however, can track motivational dips and intervene—offering a change of activity, a word of encouragement, a reminder of progress, or a gamified challenge to rekindle interest. These seemingly small interventions can make the difference between giving up and pushing through.

Personalization allows for greater autonomy—another cornerstone of successful language learning. Autonomy does not mean that the learner is left alone; rather, it means that the learner feels in control of their own journey. They choose when to learn, how to learn, and what to focus on. In this sense, mobile platforms are ideally suited to support autonomy: they offer on-demand access, diverse content types, and modular structures that let learners chart their own course. The learner becomes the agent, not the subject, of their education. Personalization is not without its challenges (Klašnja-Milićević & Ivanović, 2021). One concern is the potential for echo chambers, where learners are only exposed to what they prefer, not what they need. For example, a learner might enjoy vocabulary games but avoid speaking tasks out of discomfort. A personalized system that only rewards preference may reinforce avoidance behaviors, ultimately limiting growth. Thus, effective personalization must balance responsiveness with pedagogical guidance—encouraging learners to stretch without overwhelming them. It must be intelligent enough to differentiate between a true preference and a fear disguised as a preference.

Another challenge is privacy and data ethics. Personalized systems often rely on collecting and analyzing learner data—ranging from error patterns to usage times, from preferred content types to geolocation (Kayyali, 2025 a). While this data enables powerful insights and adaptations, it also raises critical questions about consent, transparency, and data security. Learners must be informed and empowered to control how their data is used, and designers must ensure that personalization serves the learner rather than exploiting them. There is also the risk of over-mechanization. In the quest to personalize, developers may lean heavily on algorithms, reducing the complexity of human learning to a series of binary choices and predictive analytics (Elwarraki et al, 2023). But language acquisition is not entirely logical—it is emotional, relational, and deeply human. Personalization, therefore, must retain space for spontaneity, messiness, and the kind of learning that occurs not because it was predicted, but because it was stumbled upon. Serendipity matters. Learning, at its best, includes moments of unexpected joy and discovery that no algorithm can script.

As we move further into the mobile age, the role of personalization will only grow. Artificial intelligence will become more sophisticated, interfaces more intuitive, and content more adaptive (Song et al, 2024). But the guiding principle must remain the same: personalization is not about manipulating learners into longer screen time or higher completion rates. It is about honoring the learner's uniqueness and offering them a space in which their language journey can unfold authentically, at their own rhythm, with meaningful support. Personalization enhances language acquisition not because it makes learning easier, but because it makes learning more real. It respects difference, embraces complexity, and invites learners to bring their whole selves into the process. It says, in effect: "We see you. We know you learn differently. And we've made a place for you here." And in a world where language is more than just a skill—but a gateway to belonging, to opportunity, and to self-expression—that kind of learning is not just desirable. It is essential.

DESIGNING MOBILE-BASED LEARNING PATHWAYS: PRINCIPLES AND PRACTICES

Designing a learning experience is never just about arranging content; it is about creating a journey—one that feels intentional, meaningful, and humane. In the context of language learning, this journey becomes even more intricate. Language is not learned in isolation; it is acquired in relation to purpose, emotion, culture, and personal aspiration. When we bring mobile technology into the picture, the task of designing effective learning pathways takes on both exciting possibilities and serious responsibilities. Mobile platforms can expand access, provide adaptive feedback, and offer constant engagement. But to truly serve the learner, these platforms must be built on thoughtful principles that reflect a deep understanding of how people learn, not just how they interact with screens. The term *learning pathway* suggests more than a collection of lessons (Sari et al, 2021). It implies progression, choice, flexibility, and structure. It honors the learner's agency while also providing guidance. The challenge in mobile-based environments is striking the right balance between structure and freedom. Too much structure, and the learner feels trapped in a rigid system that doesn't respond to their needs. Too much freedom, and they may feel lost, overwhelmed by options, and unsure of what comes next. A well-designed mobile-based learning pathway is one that offers a clear sense of direction while allowing space for divergence, exploration, and personalization.

The first principle in designing such a pathway is learner-centricity. That may sound obvious, but in practice, it requires a radical shift. Traditional educational content is often built around what institutions want to teach, rather than what learners need to learn (Kayyali, 2024 c). In contrast, a mobile learning pathway must begin with the learner: Who are they? What are their goals? What prior knowledge do they bring? What barriers might they face? Answering these questions isn't always easy—but mobile platforms have an advantage in that they can gather insights over time. Initial diagnostics, user inputs, behavioral data, and performance history can be used not to label learners, but to understand them better. This understanding then becomes the foundation for a dynamic, responsive design that grows with the learner rather than constraining them. Another key principle is contextual relevance. In traditional classrooms, language is often taught in decontextualized forms—fill-in-the-blank exercises, isolated vocabulary lists, and dialogues that feel artificial. Mobile pathways, however, have the capacity to bring learning into the learner's everyday world. By integrating location-based services, real-time communication tools, or cultural content tailored to the learner's environment, mobile learning becomes more than an academic exercise; it becomes lived experience. For instance, a mobile app could prompt a learner visiting a coffee shop to practice ordering in the target language, or suggest a vocabulary set related to an upcoming holiday in the learner's country. The idea is to embed language into daily life, turning the world itself into a classroom.

Progressive layering is another essential practice in pathway design. Language learning is not a one-time revelation—it's an accumulation of skills, patterns, and intuitions (Lin, 2021). A mobile-based learning path should reflect this reality by offering content that builds naturally and meaningfully. This doesn't mean simply increasing difficulty; it means spiraling learning in such a way that each new step reinforces prior knowledge while introducing new challenges. A word learned in a vocabulary module might reappear later in a listening exercise, then in a writing prompt, then in a simulated conversation. This layered approach not only strengthens retention but also mirrors the multifaceted nature of real communication. It also encourages deep learning over superficial completion, pushing learners to apply, reuse, and synthesize what they've encountered before. In designing mobile pathways, feedback deserves special attention. Feedback in traditional learning contexts is often delayed, sporadic, or generic. But mobile platforms can provide immediate, targeted, and ongoing feedback that is crucial for language acquisition. Yet it's not just about correcting mistakes; it's about cultivating awareness. A well-designed pathway doesn't just say, "You got this wrong." It might ask, "Can you spot the pattern in your errors?" or "Let's review this concept in a different context." Feedback can also be emotional as well as cognitive: a word of encouragement, a badge for consistency, a reflection prompt that helps the learner see how far they've come. The best mobile pathways treat feedback not as punishment for being wrong, but as a partner in the process of becoming more right.

Figure 2. Readiness for mobile-personalized language learning

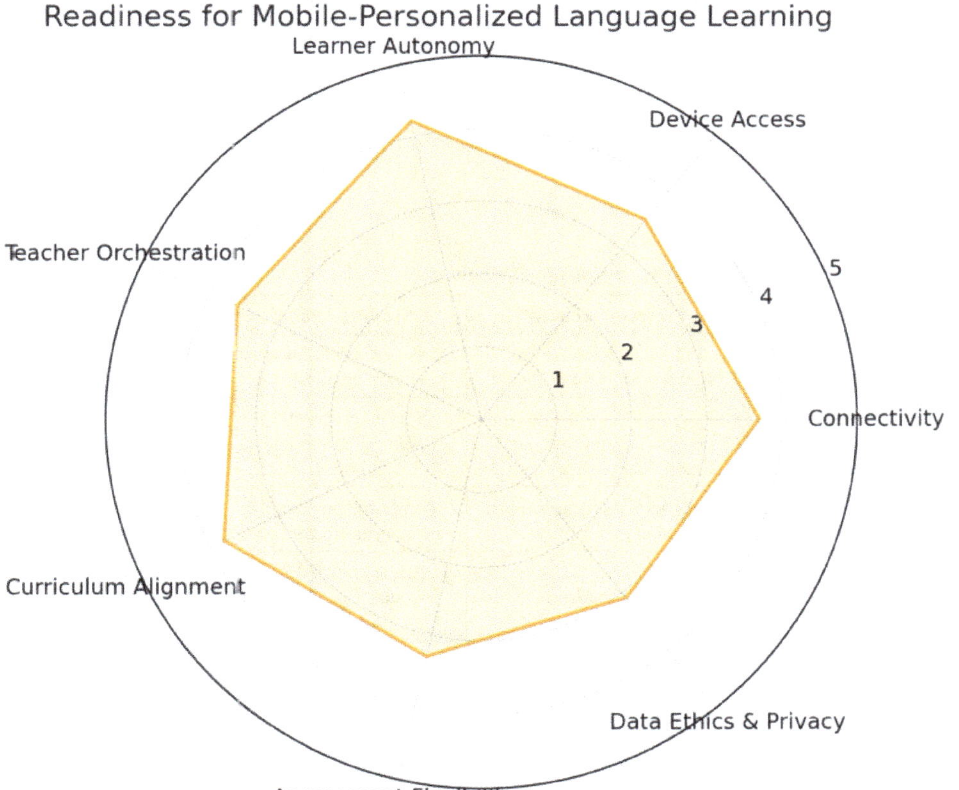

Variety and multimodality also play a crucial role. No single method works for all learners, and no single format suits all content (Lim et al, 2022). A robust mobile learning pathway should offer a range of activity types—listening, reading, speaking, writing—and a range of media formats, from text to audio to video to interactive challenges. It should also account for different levels of cognitive demand, mixing practice and play, review and discovery, repetition and creativity. Importantly, variety is not just for engagement; it's for balance. Language learning touches on so many dimensions—phonetics, grammar, pragmatics, culture—that it requires a rich ecosystem to grow (Zhang, 2022). A pathway designed with only quizzes and flashcards, no matter how polished, is like trying to teach someone to swim using only diagrams. Exposure, immersion, and active use are essential. Another principle is goal orientation—but with a human touch. Many mobile apps boast progress bars, daily streaks, or skill trees to gamify the experience (Oberoi, 2024). These can be motivating, but they must be tied to real, meaningful goals. A learner should not only see how many points they earned but also be reminded of why they're learning in the first place (Pham & Pham, 2022). The best mobile learning designs bring the learner's goals to the surface regularly. They prompt reflection: "Why are you learning this language?" "How will you use what you've learned today?" "What would you like to be able to do next?" These questions help learners connect abstract progress with personal purpose, turning routine practice into a values-driven pursuit.

Table 1. Personalized pathway design matrix

Trust Strategy	Professor Actions	Student Actions	Evidence / Artifact
Transparent Syllabus & Grading	Publish goals, rubrics, timelines; explain rationale	Review expectations; ask clarifying questions	Syllabus; rubric; timeline
Consistent Deadlines & Response Times	State and keep turnaround times; flag changes early	Plan ahead; respect timelines; communicate issues	Email/ LMS logs; update notes
Empathetic Office Hours	Invite check ins; acknowledge pressures; refer supports	Book time proactively; bring specific concerns	Office hour notes; referral records
Fair, Criteria Based Assessment	Use rubrics; blind/anonymous grading where feasible	Use rubrics; request feedback; reflect on results	Scored rubrics; moderation notes
Demonstrated Competence & Preparation	Organized sessions; responsive to questions; admit limits	Prepare; participate; give constructive feedback	Slide outlines; Q&A logs
Co created Norms & Student Agency	Offer task choices; negotiate deadlines within bounds	Select pathways; set goals; document decisions	Choice menus; learning contracts
Psychological Safety Signals	Normalize questions/errors; model vulnerability	Participate; support peers; practice civil debate	Reflection prompts; discussion guidelines

Cultural sensitivity and inclusion should never be an afterthought. Mobile learning often reaches across borders, into the hands of learners from vastly different cultural, social, and educational contexts. Pathways should reflect that diversity (Zhang & Hu, 2024). This includes offering content that is inclusive in terms of gender, ethnicity, ability, and worldview; avoiding stereotypes; and incorporating voices and perspectives from the target language culture(s) authentically and respectfully. It also means providing language options and accessibility features that allow all users—not just the tech-savvy or highly educated—to participate fully. Importantly, we must recognize that coherence matters as much as content. Learners need to feel that the journey they're on makes sense—that it's going somewhere, that there is a logic to how things are unfolding. A poorly designed pathway may have excellent materials but feel disjointed, arbitrary, or repetitive. Coherence doesn't mean predictability, but it does mean intentionality (Danilov & Mihailova, 2021). Each step should connect to the last, build on it, and gently guide the learner toward the next. Transitions between units, integration of review, and thematic continuity all contribute to a sense of forward motion and narrative arc in the learning journey. The best-designed mobile learning pathways are not static—they are living systems, capable of growing with the learner. As technology evolves, so too must our designs. This means using learner feedback not just for correction but for iteration. It means conducting research, gathering data, and adjusting the design to respond to real-world usage. It means viewing the pathway not as a product to be completed, but as a relationship to be nurtured—a collaboration between designers, educators, and learners in the shared project of language learning. In designing these pathways, we must remember that mobile technology is only as good as the thinking behind it. Flashy graphics and seamless animations can never compensate for shallow pedagogy (Lissovoy, 2022). What matters is whether the pathway honors the learner's humanity—whether it offers dignity, challenge, support, and opportunity. Designing mobile-based language learning pathways, then, is not a technical task. It is a moral one. It is the task of making sure that in a world increasingly shaped by algorithms, learners still feel seen, guided, and inspired. And that is the quiet power of good design: it does not shout, but it listens. It does not dominate, but invites. It does not impose, but accompanies. And when it is done with care, it can turn even the smallest device into a doorway—not just to a new language, but to a larger life.

LEARNER AUTONOMY, MOTIVATION, AND ENGAGEMENT THROUGH MOBILE INTEGRATION

The intersection of learner autonomy, motivation, and engagement represents one of the most delicate and essential threads in the tapestry of language acquisition. These three elements—distinct yet intimately interconnected—form the emotional and cognitive infrastructure upon which successful language learning is built. While mobile integration in education is often celebrated for its scalability, accessibility, and technological sophistication, its real power lies in its potential to empower learners: to cultivate autonomy, spark genuine motivation, and sustain long-term engagement. But these are not qualities that can be manufactured or artificially imposed. They must be nurtured—thoughtfully, intentionally, and with deep respect for the learner as a human being navigating the challenges and joys of learning a new language. Autonomy, at its heart, is about agency. It is the learner's ability to make informed, independent choices about what they learn, how they learn it, and why they do so (Hase & Blaschke, 2021). This does not imply learning in isolation or without support; rather, it reflects a shift in the locus of control from teacher to learner. In traditional language classrooms, learners have often been positioned as passive recipients—expected to absorb rules, mimic structures, and regurgitate vocabulary on command. This model, while efficient in delivering standardized curricula, often fails to ignite ownership. Autonomy challenges this paradigm. It invites learners to step forward, to define their goals, reflect on their progress, and take responsibility for their development (Bakar, 2021).

Mobile learning platforms, when designed well, can be powerful allies in fostering autonomy. They offer flexibility in time and place—allowing learners to practice in the rhythm of their own lives, rather than according to the bell schedule of an institution. They allow learners to revisit materials as often as needed without fear of judgment. They provide multiple pathways, letting users explore topics that interest them, skip what they already know, or dive deeper into areas of difficulty. This flexibility doesn't just cater to convenience—it affirms the learner's agency. It communicates a message that says: "Your learning is yours." But autonomy does not thrive in a vacuum. It depends on motivation—a deeply personal and often fluctuating force that drives learners to persist, to explore, to care. Motivation in language learning is not static; it ebbs and flows, influenced by context, success, failure, interest, fatigue, and a myriad of external and internal factors (Nikoletta, 2023). In the early stages of language learning, motivation may be high—fueled by curiosity, travel plans, or the desire to connect with others. Yet as learners encounter difficulty, plateau in their progress, or become overwhelmed by the demands of grammar and syntax, motivation can wane. Mobile integration, when thoughtfully applied, can serve as a buffer against these motivational declines. Unlike traditional classrooms, which rely heavily on scheduled momentum and external encouragement, mobile platforms can provide personalized nudges, adaptive content, and immediate feedback—tools that can help sustain the learner's energy even in solitary moments. They can remind learners of their goals, celebrate their milestones, and reframe failure as part of the learning process rather than as a judgment of ability.

More importantly, mobile learning can tap into what psychologists refer to as intrinsic motivation—the internal satisfaction that comes from doing something for its own sake. Intrinsic motivation flourishes when learners feel competent, autonomous, and connected. A well-designed app might create a sense of competence by offering scaffolded challenges that gradually increase in difficulty (Bacigalupo et al, 2024). It may offer autonomy by letting users choose when and how they engage. It might create a sense of connection through community features—forums, chats, peer feedback—that remind learners they are not alone. These features, when used not as gimmicks but as genuine support systems, can turn the solitary act of mobile learning into a shared, motivating experience. Still, motivation alone is not enough. Engagement is the engine that turns motivation into action. It is not merely about "using the app" or "completing a task." Engagement is about being mentally and emotionally present—curious, focused, and invested in what one is doing. It is the difference between clicking through flashcards while half-watching television and leaning in with interest during a conversational exercise. And in a digital environment full of distractions, designing for true engagement is both a technical and human challenge.

Mobile integration offers a rich palette of tools to enhance engagement—but it must be done with care. Gamification, for example, can add excitement and structure through points, badges, and leaderboards (Saleem et al, 2022). But if overused or poorly designed, it can reduce language learning to a trivial game, encouraging surface-level interactions instead of deep learning. Engagement that lasts is not about dopamine hits—it's about relevance, authenticity, and the sense that what one is doing matters. To foster this kind of engagement, mobile platforms must offer content that connects with the learner's life. Lessons about shopping, travel, or work should feel grounded in realistic scenarios, not robotic dialogues. Activities should invite reflection, creativity, and even error—because true engagement means feeling safe enough to try and fail. Learners should be able to see how what they are learning relates to who they are and who they want to become. When this connection is made—when the learner sees themselves in the learning experience—engagement becomes a natural outcome, not a forced one.

Figure 3. Shift in learning time with mobile personalization, 2021–2026

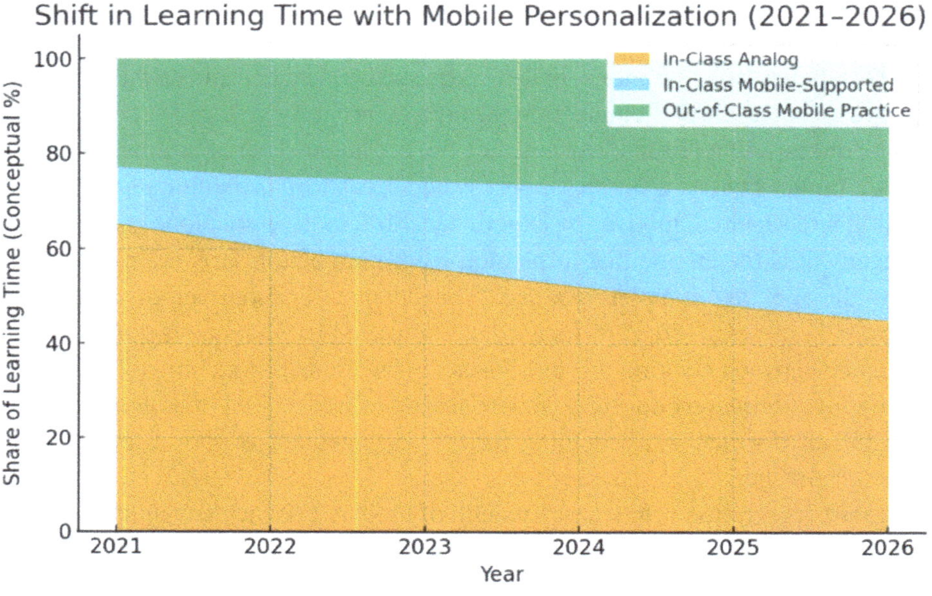

One of the most beautiful aspects of mobile learning is its potential to align with the rhythms of human life (Fijačko et al, 2021). In contrast to classroom models, which operate on fixed schedules, mobile integration respects the fragmented, nonlinear, and deeply personal nature of adult learning. A learner might practice verb forms on the subway, listen to a dialogue while cooking, or review vocabulary before sleep. These micro-moments of engagement, while individually small, accumulate into meaningful progress over time. And they reinforce the idea that language learning is not something one does in a classroom—it is something one lives. Yet, while mobile platforms enable autonomy, motivation, and engagement, they cannot guarantee them. These are human qualities, and they require more than clever design. They require empathy, nuance, and the recognition that learners are not data points—they are people. People with anxieties, families, dreams, and distractions. A truly supportive mobile learning experience is one that acknowledges these realities and offers tools that are not only intelligent but compassionate. For example, offering reminders that are encouraging rather than guilt-inducing, designing challenges that invite curiosity rather than shame, and providing space for reflection as well as achievement.

Moreover, we must be cautious not to confuse activity with progress. A learner might open a mobile app every day and still feel stuck. Autonomy must be paired with guidance; motivation must be nourished with meaningful experiences; engagement must be deepened with relevance and support (Aldosari & Alsager, 2023). This is why mobile integration in education should not be seen as a replacement for human connection, but as a complement to it. Teachers, mentors, and peer communities still play a vital role in helping learners set goals, process setbacks, and celebrate breakthroughs. The most powerful mobile learning experiences are those that blend the best of both worlds—the accessibility and flexibility of technology with the care and insight of human support. It is worth remembering that learner autonomy is not only a pedagogical goal—it is a democratic one. Empowering learners to take charge of their education is a form of respect. It says: You are capable (Kayyali, 2024 d). You know yourself. You have the right to learn in ways that work for you. In this way, mobile integration is not just a technological evolution—it is an ethical stance. It reflects a belief in the dignity of the learner and the value of their journey. As we continue to integrate mobile technologies into language learning, our responsibility is not simply to make learning more efficient or more convenient. Our responsibility is to make it more humane. To design experiences that honor the learner's autonomy, fuel their motivation, and sustain their engagement not just for a moment, but for the long and beautiful path of language acquisition. When we do this well, we do more than teach a language. We invite people into a fuller expression of themselves—and that, ultimately, is what education is meant to do.

CHALLENGES, OPPORTUNITIES, AND THE ROAD AHEAD FOR MOBILE-ASSISTED LANGUAGE LEARNING

The story of mobile-assisted language learning (MALL) is still being written. What began as an experiment in portability and convenience has evolved into a full-fledged movement reshaping how people around the world learn, use, and relate to languages (Zain & Bowles, 2021). The dream of personalized, flexible, and engaging language learning has never been more tangible. Yet, as with all transformations, the path forward is not without friction. The road ahead is marked by both promise and complexity—an intricate landscape where technological innovation must meet pedagogical wisdom, and where accessibility, ethics, and human connection must not be left behind.

One of the most pressing challenges in the field today is the illusion of progress that mobile learning can sometimes create. With the rise of gamified platforms, learners are often rewarded with points, streaks, badges, and animations—metrics that signal activity, but not always mastery. Tapping through vocabulary decks or completing lessons at a rapid pace can feel productive, yet the depth of understanding and retention may be shallow (Bastawros, 2024). This surface-level engagement, while initially motivating, can leave learners disillusioned over time when they realize that despite all the progress bars filled and achievements unlocked, they struggle to speak with fluency or comprehend real-world conversations. This disconnect arises not from malice or negligence, but from a mismatch between design priorities and educational goals. Many mobile learning tools are built with a consumer mindset: optimize for engagement, retention, and daily active users. While these metrics make sense in the tech world, they don't always translate to meaningful learning. The road ahead demands a reevaluation of what we value in educational design. We must ask: Are we helping learners grow in competence, confidence, and communicative ability—or are we just keeping them busy?

Another challenge is equity of access. Mobile technology is often celebrated for democratizing education—and rightly so. It has brought language learning into rural communities, refugee camps, busy homes, and remote classrooms (Tobin & Hieker, 2021). It has enabled people with disabilities, caretaking responsibilities, or full-time jobs to engage with education on their own terms. Yet, access is never as simple as owning a device. Digital divides persist not only in terms of hardware and connectivity, but in digital literacy, app usability, language availability, and contextual relevance. A beautifully designed app in English may be unintelligible to a learner whose primary language is Amharic or Urdu. A high-bandwidth platform may be unusable in regions with spotty internet. Designing for access, therefore, is not just about scaling reach—it is about radical inclusion (Pineda, 2024). It means building tools that work offline, interfaces that are intuitive for first-time users, and content that respects linguistic and cultural diversity. It means involving learners from marginalized communities in the design process—not just as recipients, but as co-creators. It means acknowledging that mobile learning cannot be a one-size-fits-all solution, but must adapt to a multiplicity of realities, devices, and life conditions. Closely related to this is the question of language ideologies embedded in mobile learning tools. Whose language is being taught? Whose accent is being prioritized? What norms, dialects, and cultural references are presented as "standard" or "correct"? Too often, mobile platforms replicate colonial or elitist hierarchies—privileging dominant languages, urban speech patterns, and Western-centric examples. A truly inclusive vision of MALL must interrogate these choices and push for more diverse linguistic representations (Robinson-Jones, 2024). Learners should be exposed to multiple varieties of the target language and taught to navigate linguistic plurality, not just conformity.

There is also the question of teacher roles in the mobile learning ecosystem. As technology takes on more functions—providing feedback, structuring lessons, offering explanations—some have wondered whether the teacher becomes obsolete. This is not only a misguided fear, but a dangerous one. No app, however advanced, can replicate the nuance, empathy, and insight that a skilled teacher brings to the learning process. The real opportunity lies in rethinking the relationship between teachers and technology, not replacing one with the other. Mobile tools can free up teachers to do what they do best—facilitate conversations, personalize support, and nurture motivation—while automating repetitive or data-heavy tasks (Tyagi, 2023). The road ahead must include professional development that helps educators integrate mobile tools thoughtfully, not as gimmicks or replacements, but as extensions of their pedagogical care. Perhaps the most profound challenge ahead is the question of sustainability—not just in the environmental sense, though that matters too, but in terms of emotional, cognitive, and ethical sustainability. What kind of learning culture are we building when learners are expected to engage with screens every day, often in solitary environments, mediated by algorithms? Are we cultivating curiosity and resilience—or dependency and burnout? Are we preparing learners to interact with real human beings, or only with neatly packaged user interfaces? These are not questions that can be answered with more code or faster processors. They require reflection, humility, and a return to the core purpose of language learning: to connect, to express, to understand, to belong.

Table 2. Barriers, solutions, supports

Domain	Typical Barrier	Practice / Intervention	Support Needed	Metric / Evidence
Communication	Ambiguous expectations	Transparency framework; Q&A forum	LMS tools; time to clarify	Fewer clarification emails; survey clarity scores
Assessment	Perceived grading bias	Criteria; exemplars; moderation	Calibration time; co grading	Reduced appeals; rubric reliability
Classroom Climate	Fear of speaking up	Norms; think pair share; anonymity options	Facilitation training	â†‘ participation; climate survey scores
Equity & Inclusion	Unequal voice/ opportunity	Rotating roles; inclusive materials	DEI resources; community partners	Representation audit; participation parity
Support & Wellbeing	Untended stress & overload	Check ins; referral pathways; flexibility windows	Counseling services; wellbeing policies	Referral uptick (early); absenteeism down
Academic Integrity	Mistrust from misconduct cases	Preventive education; restorative responses	Integrity office; clear process	Lower incidents; learning reflections
Institutional Alignment	Misaligned policies/ resources	Shared rubrics; faculty PD; student feedback loops	Leadership backing; data dashboards	Student satisfaction; retention/ engagement

And yet, despite these challenges, the opportunities for mobile-assisted language learning remain vast and inspiring. We are living in a time when the walls between formal and informal learning have been all but dismantled. Learners can start a sentence in a textbook and finish it in a chat with a native speaker (Rasyid et al, 2023). They can move from theory to practice, from classroom to real world, in a matter of minutes. They can document their learning in real time, share it across networks, and receive feedback from communities around the globe. This kind of permeability—the ability to move seamlessly between contexts—is one of the greatest gifts that mobile learning offers. It brings language to life. There is also immense potential in the integration of emerging technologies. Augmented reality (AR) and virtual reality (VR) can place learners in immersive environments where they practice language in context—ordering food in a virtual café, asking for directions on a digital street, or participating in a simulated job interview (Kayyali, 2021). Artificial intelligence can provide real-time pronunciation coaching, adaptive grammar exercises, or conversation simulations that respond naturally to input. These tools, when used ethically and pedagogically, can make language learning more dynamic, personalized, and effective than ever before.

But technology alone will never be enough. The real opportunity lies in cultivating a human-centered vision of mobile learning—one that sees technology not as a solution, but as a companion (Kayyali, 2024 e). One that puts the learner's dignity, agency, and well-being at the center. This means designing not just for outcomes, but for experience. Not just for completion rates, but for connection. Not just for efficiency, but for meaning. It also means embracing pluralism and plurality. The future of mobile language learning is not monolithic. It is diverse, layered, and context-sensitive. In one country, mobile learning may be used to preserve endangered languages through community storytelling. In another, it may be a lifeline for migrant workers trying to navigate healthcare or employment. In yet another, it may serve as a bridge between generations, helping parents and children share language in ways that bring them closer. The road ahead must make space for all of these pathways—not as fringe projects, but as core expressions of what language learning can be.

There is, finally, a profound opportunity in the ethical reinvention of learning technologies. As awareness grows around data privacy, algorithmic bias, and tech fatigue, there is a chance to build platforms that are transparent, accountable, and learner-driven (Ismaili, 2024). Imagine a mobile app that tells you not just what you got wrong, but why it thinks you got it wrong—and how sure it is. Imagine a system where learners can access their data, challenge its interpretations, and co-author their learning trajectory. Imagine technologies designed not to hook users into endless engagement, but to empower them to step away once their needs are met. The future of mobile-assisted language learning will not be written by technology alone. It will be shaped by the choices we make now—about design, access, ethics, pedagogy, and care. It will be shaped by the courage to question trends, the humility to listen to learners, and the imagination to envision learning environments that are not only smart, but just. The journey ahead will not be easy. It will be uneven, contested, and at times, uncertain. But if we move forward with integrity, creativity, and a relentless focus on the human beings behind the screens, we may just create something extraordinary. Not just better apps. Not just smarter systems. But better learners. Smarter communities. And a more connected, compassionate world.

CONCLUSION

As we conclude this exploration of personalized language learning through mobile devices, it becomes clear that we are not merely discussing a set of tools or a trend in educational technology. We are witnessing the unfolding of a pedagogical transformation—one that repositions the learner at the center of the experience and reshapes the contours of language acquisition in profoundly human ways. The integration of mobile devices into language education is not just about making learning portable or convenient. It is about making learning personal, adaptive, and continuous. It is about creating a space where learners are not passive recipients of instruction, but active navigators of their own linguistic development. Throughout this chapter, we have examined how mobile technology enables learning environments that are increasingly responsive to the diverse needs, goals, and contexts of learners. From AI-driven apps that adjust to learner performance, to mobile platforms that allow for culturally relevant and real-time communication practices, the ecosystem of mobile-assisted language learning is rich with promise. But perhaps more importantly, this shift reflects a broader cultural and educational reckoning: a growing acknowledgment that education must honor the complexity of human learning—its pace, its variability, and its deeply personal character.

Yet, with all its potential, this transition demands responsibility. Educators, developers, and policymakers must ask not only what technology can do, but what it should do. Personalization, when reduced to data points and algorithmic feedback, risks dehumanizing the very learners it aims to support. True personalization must go beyond convenience or customization—it must engage with learners' motivations, identities, and aspirations. It must understand that language is not simply a system of rules to be mastered, but a living, evolving practice shaped by culture, emotion, and experience. Therefore, technology must serve the learner, not the other way around. In practical terms, this means designing mobile learning systems that promote reflection, encourage real-world application, and foster a sense of progress that is both measurable and meaningful. It means embracing learner autonomy without abandoning the guiding presence of teachers and mentors. It means making room for failure, for play, for experimentation—acknowledging that language learning is as much about exploration as it is about mastery. And it means ensuring that access to these tools is equitable, so that the benefits of mobile-enhanced personalization do not become privileges reserved for a few, but rights extended to all.

As we look to the future, the challenge will not be merely technical—it will be deeply pedagogical and ethical. How can we design learning systems that are not only smart, but wise? Not only efficient, but empathetic? The answers will require collaboration across disciplines, cultures, and contexts. They will require listening to learners, valuing their voices, and recognizing the nuanced ways they engage with language in everyday life. The value of personalized learning pathways through mobile devices lies not just in their capacity to enhance acquisition, but in their ability to cultivate confidence, curiosity, and connection. When language learning becomes a journey shaped by the learner, supported by technology, and enriched by meaningful interaction, we move closer to realizing a vision of education that is not only effective—but profoundly human.

REFERENCES

Aldosari, M. S., & Alsager, H. N. (2023). A step toward autonomy in education: Probing into the effects of practicing self-assessment, resilience, and creativity in task supported language learning. *BMC Psychology*, *11*(1), 434. DOI: 10.1186/s40359-023-01478-8 PMID: 38062504

Alexandra, T. (2024). Foreign Language Acquisition Difficulties: Diagnosing Individual Differences.

Bacigalupo, M., Bekh, O., Binasco, A., Israel, H., & WEIKERT, G. L. (2024). " Scaffold", a deck of cards to design competence-oriented learning experiences (No. JRC136622). Joint Research Centre.

Bakar, S. (2021). Investigating the dynamics of contemporary pedagogical approaches in higher education through innovations, challenges, and paradigm shifts. *Social Science Chronicle*, *1*(1), 1–19. DOI: 10.56106/ssc.2021.009

Bastawros, H. (2024). Enhancing Productive Vocabulary of ESL Learners: A Qualitative Case Study.

Caldwell-Harris, C. L., & MacWhinney, B. (2023). Age effects in second language acquisition: Expanding the emergentist account. *Brain and Language*, *241*, 105269. DOI: 10.1016/j.bandl.2023.105269 PMID: 37150139

Chen, Y. M. (2022). Understanding foreign language learners' perceptions of teachers' practice with educational technology with specific reference to Kahoot! and Padlet: A case from China. *Education and Information Technologies*, *27*(2), 1439–1465. DOI: 10.1007/s10639-021-10649-2

Danilov, I. V., & Mihailova, S. (2021). Neuronal coherence agent for shared intentionality: A hypothesis of neurobiological processes occurring during social interaction. *OBM Neurobiology*, *5*(4), 1–28. DOI: 10.21926/obm.neurobiol.2104113

EFTHYMIA-AGAPI. K. A. R. A. T. H. A. N. O. U. (2024). UNLOCKING VOCABULARY ACQUISITION: THEORETICAL FOUNDATIONS AND PRACTICAL APPROACHES FOR AUTISTIC EFL STUDENTS (Doctoral dissertation, Aristotle University of Thessaloniki).

El Zahraa, F. (2025). Leveraging Artificial Intelligence and Digital Technologies to Enhance Sociolinguistic Competence and Arabic Language Skills. In *Proceeding of the International Conference on Religious Education and Cross-Cultural Understanding* (Vol. 1, No. 1, pp. 33-49).

Elwarraki, O., Aammou, S., Jdidou, Y., & Lahiassi, J. (2023). Toward the future of personalized learning: Emerging trends and challenges in recommendation systems. ICERI2023 Proceedings, 7226-7235.

Fijačko, N., Masterson Creber, R., Gosak, L., Štiglic, G., Egan, D., Chaka, B., Debeljak, N., Strnad, M., & Skok, P. (2021). Evaluating quality, usability, evidence-based content, and gamification features in mobile learning apps designed to teach children basic life support: Systematic search in app stores and content analysis. *JMIR mHealth and uHealth*, *9*(7), e25437. DOI: 10.2196/25437 PMID: 34283034

George, A. S. (2024). Technology Tension in Schools: Addressing the Complex Impacts of Digital Advances on Teaching, Learning, and Wellbeing. *Partners Universal Multidisciplinary Research Journal*, *1*(3), 49–65.

Hamlin, M. D. (2022). Developing self-directed learning skills for lifelong learning. In *Self-directed learning and the academic evolution from pedagogy to andragogy* (pp. 209–234). IGI Global Scientific Publishing. DOI: 10.4018/978-1-7998-7661-8.ch012

Hase, S., & Blaschke, L. M. (2021). *Unleashing the power of learner agency*. Unleashing the Power of Learner Agency.

Iftikhar, N. (2025). MOBILE-ASSISTED LANGUAGE LEARNING (MALL): REVOLUTIONIZING SECOND LANGUAGE ACQUISITION. [JALT]. *Journal of Applied Linguistics and TESOL*, *8*(1), 1038–1045.

Ihnatova, O., Poseletska, K., Matiiuk, D., Hapchuk, Y., & Borovska, O. (2021). The application of digital technologies in teaching a foreign language in a blended learning environment. Linguistics and Culture Review, 114-127.

Ismaili, Y. (2024). AI Literacy and Adaptive Learning in Moroccan Education: Advancing Critical Thinking and Personalized Learning. *Research on Education and Media*, *16*(2), n2. DOI: 10.2478/rem-2024-0017

Kayyali, M. (2021). Positive impact of high technology on higher education.

Kayyali, M. (2024 a). Global Perspectives on Personalized Learning: Case Studies and Best Practices. In Transforming Education for Personalized Learning (pp. 66-84). IGI Global.

Kayyali, M. (2024 b). Immersive Technologies: Virtual and Augmented Reality in Higher Education. In Reshaping learning with next generation educational technologies (pp. 99-114). IGI Global.

Kayyali, M. (2024 c). Future directions towards sustainable educational equity: Policies and practices. In *Exploring Educational Equity at the Intersection of Policy and Practice* (pp. 81–100). IGI Global. DOI: 10.4018/979-8-3693-1614-6.ch005

Kayyali, M. (2024 d). Inclusive Practices in Higher Education and Workplaces Strategies for Empowering Women. In A Cross-Cultural Examination of Women in Higher Education and the Workplace (pp. 237-262). IGI Global. DOI: 10.4018/979-8-3693-0102-9.ch012

Kayyali, M. (2024 e). Future possibilities and challenges of AI in education. In *Transforming education with generative AI: Prompt engineering and synthetic content creation* (pp. 118–137). IGI Global Scientific Publishing. DOI: 10.4018/979-8-3693-1351-0.ch006

Kayyali, M. (2025 b). E-Learning Platforms and Their Influence on Lifelong Learning. In Impact of Digitalization on Education and Social Sustainability (pp. 155-176). IGI Global.

Kayyali, M. (2025 b). The Impact of Personalized Learning on Student Engagement and Achievement in STEAM. In *Integrating Personalized Learning Methods Into STEAM Education* (pp. 51–78). IGI Global Scientific Publishing. DOI: 10.4018/979-8-3693-7718-5.ch003

Klašnja-Milićević, A., & Ivanović, M. (2021). E-learning personalization systems and sustainable education. *Sustainability (Basel)*, *13*(12), 6713. DOI: 10.3390/su13126713

Kumar, P., Gupta, A., & Budarina, A. O. (Eds.). (2023). *Innovative Teaching Learning Practices: A Paradigm Shift*. GLOBAL ACADEMY YAYINCILIK VE DANIŞMANLIK HİZMETLERİ SANAYİ TİCARET LİMİTED ŞİRKETİ.

Lăpădat, M. M. (2023). Harnessing technological transformation: An exploration of the impact of mobile assisted language learning on English vocabulary acquisition. Analele Universităţii din Craiova, Seria Ştiinţe Filologice. *Limbi Străine Aplicate*, (1), 266–279.

Lim, F. V., Toh, W., & Nguyen, T. T. H. (2022). Multimodality in the English language classroom: A systematic review of literature. *Linguistics and Education*, *69*, 101048. DOI: 10.1016/j.linged.2022.101048

Lin, T. H. (2021). Revelations of service-learning project: Multiple perspectives of college students' reflection. *PLoS One*, *16*(9), e0257754. DOI: 10.1371/journal.pone.0257754 PMID: 34555105

Lissovoy, N. D. (2022). Capitalism, pedagogy, and the politics of being.

LIU. Y. (2024). Towards a Contextualised Interpretation of Chinese University Student Engagement with Teacher Written Feedback (Doctoral dissertation, University of Hong Kong).

Nikoletta, V. (2023). The effects of EFL variables on motivation and the learning outcome.

Oberoi, Z. (2024). Using gamification to enhance development.

Park, J. J., Kim, B. H., Wong, N., Zheng, J., Breen, S., Lo, P., & Poon, O. A. (2023). Inequality beyond standardized tests: Trends in extracurricular activity reporting in college applications across race and class. *American Educational Research Journal*, •••, 00028312241292309.

Permatasari, N. P. I., & Wulandari, C. I. S. A. (2025). English Speaking Fluency and Confidence for Global Competitive Work: The Highlight of Authentic Role-play Activity for University Students. *Jurnal Multidisiplin Sahombu*, *5*(02), 400–411.

Pham, A. T. V., & Pham, H. T. (2022, September). Developing the habit of learning English using Duolingo application. In *Proceedings of the 6th International Conference on Digital Technology in Education* (pp. 131-136). DOI: 10.1145/3568739.3568763

Pineda, V. S. (2024). Radical Inclusion: The Key to Urban Transformation. In *Inclusion and Belonging in Cities of Tomorrow: Governance and Access by Design* (pp. 1–21). Springer Nature Singapore. DOI: 10.1007/978-981-99-3856-8_1

Procel, G. J. O., Medina, M. L. F., Sotomayor, D. J., & Sanchez, M. A. (2024). Using Technology in English Teaching. *Journal of Environmental Research and Public Health*, *17*(9), 9.

Rasyid, F., Ibna Seraj, P. M., Ghofur, A., & Asrifan, A. (2023). Students' Perception toward Teaching Strategies of Native and Nonnative English-Speaking Teachers: A Case Study in Indonesia. *Education Research International*, *2023*(1), 7827917. DOI: 10.1155/2023/7827917

Robinson-Jones, C. (2024). Tension in the linguistic landscape: The implications of language choices for diversity and inclusion in multilingual museums representing minorities. *International Journal of Multilingualism*, *21*(2), 881–905. DOI: 10.1080/14790718.2022.2135710

Saleem, A. N., Noori, N. M., & Ozdamli, F. (2022). Gamification applications in E-learning: A literature review. Technology. *Knowledge and Learning*, *27*(1), 139–159. DOI: 10.1007/s10758-020-09487-x

Sari, S. N., Aminatun, D., Sari, S. N., Aminatun, D., Sari, S. N., & Aminatun, D. (2021). Students' perception on the use of English movies to improve vocabulary mastery. *J. English Lang. Teach. Learn*, *2*(1), 16–22. DOI: 10.33365/jeltl.v2i1.757

Song, T., Li, X., Wang, B., & Han, L. (2024). Research on intelligent application design based on artificial intelligence and adaptive interface.

Tobin, E., & Hieker, C. (2021). What the EdTech experience in refugee camps can teach us in times of school closure. Blended learning, modular and mobile programs are key to keeping disadvantaged learners in education. *Challenges*, *12*(2), 19. DOI: 10.3390/challe12020019

Tyagi, A. K. (Ed.). (2023). *Automated secure computing for next-generation systems*. John Wiley & Sons.

Yangambi, M. W. (2024). Culturally Relevant Teaching in the Digital Age: Student-Centered and Personalized Learning in Developing Countries. *Creative Education*, *16*(1), 71–89. DOI: 10.4236/ce.2025.161005

Yılmaz, A., & Söğüt, S. (2022). Language education for social justice: Reproductions or disruptions through technology. *Computers & Education*, *187*, 104535. DOI: 10.1016/j.compedu.2022.104535

Zain, D. S. M., & Bowles, F. A. (2021). Mobile-assisted language learning (MALL) for higher education instructional practices in EFL/ESL contexts: A recent review of literature. *Computer-Assisted Language Learning Electronic Journal*, *22*(1), 283–307.

Zhang, W. (2022). Language, Culture, and Ecology: An Exploration of Language Ecology in Pragmatics. *English Language Teaching*, *15*(6), 80–87. DOI: 10.5539/elt.v15n6p80

Zhang, X., & Hu, J. (2024). A study on the learning behaviors and needs of design-maker communities of practice in the era of mobile learning. *Library Hi Tech*, *42*(2), 580–606. DOI: 10.1108/LHT-12-2021-0486

KEY TERMS AND DEFINITIONS

Adaptive Feedback: Immediate, targeted guidance (e.g., phoneme cues, vocabulary resurfacing) driven by mobile analytics.

Data Ethics: Transparent, consensual use of learner data for personalization without compromising privacy or agency.

Diagnostics & Profiling: Early and ongoing checks (usage, errors, pace) that inform how a mobile system tailors content and support.

Learner Autonomy: The ability to choose what/when/how to learn, amplified by on-demand mobile activities and reflection.

Microlearning: Short, focused, in-the-moment tasks (e.g., voice-note drills, QR prompts) that compound into durable gains.

Mobile-Assisted Language Learning (MALL): Using handheld devices to deliver adaptive, anytime language practice that bridges formal and informal learning.

Personalized Pathway: A learner-specific sequence of mobile tasks that adapts to goals, performance, preferences, and context.

Chapter 4

An Analysis of Babbel Apps in Reading: Benefits and Drawbacks From the User Perspective

Sholihatul Hamidah Daulay
https://orcid.org/0000-0002-7660-8763
Universitas Islam Negeri Sumatera Utara, Indonesia

Muhammad Dalimunte
https://orcid.org/0000-0003-2089-514X
Universitas Islam Negeri Sumatera Utara, Indonesia

Hotma Berutu
https://orcid.org/0009-0002-8813-0979
Universitas Islam Negeri Sumatera Utara, Indonesia

Eka Apriani
https://orcid.org/0000-0002-8427-2817
Institut Agama Islam Negeri Curup, Indonesia

Maslathif Dwi Purnomo
Universitas Islam Negeri Sumatera Utara, Indonesia

Shaumiwaty Shaumiwaty
Institut Agama Islam Negeri Takengon, Indonesia

Julisah Izar
Universitas Jambi, Indonesia

ABSTRACT

This study aimed to investigate the benefits and drawbacks of the Babbel App based on users' perspectives on the Google Play Store. This study used interpretative qualitative as the research methodology. The data source used is screenshots of users' feedback on the Google Play Store for the Babbel App. The study came up with the finding that Babbel is a good tool for assisting users in learning a new language because it provides language learning classes in 14 different languages. Therefore, the users around the world benefit from the presence of Babbel. However, users also

DOI: 10.4018/979-8-3373-3366-3.ch004

revealed some drawbacks of Babbel App such as (1) the tricky use of Babbel app (2) the limited language choices, especially for the 'native language' choice. (3) users are required to understand English first before utilizing the app, and (4) it requires a high payment when the users want to jump to the next level.

INTRODUCTION

Technology gives fundamental changes on how we teach, learn and acquire second language L2) (Chapelle, 2007; Motteram & Sharma, 2009; Otto, 2017). Instructional context which involve technology in it offers lots of benefits to L2 learners, including: providing educational material based on its authenticity (Larsen-Freeman & Anderson, 2011), allowing for independent, self-paced learning (Pim, 2013), providing opportunities for developing intercultural competence through communication with native and other non-native speakers (Whyte, 2011), and increasing motivation (Baleghizadeh, 2015). Language teachers can also use technological innovations to create a more exciting and interactive classroom environment by incorporating materials from the outside world into the classroom (Stanley, 2013), provide learners with multimodal feedback (Elola & Oskoz, 2016), and assist students with special learning needs (Roblyer & Doering, 2010). It would not be exaggerating to say that technology today supports every element of language learning and teaching.

The advancement of technologies in employing devices like mobile phone to teach and learn L2 has piqued the interest of many scholars (Viberg & Grönlund, 2012). In reality, Mobile Assisted Language Learning (MALL), a subset of M(-mobile)-learning, is a fast emerging field of study with interesting implications for second language learning and instruction. (Pachler et al., 2010). Smartphones, tablet computers, laptops, MP3 and MP4 players, and other hand-held mobile devices appeal to consumers for a variety of reasons, including timelessness, accessibility, immediacy, and interaction. (Zheng et al., 2005). According to Thornton and Houser (2005), " Mobile devices can be useful instruments for a variety of educational tasks." Furthermore, new mobile technologies (e.g., Duolingo, Edmodo, Kahoot, Elsa, etc.) that promise to aid language learning are being created. Chen (2106) concludes that while lots of mobile English learning apps provide affordances for adult learners, " no single language learning app can provide a one-size-fits-all solution to fulfill the language learning demands of adult learners.,".

Crucially, most MALL-applications developed by industries are based on established learning principles from fundamental memory research. For instance, retrieval-based learning benefits learning over re-studying (Agarwal et al., 2021; Roediger & Butler, 2011; Yang et al., 2021). The increased interest in MALL and its uses should not be accepted at face value. According to Nushi and Jenabzadeh

(2016), many "mobile language applications" have been created by people who do not work in the field of second language instruction, and their efficacy cannot and should not be ignored." Viberg and Grönlund (2012) also note that "there are few empirical studies that show how the usage of mobile devices can improve individual language learning outcomes." Given those cautionary notes, it is important to critically examine the available language learning applications so that language learners and teachers are aware of their potential benefits and drawbacks, particularly Babbel app, and can make informed decisions about whether or not to include them in language learning and teaching packages. Based on the explanation stated previously, this paper is aimed to investigate the users' perspective related to benefits and drawbacks of Babbel app in learning new languages.

The Issue of App Effectiveness

The efficacy of Mobile-Assisted Language Learning (MALL) applications is currently a prominent area of inquiry within the educational landscape (Hirsh-Pasek et al., 2015), characterized by a variety of designs that aim to improve educational outcomes. These educational applications fulfill multiple roles, including the facilitation of knowledge acquisition (Hannon, 2017), the provision of instantaneous feedback, and the engagement of learners in contextually situated learning via augmented reality (Bower et al., 2014). Elements such as learner engagement, cognitive involvement, goal-directed guidance, alongside scaffolded and interactive learning experiences, significantly contribute to their overall effectiveness (Hirsh-Pasek et al., 2015). Moreover, it is essential to consider additional factors pertinent to older learners, including their motivations, expectations, and prior experiences. Similarly, mechanisms such as prompting questions, automated immediate feedback, and push notifications have been demonstrated to effectively sustain learner engagement (Kudo et al., 2015; Sung et al., 2016). Furthermore, factors rooted in educational psychology, such as the spacing effect and testing effect, are integral to the design of MALL applications, enabling students to revisit and consolidate their knowledge (Ebbinghaus, 2013).

As the demand for personalized learning experiences escalates among students (Wanner & Palmer, 2015) and personal mobile devices become increasingly prevalent, it is imperative to harness mobile application technologies to tailor educational experiences, a point underscored by Supyan Hussin's research (2017) regarding effective mobile applications that incorporate fundamental pedagogical strategies and features such as individualized instruction and reciprocal learning.

Employing Illeris's three-dimensional learning theory (2009) in conjunction with the empirical case study of Mobile-Assisted Language Learning (MALL) applications, while also considering pertinent scholarly literature (Liang et al., 2018), this analysis systematically investigates language learning applications through six critical dimensions: cognition, emotion, society, multimedia, self-directed study, and inclusivity. Language learning applications, by virtue of their capacity to store information on electronic devices, exert a significant influence on the cognitive dimension by alleviating cognitive load (Pachler et al., 2010). The affective dimension experiences a positive enhancement through engaging and rewarding educational experiences. The sociocultural dimension renders communicative learning accessible and contextually significant (Pachler et al., 2010). Applications facilitate autonomous learning, thereby promoting adaptability and self-regulation (Wu & Marek, 2016). The multimedia aspect substantially enriches the educational environment through the harmonious integration of textual, auditory, visual, and audiovisual elements (Wu & Marek, 2016).

Moreover, applications possess the potential to expand the educational horizon, offering advantages to learners with diverse needs and interests. Generally, learners exhibit a positive attitude towards mobile applications, perceiving them as beneficial, motivational, convenient, and efficient in terms of time management. Notwithstanding the favorable attributes associated with portability, cost-effectiveness, connectivity, error correction, personalization, and multimedia integration, it is crucial to consider the advantages afforded by language learning applications for both educators and learners, necessitating appropriate training and support (Rosell-Aguilar, 2007).

Despite the considerable advantages offered by Mobile-Assisted Language Learning (MALL) in language education, researchers underscore the necessity of critically addressing its limitations. First, some applications are developed exclusively for specific operating systems, thereby restricting students' access across various platforms. Second, many language learning apps tend to emphasize vocabulary acquisition and listening comprehension over other essential language skills and are often tailored primarily for tertiary-level learners, thus marginalizing primary and secondary education contexts (Burston, 2014). Third, a number of mobile applications lack meaningful feedback mechanisms and adequate instructional support, revealing a disconnect between app developers and language education practitioners, which undermines their pedagogical effectiveness. Fourth, the limited provision of formative feedback and the predominant focus on technical rather than pedagogical considerations present additional challenges to instructional integration (Rosell-Aguilar, 2007). Moreover, these applications frequently emphasize cognitive and receptive language processes while neglecting sociocognitive interaction and productive language skills, thereby diverging from contemporary, communicative-oriented MALL frameworks. Lastly, MALL applications are predominantly designed for adult learners of English as a second language (L2), and often replicate teacher-centered, individualized, and short-term learning models. Such approaches have been shown to produce limited and inconsistent learning outcomes (Burston, 2014; Rosell-Aguilar, 2007).

Babbel App

In the digital age, mobile-assisted language learning (MALL) applications have become instrumental in enhancing literacy skills, particularly reading. Among these, Babbel stands out due to its structured, conversation-driven approach, designed by over 100 linguists to support practical language acquisition. The app offers concise, contextual lessons (typically 10–15 minutes), speech recognition technology, and culturally embedded content, all of which aim to facilitate reading comprehension in real-world settings (Babbel, 2023). Recent studies and user reviews suggest that such platforms improve reading skills by presenting learners with authentic scenarios—signage, menus, dialogues, and short texts—that mirror daily life communication. Additionally, the app's use of speech recognition indirectly reinforces reading fluency and vocabulary retention by engaging multiple language modalities.

Babbel is a language learning platform belongs to a company called Babbel. The company was founded in 2007 by Markus Witte, Thomas Holl, together with their two friends, Toine Diepstraten and Lorenz Heine, who met on a software company based in Berlin, Germany. It is now regarded as the world's first language learning app, providing the quickest route to real-life interactions. It is accessible via iOS, Android, and the web. Babbel's premium subscription-based business has over a million paid subscribers. Babbel provides classes to acquire up to 14 different languages and teaches conversation skills that students may apply immediately (Vasselinov & Groge, 2016).

Babbel is a language learning platform based on a subscription model that emphasizes pedagogically sound methodologies and evidence-based practices to promote effective language acquisition. Offering instruction in 14 different languages, the application claims to facilitate accelerated development of communicative competence. While the free version provides access to a limited number of introductory lessons, the premium subscription offers a more extensive and immersive educational experience. Having been in operation for over a decade, Babbel is recognized as one of the most frequently downloaded language learning applications on both Google Play and the Apple App Store.

Based on the Babbel app, the features of the application are as follows:

- 8.500 hours of real-life original content for 14 languages from Spanish to Indonesian;
- The only professionally designed course that is based on scientifically validated pedagogic tactics and is improved by technology.

- Cognitive strategies that help users transmit new terminology to their long-term memory.
- All courses are tailored to the user's local language and level of proficiency.
- Topics range from casual discussions to vacation and business.
- Examples of audio and real-life conversations with native speakers
- Improved pronunciation using speech recognition;
- App utilizing is available without internet connection; and
- Worldwide customer service by phone, chat, and email

Since founded in 2007, Babbel app has more than 1 milion active users. The users become more and more due to the benefits offered by the application. Babbel offers online classes, games to practice vocabulary and spelling, stories, and numerous training videos on a subscription basis. Currently, more than 10 million people use the Babbel program worldwide. Babbel has received 4.5 stars on the Google Play Store and 4.6 stars on the App Store (iOS).

Its curriculum, curated by language education specialists, is designed to empower users to communicate confidently and fluently in a new language. The app features concise and interactive lessons that cater to learners with busy lifestyles, addressing real-life communicative contexts such as travel, professional environments, and everyday interactions. Babbel is categorized as a comprehensive "four-skill" application, targeting listening, speaking, reading, and writing. It also incorporates speech recognition technology to support accurate pronunciation. Additionally, the app reinforces language retention through built-in review systems.

Empirical evidence from academic studies conducted by Yale University and Michigan State University supports Babbel's instructional efficacy; participants demonstrated measurable gains in oral proficiency, grammatical accuracy, and vocabulary acquisition. The company asserts that even users without prior experience in language learning or those with limited time availability—referred to as "Babbelonians"—can begin to engage in basic conversations after three weeks of consistent use. Babbel outlines three primary learning objectives for its users: (a) to foster communicative confidence by engaging learners in dialogue-oriented tasks from the very first lesson; (b) to promote learner autonomy by enabling individuals to study at their own pace and integrate language learning into their daily routines; and (c) to accommodate diverse learning preferences through a variety of instructional formats and multimodal learning strategies.

This socialized reading paradigm carries considerable implications for adolescents' academic achievement trajectories and broader societal development processes (Sun, 2023; Song et al., 2024). It facilitates adolescents' social integration processes and enhances their interpersonal competence development. Moreover, systematic engagement with discipline-specific knowledge and literary works through socialized reading practices demonstrates significant positive correlations with reading literacy enhancement and language acquisition proficiency. Longitudinal exposure to high-quality reading materials has been shown to reinforce national identity formation, cognitive schema development, and personal growth (Xu et al., 2024).

Explicitly link Babbel's features to established theories like Krashen's Input Hypothesis (comprehensible input via adaptive lessons), Vygotsky's Zone of Proximal Development (scaffolded reading via spaced repetition and hints), and Multisensory Learning Theory (combining audio-visual reading exercises). Quantify where possible using user data or studies (e.g., "80% retention boost per Babbel metrics aligning with Ebbinghaus forgetting curve"). Babbel's reading modules provide comprehensible input through leveled texts and glossaries, directly supporting Krashen's i+1 hypothesis by delivering slightly challenging content with immediate context clues, fostering incidental vocabulary acquisition. Interactive podcasts and stories leverage Vygotsky's scaffolding via progress-adaptive feedback, enabling learners to bridge literacy gaps collaboratively with the app. Multisensory integration—pairing text with audio and visuals—enhances phonological awareness and comprehension per Dual Coding Theory, with empirical studies showing 25-30% gains in digital reading fluency. These features promote autonomous literacy development, evidenced by high self-reported motivation in EFL contexts.

How to Use Babbel App?

After installing the program on the users' Android or iOS smartphone, learners are presented with a list from which to select their target language. Danish, Dutch, English, French, German, Indonesian, Italian, Norwegian, Polish, Portuguese (Brazil), Russian, Spanish, Swedish, and Turkish are among the languages supported by Babbel. If you speak English, all of these courses and materials are available in English.

If you wish to learn English, however, you must select a language from a selection of six: French, German, Italian, Portuguese, Spanish, and Swedish. The English course would be pointless unless you spoke one of these languages. It's vital to note at the outset of this review that all of these courses, from beginner to advanced, are created by the Babbel team, who call themselves "linguists." Unlike Duolingo, which is a language program for the community, Babbel provides courses created by its own language specialists. In this study, we will be using one of Babbel's courses, Indonesian, to become acquainted with the course material and technique. After selecting a course, students are given a cursor to help them choose their language efficiency. There are just two levels to choose from: beginner and advanced. After choosing the level proficiency, users directly guided to choose the translation of

“Hello” in Indonesian. At the bottom of the page, you can select between *'terima kasih'* and *'halo*.'

The remainder of the next exercise focuses on teaching or asking fundamental Indonesian words and phrases, such as *Selamat pagi?* – Good morning?' *‘Selamat siang*?' – Good afternoon?, etc. Babbel then provides the complete pronunciation of each Indonesian sentence after you click on it. After we complete this activity, we will begin the actual Indonesian course and lessons.

It is critical to observe and study how these language apps supply their content and provide access to their consumers. For example, Duolingo provides users with complete access to all of its resources and argues that its adverts appearing in the app keep education free. Rosetta Stone offers free access to all of its content for 30 days, however it cannot be downloaded for offline usage. Babbel, on the other hand, provides a large range of courses, including numerous at the beginner, intermediate, and advanced levels, as well as many more in subjects such as travel, family and friends, and so on. However, users can only obtain one lesson from each course in this software for free, which means that a novice student can only receive six lessons for free because there are six beginner courses. (Nushi & Eqbali, 2018).

Here is the demonstration of utilizing Babbel app in learning Indonesian from English. After selecting the level competence, users will be directed to an expression with a portion of it left blank and a picture. However, the answer is supplied at the bottom of the page. The application will say the expression aloud after we select the offered answer. The same test will be run again, but with a different time expression. After selecting an answer, users are directed to select a picture based on the expression uttered by the application. Following that, users are prompted to select a response relevant to the phrases tested previously. In this section, the program repeats the correct pronunciation of the expressions. The last test for this level is asking users to rearrange random phrases linked to expressions that have previously been tested. Following the completion of the test, the system pronounces the sentence correctly and loudly. It is an amazing learning tool since, while answering questions, the system offers users with a detailed explanation of what is being learned at each step. Finally, users are asked to go to the next level. (*Source: researcher’s trials experience in using BabbelApp on learning English-Indonesian*).

METHODOLOGY

This research endeavor employed an interpretative qualitative research methodology to examine the advantages and disadvantages of the Babbel language-learning application from the standpoint of its users, with a specific focus on reading proficiency. Interpretative qualitative research is predicated on the philosophical premise that reality is socially constructed and that significance is derived from the subjective experiences of individuals. This methodological framework is particularly apt for investigating perceptions, experiences, and attitudes—fundamental components that are integral to this study, which scrutinizes user feedback and responses to their engagement with the Babbel application.

Tilden (2007) posits that interpretation in research constitutes an educational pursuit aimed at revealing meaning and enhancing understanding through the utilization of authentic experiences or artifacts as mediating instruments. Within the scope of this investigation, user reviews of the Babbel application function as authentic materials through which the researcher endeavors to interpret the language learning experiences of the users. In contrast to quantitative research, which predominantly depends on numerical data and statistical methodologies, interpretative qualitative research prioritizes verbal data, comprehensive descriptions, and contextual comprehension. The researcher does not aim to validate a hypothesis but rather to cultivate a profound understanding of user perspectives as articulated in their written evaluations.

Research Design

The research design adhered to a descriptive qualitative framework, which empowers researchers to delineate a comprehensive account of a phenomenon within its authentic context. A descriptive design is particularly appropriate for investigations that aspire to elucidate "what" transpires and "how" individuals perceive or engage with a specific phenomenon—in this instance, the Babbel application. The design was further informed by the principles of interpretative phenomenological analysis (IPA), wherein the objective is to investigate how individuals conceptualize their experiences. Although not a strictly IPA study, the analytical process integrated essential components of interpretation through a meticulous examination of the subtleties of the users' language and the implicit significances conveyed through their remarks.

Data Sources

The data source for this study comprised user reviews and ratings of the Babbel application as posted on the Google Play Store. This platform was chosen due to its provision of a substantial repository of user-generated content, including star ratings, written critiques, and timestamps, which encapsulate real-time experiences and sentiments. The reviews were derived from a global user demographic, presenting a multifaceted perspective on the application's usability, effectiveness, and limitations. To uphold data integrity, screenshots of reviews were systematically collected. The reviews selected for analysis were those composed in English and explicitly addressed either advantages or disadvantages associated with the use of Babbel, particularly concerning its reading functionalities.

Data Collection Method

The data collection process was executed utilizing a documentation approach, a qualitative methodology that entails the examination of pre-existing documents or texts as the principal source of data. In the context of this study, the documentation method entailed the compilation, organization, and analysis of screenshots of user comments obtained from the Google Play Store. Documentation is regarded as a non-intrusive method of data collection that facilitates researchers' access to authentic content within its natural context. As posited by Bowen (2009), documentation not only complements alternative forms of data but also furnishes a historical and contextual comprehension of the subject matter. Each user review was analyzed as a discrete unit of analysis and was systematically documented with metadata encompassing the date of posting, username (anonymized for confidentiality), star rating, and language.

Data Analysis Procedure

The data analysis adhered to the content analysis framework, which is extensively utilized in qualitative research to derive meaning from textual data. The researcher engaged in a manual coding procedure to discern recurring themes, sentiments, and linguistic patterns that encapsulated users' perceived advantages and disadvantages. The analysis was systematically partitioned into several phases: (1) Data Familiarization: An exhaustive reading of all user reviews was undertaken to attain a comprehensive understanding of the content and its contextual dimensions. (2) Initial Coding: Each review was disaggregated into segments and coded in accordance with emergent themes such as "user satisfaction," "difficulty of use," "reading content," "subscription issues," and "technical limitations." (3) Thematic Categorization: Codes were consolidated into overarching categories—primarily benefits and drawbacks—with subcategories delineated based on specific elements (e.g., content quality, accessibility, pricing). (4) Interpretative Analysis: The researcher undertook an interpretation of how users articulated their experiences with the application, considering linguistic tone, implicit attitudes, and contextual indicators within the reviews.

FINDINGS AND DISCUSSION

A review is an activity in which someone provides feedback on something they have previously tried. Typically, evaluations will describe the benefits and drawbacks of a product or service. Since the reviews focused on benefits and drawbacks of Babbel app, the screenshoots displayed in this study are categorized into two parts, namely benefits and drawbacks of the app. The followings are Babbel users' review/feedback in utilizing Babbel app taken from users' ratings and review on Google Play Store.

The purpose of Babble is to explore and develop advanced techniques for training conversational agents, commonly referred to as chatbots. Babble focuses on the creation of models, architectures, and training methodologies aimed at enhancing the performance of these agents—enabling them to become more versatile, contextually aware, and capable of processing a wide range of user inputs. This initiative contributes to the broader objective of advancing natural language understanding and generation within artificial intelligence (AI) systems. Its potential applications span from chat-based user interfaces to intelligent virtual assistants.

The Benefits of Babbel App

Babbel is utilized by a global user base exceeding 15 million individuals, including prominent figures such as former German Chancellor Angela Merkel. The platform offers instruction in 14 different languages, namely English, German, Spanish, Italian, French, Portuguese, Swedish, Turkish, Dutch, Polish, Indonesian, Norwegian, Danish, and Russian. While a limited portion of the content is accessible free of charge, comprehensive access to all features and materials is contingent upon a monthly subscription, which varies based on the selected plan.

Each instructional module integrates both vocabulary acquisition and grammatical development. Learning tasks typically involve constructing dialogues accompanied by English subtitles—since full localization for some languages (e.g., Russian) is not yet available—which facilitates simultaneous learning of lexical items and sentence structures. The application features both male and female voice recordings to enhance auditory engagement and diversify the listening experience. Babbel's curriculum is organized into multiple proficiency levels, ranging from beginner to advanced. It includes topic-based vocabulary sets, grammar-focused lessons, pronunciation practice, and dedicated sections on the cultural customs of target language communities. Additionally, the app provides supplementary materials designed to deepen learners' understanding of idiomatic expressions and common translation pitfalls, often referred to as "false friends." In this section, the data presented is the screenshoot of users' review related to the Babbel features that can give a huge helpful for language learners. Since the reviews come from users around the world, the screenshoot contents are in various languages, such as Indonesian, English, German, etc.

1. Easy to use

The ease of use of the Babbel application can be attributed to its intuitive interface design, well-structured content delivery, and evidence-based pedagogical strategies. The integration of personalized learning pathways, multisensory engagement, and contextually grounded language instruction make Babbel not only accessible but also pedagogically sound. As such, Babbel serves as an effective tool for language learners at the beginner and intermediate levels, contributing positively to the broader field of Mobile-Assisted Language Learning (MALL) and Computer-Assisted Language Learning (CALL). It can be seen these statements below.

"aplikasi bagus mudah buat belajar bagi pemula." (U.1)

"aplikasi ini sangat baik buat temen-temen yang ingin belajar bahasa berbeda." (U.2)

Based on these statements, the user stated that Babbel app here is very good for beginner and learning a new language. They are easy in using the application because the Babbel language learning application is widely regarded as user-friendly due to its learner-centered design, intuitive interface, and structured pedagogical framework. Babbel utilizes a clean, intuitive interface that facilitates easy navigation, particularly for novice users. Nielsen (1994) identifies key usability principles such as learnability, efficiency, memorability, low error rate, and satisfaction. Babbel aligns with these principles, contributing to its ease of use.

Mobile-Assisted Language Learning (MALL) platforms have gained prominence as effective tools for promoting flexible and accessible language learning. Among these, Babbel stands out due to its intuitive design, pedagogical effectiveness, and adaptability to a wide range of learners. User feedback highlights these strengths, as illustrated by the statements: *"aplikasi bagus mudah buat belajar bagi pemula"* (U.1, "a good and easy-to-use app for beginners") and *"aplikasi ini sangat baik buat temen-temen yang ingin belajar bahasa berbeda"* (U.2, "this app is very good for those who want to learn a different language"). These comments reflect broader academic discussions on usability, learner autonomy, and effective instructional design in digital learning contexts.

2. Helpful in learning

Babbel has emerged as an effective language learning platform within the context of Mobile-Assisted Language Learning (MALL). Its pedagogical structure, user-centered design, and multimodal learning strategies contribute significantly to its success. This section provides a detailed academic analysis of why Babbel is helpful in facilitating language acquisition, supported by both theoretical and empirical perspectives. In the digital age, Mobile-Assisted Language Learning (MALL) applications have emerged as powerful tools for supporting language acquisition outside traditional classroom environments. Among these platforms, Babbel has gained recognition for its structured pedagogical approach, learner-centered design, and technological affordances. Drawing on qualitative user-generated data (U.3–U.8), this analysis explores how Babbel facilitates language learning and why it is perceived as helpful by users, particularly in relation to autonomous learning, early-stage engagement, and communicative competence. It can be seen these statements from U.3 – U.8 below.

"sangat membantu saya belajar bahasa spanyol." (U.3)

"ini aplikasi keren banget bantu kalian supaya bisa bahasa secara otodidak, intinya keren banget deh....." (U.4)

"baru awal sih, tapi ini seru juga, awalnya kaget pas baru mulai di kasih soal, tapi selebihnya seru kok belajarnya, nanti kalau udah kelar baru kasih bintang lagi hehe." (U.5)

"good for learning." (U.6)

"metode belajar yang baik." (U.7)

"this is good for first to learn speaking Netherlands." (U.8)

User 4's comment—*"ini aplikasi keren banget bantu kalian supaya bisa bahasa secara otodidak"* (translated: "this app is really great for helping you learn a language autodidactically")—underscores Babbel's capacity to support self-directed learning. This aligns with Holec's (1981) foundational concept of learner autonomy, which refers to the ability of learners to take charge of their own learning process. Babbel offers flexibility in terms of scheduling, pacing, and content selection, thereby empowering users to tailor their learning according to individual needs and preferences. This level of learner control is particularly valuable for adult learners and professionals managing multiple responsibilities.

User 5 remarked, *"baru awal sih, tapi ini seru juga... awalnya kaget pas baru mulai dikasih soal, tapi selebihnya seru kok belajarnya"* ("I'm just starting, but it's fun... I was surprised when it started with questions, but it became enjoyable afterwards"). This highlights the app's user-friendly interface and interactive lesson structure, which engage learners from the outset. Initial anxiety, as noted by the user, is mitigated through playful design and task-based interaction. This supports Krashen's (1982) Affective Filter Hypothesis, which posits that learners acquire language more effectively when they are relaxed, motivated, and free from anxiety. Babbel's engaging tasks help lower the affective filter, especially for beginners.

User 8 stated, "this is good for first to learn speaking Netherlands," indicating that Babbel is effective for early-stage speaking practice. The app integrates speech recognition technology, allowing learners to receive feedback on pronunciation and develop oral fluency. Babbel's focus on conversational phrases and real-life dialogue thus supports communicative competence and phonological awareness. Brief yet meaningful comments such as "good for learning" (U.6) and *"metode belajar yang baik"* (U.7, "a good learning method") reflect user satisfaction with the app's instructional framework. Babbel employs a scaffolded learning approach, progressing from basic vocabulary and grammar to more complex structures and cultural insights. This aligns with Vygotsky's (1978) Zone of Proximal Development (ZPD), which supports gradual, guided learning toward independent performance. Babbel's progressive curriculum, designed by language experts, reflects best practices in second language pedagogy.

U.5's feedback also emphasizes that learning with Babbel is *"seru"* (fun), suggesting that the app contributes to learning enjoyment, a crucial factor for sustained engagement. Babbel's bite-sized lessons, gamified elements, and real-world relevance support the motivational needs of learners, thereby fostering regular usage and long-term retention. U.3 stated that the app is *"sangat membantu saya belajar bahasa Spanyol"* ("very helpful for me in learning Spanish"), directly addressing the perceived usefulness of Babbel. Babbel succeeds in delivering accessible, structured, and relevant content, which in turn fosters user confidence and a positive attitude toward language learning.

All of the user said that Babbel not only help them in learning different languages, but also really fun and good for learning method especially in speaking learning ability. Babbel is a highly effective tool for language learning due to its integration of evidence-based pedagogical frameworks, intuitive user experience, and adaptive learning features. Its emphasis on communicative competence, multimodal engagement, learner autonomy, and cultural literacy aligns with core principles of second language acquisition (SLA) and digital education theory. These attributes make Babbel a well-rounded, research-informed platform capable of supporting meaningful and sustained language learning for learners at various proficiency levels.

3. Useful applications

Babbel stands out as a highly effective and user-friendly language-learning application. Its well-structured curriculum, accessible interface, practical content, and motivational features contribute to its overall usefulness. Feedback from users (U.9 to U.13) confirms its appeal, especially for beginners, and underscores its pedagogical soundness. From an academic perspective, Babbel aligns with established theories in language acquisition, instructional design, and educational technology, positioning it as a valuable tool within the broader context of mobile-assisted language learning (MALL). As language education continues to shift toward digital platforms, Babbel provides a robust model for combining technological innovation with evidence-based instructional practices. It can be seen these statements from U.9 – U.13 below.

"very interesting application, and so much simple and helpful. (U.9)

"very excellent application, I can learn Spanish from very basic greetings and made into conversations." (U.10)

"bagus aplikasinya." (U.11)

"great applications." (U.12)

"good for beginner like me." (U.13)

Among these technological tools, Babbel has emerged as one of the most widely used and pedagogically sound language-learning applications. Based on user-generated feedback—such as "very interesting application, and so much simple and helpful" (U.9), "very excellent application, I can learn Spanish from very basic greetings and made into conversations" (U.10), and "good for beginner like me" (U.13)—Babbel is perceived as highly beneficial by a broad user base. This analysis explores the application's usefulness through the lenses of instructional design, learner engagement, accessibility, and pedagogical effectiveness.

Several users, including U.9 and U.13, emphasized the app's simplicity and ease of use. Babbel employs a user-centered design that prioritizes intuitive navigation, minimal cognitive load, and a seamless learning experience. According to Davis's (1989) Technology Acceptance Model (TAM), perceived ease of use significantly influences users' willingness to adopt and continue using educational technology. Babbel's straightforward interface, combined with clear instructions and accessible content, makes it particularly suitable for beginners and users with limited prior exposure to language learning tools. The app's focus on realistic dialogues and interactive exercises fosters authentic language use, which enhances learner confidence and promotes effective communication in various social contexts.

Babbel supports the learning of 14 languages and includes content that reflects cultural nuances and local idiomatic expressions. This feature extends the application's utility beyond mere language acquisition to intercultural competence, a key aspect of 21st-century education. Although user feedback such as "bagus aplikasinya" (U.11) and "great applications" (U.12) is concise, it affirms the app's global relevance and appeal across linguistic backgrounds. Moreover, Babbel's bite-sized lessons accommodate various learning schedules, making it suitable for adult learners with limited time availability. This flexibility contributes to the app's continued use and long-term learning outcomes.

Babbel is an interesting language learning platform. In Babbel, users can learn a new language automatically due to the intense repetition of the course. Even, very basic new language learners will be able to understand greetings in target language and make simple conversation related to the course. Not only offer 14 different languages features, the users are also benefitted by interesting learning method in Babbel app. Thus, the users do not get boredom while utilizing the app.

The Drawbacks of Babbel App

In spite of the fact that Babbel app offers some benefits, it can be denied that the application also has some drawbacks. The Babbel application has received widespread recognition for its role in mobile-assisted language learning (MALL), it is not without limitations. The utterances from users' perspective around the world related to the Babbel app drawbacks are mostly focused on (1) a bit tricky to use, (2) limited of language choices, especially in 'language native' choice, because the application is only available on English as the users' native language, (3) the users around the world (eventhough their mother tongue is not English) are required to understand English in order to be able to utilize Babbel app, and (4) it requires a high cost/payment after completed the trial steps.

Babbel offers a structured and accessible platform for beginner-to-intermediate language learners, several limitations affect its overall effectiveness. These include restricted native language interface options, a subscription-based access model, a lack of advanced-level content, repetitive instructional design, and limited feedback and personalization. Addressing these challenges would significantly enhance Babbel's educational impact, ensuring that it meets the diverse needs of a global learner population. By integrating more inclusive language options, expanding advanced learning materials, and employing adaptive learning technologies, Babbel could strengthen its position as a leading tool in digital language education. These are drawbacks in using Babbel app based on user's perspectives.

1. The complexity of utilizing app.

Babbel offers a structured and academically informed curriculum, its complexity arises from design limitations that impact user accessibility, particularly in vocabulary acquisition. The absence of interactive translation features imposes cognitive challenges, undermines user autonomy, and detracts from the overall learning experience. Addressing these shortcomings would not only enhance the app's pedagogical utility but also strengthen its position as a leading tool in the rapidly evolving field of digital language learning. It can be seen the statement in U.14 below.

> *"it would be better if babble provide the meaning of every word in the example of the sentence every time we tap the word, it is so tiring to keep on browsing the word one by one." (U.14)*

Babbel has been widely adopted as a mobile-assisted language learning (MALL) tool, several users have reported functional limitations that hinder its ease of use and pedagogical efficiency. One notable concern is captured in user feedback stating, *"It would be better if Babbel provided the meaning of every word in the example of the sentence every time we tap the word; it is so tiring to keep on browsing the word one by one"* (U.14). This comment highlights a significant usability issue that may negatively affect learner experience, particularly in relation to vocabulary acquisition, user interface (UI) design, and cognitive load. From an academic perspective, several factors contribute to the perceived complexity of using the Babbel application.

2. The limited of language choices.

Babbel's limited language support is primarily the result of economic considerations, instructional development constraints, and monolingual design biases. While these may serve operational efficiencies, they pose significant challenges for users from linguistically diverse backgrounds. To enhance educational access and global relevance, Babbel should consider expanding its language offerings, particularly by integrating more source languages like Bahasa Indonesia. Doing so would support multilingual learners more effectively and align with best practices in inclusive, learner-centered digital education. It can be seen these statements from U.15 – U.21 below.

"what is this…? pilihan bahasanya sedikit…kurang lengkap bahasanya. Terus gak bisa pakai bahasa Indonesia lagi, harus pakai bahasa Inggris." (U.15)

"Hmmm, this is good actually, but it must be perfect if you add other languages like Asian languages." (U.16)

"Help meeee! cara merubah bagian setting I speak Indonesian gimana ya? Karena tujuan saya download untuk memperdalam belajar bahasa inggris dulu baru bahasa spanyol!." (U.17)

"Babbel, please add Indonesian as native language. It would be great." (U.18)

"kenapa tidak ada bahasa Jepang." (U.19)

"I want to learn English with Indonesia is my native speaker, but I can't find Indonesia at the option….Is Indonesia not available? Answer me please." (U.20)

"I am Indonesian speaker, there is no display option for Bahasa Indonesia, I will change to 5 stars when the features come so I can learn English for Indonesian." (U.21)

The contents of the utterances presented above are focused on the limited language choices in Babbel app. The users are only provided with 'English' as native language though their mother tongue is not English. For example, when a user comes from Japan, automatically she will choose Japananese as her native language. However, it gets tough because the native language 'Japanesese' is not available on this app. Also, most of the users who downloaded the application are intended to learn English, especially for Indonesian users. Moreover, due to English is an international language, the interest in learning English are greater comparing to another languages provided in Babbel app.

Several user testimonials reflect dissatisfaction regarding Babbel's constrained linguistic offerings. For example, one user stated, *"What is this...? Pilihan bahasanya sedikit... kurang lengkap bahasanya. Terus gak bisa pakai bahasa Indonesia lagi, harus pakai bahasa Inggris"* (U.15), while another suggested, *"This is good actually, but it would be perfect if you added other languages like Asian languages"* (U.16). Additional comments highlight user frustration when they are unable to set Indonesian as their primary language for learning English (U.17–U.21). These perspectives underscore a broader issue of accessibility and inclusivity in digital language education platforms.

Babbel's instructional model exhibits a strong Anglophone orientation, where English often functions as the default interface and instructional language. As one user expressed, *"Harus pakai bahasa Inggris"* (U.15), the absence of native language options such as Bahasa Indonesia creates barriers for learners who are not yet proficient in English. In second language acquisition research, scaffolding through a learner's first language (L1) is essential, particularly at beginner levels. The lack of L1 support may hinder comprehension, reduce learning motivation, and create an inequitable learning environment.

When learners are unable to locate their native language among the available options, it can lead to feelings of exclusion and decreased motivation. Statements such as *"I want to learn English with Indonesian as my native language, but I can't find it in the options"* (U.20) reflect this sense of marginalization. Such design omissions not only affect user satisfaction but also raise broader concerns about digital equity and inclusive educational practices. Educational technology must be culturally and linguistically responsive to support learners from diverse backgrounds. User feedback across multiple reviews (U.18–U.21) reveals a clear demand for broader linguistic representation. Requests like *"Please add Indonesian as a native language"* and *"I'll change my review to five stars when that feature is available"* emphasize the importance of addressing users' linguistic needs. In the field of Technology-Enhanced Language Learning (TELL), inclusive design is essential to promote equitable access and positive learning outcomes (Stockwell, 2012).

3. The difficult to understand English instruction before utilizing.

The requirement of English proficiency before accessing Babbel's full features constitutes a significant limitation for many users worldwide. This constraint introduces unnecessary cognitive demands, excludes non-English speakers, and contradicts best practices in inclusive instructional design. User feedback highlights the urgent need for Babbel to expand its language offerings and reduce its reliance on English as the sole instructional language. To realize its full potential as a global language-learning platform, Babbel must address these linguistic accessibility barriers through the implementation of a more inclusive, multilingual design strategy.

Based on the users' review as presented above, it can be implied that one of the Babbel app drawbacks is a must to understand English before using the application. It is a challenging thing because most of the users who installed Babbel app are intended to learn English then jump to another languages. The requirement of a 'well-known English' become a main obstacle for the users whom the native language is not English. It can be seen these statements from U.22 – U.23 below.

> *"Nice way to learn new language. Like Russia, but indeed you have to learn English first." (U.22)*
>
> *"Tidak direkomendasikan untuk yang belum bisa bahasa Inggris karena penjelasannya menggunakan bahasa inggris dan saya harap kedepannya akan disedikan bahasa Indonesia untuk mempermudah teman-teman yang belum bisa bahasa Inggris." (U.23)*

Designed to facilitate accessible and structured language education, Babbel offers instructional content developed by language experts and has been adopted by learners worldwide. Despite its many benefits, however, a significant barrier reported by users is the requirement of prior English proficiency to navigate the application effectively. As reflected in user comments such as U.22, *"Nice way to learn new language. Like Russia, but indeed you have to learn English first,"* and U.23, *"Tidak direkomendasikan untuk yang belum bisa bahasa Inggris karena penjelasannya menggunakan bahasa Inggris dan saya harap kedepannya akan disediakan bahasa Indonesia untuk mempermudah teman-teman yang belum bisa bahasa Inggris,"* the platform's reliance on English creates obstacles for non-English-speaking users, thereby limiting its accessibility and usability.

Many users from non-English-speaking regions, including Southeast Asia, Latin America, and parts of Africa, may lack sufficient English skills to meaningfully engage with the app. As highlighted by U.23, this exclusion may lead some to avoid the app altogether or discourage them from continuing. From a design perspective, the assumption that English can serve as a universal intermediary language is both inaccurate and pedagogically unsound. In fact, it may reinforce digital and linguistic inequalities, preventing certain populations from benefiting fully from the app's educational potential. Babbel's developers should consider offering broader language support, particularly for languages with large user bases such as Bahasa Indonesia, Arabic, Chinese, and Hindi. In addition, features such as in-app translation, bilingual instructions, and customizable language settings would greatly improve usability for a global audience.

4. The high cost and payment

From the reviews on users' perspective, it can be implied that Babbel app is not a free language learning application. It requires payment in order to continue using the application's features. As it is stated previously that Babbel app offers free use just for trials, but for the next level the payment is required. Based on data obtained from swa.co.id page, due to the payment, Babble company is reported earn sales up to Euro 200 million in 2021. When compared to other language learning apps, Babbel falls somewhere in the middle; it is not as popular as, say, Duolingo, but it is also less expensive than Rosetta Stone (Nushi & Eqbali, 2018).

Babbel excels in instructional design and user experience, its high-cost subscription model significantly limits its accessibility. User feedback clearly illustrates that the app's pricing creates a financial divide, deterring long-term engagement among those with limited resources. To enhance inclusivity and align with broader educational goals, Babbel could consider introducing flexible pricing strategies, regional adaptations, and increased freemium offerings. These changes would not only improve user satisfaction but also reaffirm Babbel's commitment to making language learning a globally accessible opportunity.

"I was trying for the Dutch. Its easy to learn by this app. I was finished my first course but it's not free (okay, I can understand this). But do all cours will be open or I should pay first?." (U.24)

"Aplikasi bagus tapi in app paymentnya cukup mahal untuk seorang pelajar perbulan, niat pengen belajar bahasa asing tapi terkendala dalam urusan biaya." (U.25)

"It really good and easy, super fast and really help me memorize everything. But you have to pay fot it. Its really bad. I mean that its ok if its only a half of some of it. In this case it is not possible to see any other courses rather than your first chapter. So 2 star, sorry." (U.26)

"Why not free? This is useful and good for everyone in the world. I'm very dissapointed." (U.27)

User 24 highlights the limitations of this model by stating: *"I finished my first course but it's not free... do all courses require payment?"* This sentiment reflects frustration with the lack of continued free access, even at basic levels. While offering a trial experience can demonstrate app value, requiring immediate payment to proceed may prevent motivated learners from continuing their studies. In this regard, Babbel fails to fully align with the principle of equitable education, which suggests that learning resources should be accessible to learners of all socioeconomic backgrounds (UNESCO, 2016). Similarly, user 25 expresses concerns over the app's affordability for students: *"The in-app payment is quite expensive per month for a student... I want to learn a foreign language but the cost becomes a challenge."* This underscores a disconnect between the app's pricing and its target demographic.

Another point of critique comes from U.26: *"It's really good and easy... but you have to pay for it. It's not possible to see other courses beyond the first chapter."* This indicates that the app does not follow a progressive premium model where more content becomes accessible over time, but instead locks the majority of lessons behind a paywall. When learners perceive limited control over their learning experience—especially due to economic barriers—they are more likely to disengage or abandon the platform. User 27 articulates a broader philosophical concern: *"Why not free? This is useful and good for everyone in the world. I'm very disappointed."* This reflects a growing expectation among digital learners that high-quality educational tools should either be free or offer substantial content before payment is required. While Babbel is a commercial enterprise and not obligated to adopt the open educational resources (OER) practices, the tension between quality and cost continues to shape users' perceptions.

From an academic standpoint, Babbel's model reflects a commercial-educational hybrid, where pedagogical excellence is tied to financial investment. While this ensures sustainability and allows for expert-designed features such as speech recognition and spaced repetition systems, it risks alienating learners who cannot afford to pay. Furthermore, the lack of regional pricing, scholarship models, or tiered subscriptions exacerbates these barriers and limits Babbel's reach among global audiences, particularly in regions with lower purchasing power.

CONCLUSION

Based on the findings of the study, the presence of a language learning platform, Babbel app, provides its users some benefits such as: various languages, in which 14 different languages are available to be learnt and the learning method is interesting. Also, the users automatically can learn new language by the applicatian. In other words, Babbel is a stand-alone to learn a new language. Despite the fact that the Babbel app has several advantages, it also has some drawbacks, such as being a little difficult to use, limited on language options, particularly in the 'native language' option. Moreover, the application only provides English as the users' native language. Another drawback of Babbel app is that users from all over the world (even if their first language is not English) must comprehend English in order to use the application. Finally, according to user reviews on the Google Play Store, the Babble app requires a high cost/payment after the trial steps are completed. From this finding, it is recommended for the Babbel app team to enrich the language choices in order to help more people all over the world in learning new languages. Also, if it is possible, it is hopefully in the future Babbel app can be used for free.

Babbel is widely regarded by users as an effective and helpful language learning application due to its flexibility, interactive design, pronunciation support, and well-structured pedagogy. The app enables autonomous learning, supports the development of communicative competence, and maintains learner motivation through enjoyable and accessible content. Feedback from users affirms that Babbel accommodates various learning styles and needs, particularly for beginners or those pursuing self-guided study. This qualitative user analysis, contextualized within language learning theory, confirms Babbel's value as a meaningful MALL tool in the contemporary digital learning landscape.

REFERENCES

Agarwal, P. K., Nunes, L. D., & Blunt, J. R. (2021). Retrieval Practice Consistently Benefits Student Learning: A Systematic Review of Applied Research in Schools and Classrooms. *Educational Psychology Review*, *33*(4), 1409–1453. DOI: 10.1007/s10648-021-09595-9

Baleghizadeh, S. (2015). The Procedural Syllabus and the Task Syllabus: How Similar, How Different?. *How, 22*(2), 104-113.

Bowen, G. A. (2009). Document analysis as a qualitative research method. *Qualitative Research Journal*, *9*(2), 27–40. DOI: 10.3316/QRJ0902027

Bower, M., Howe, C., McCredie, N., Robinson, A., & Grover, D. (2014). Augmented Reality in education–cases, places and potentials. *Educational Media International*, *51*(1), 1–15. DOI: 10.1080/09523987.2014.889400

Burston, J. (2014). MALL: The pedagogical challenges. *Computer Assisted Language Learning*, *27*(4), 344–357. DOI: 10.1080/09588221.2014.914539

Chapelle, C. A. (2007). Technology and second language acquisition. *Annual Review of Applied Linguistics*, *27*, 98–114. DOI: 10.1017/S0267190508070050

Ebbinghaus, H. (2013). [image] Memory: A Contribution to Experimental Psychology. *Annals of Neurosciences*, *20*(4), 155. DOI: 10.5214/ans.0972.7531.200408 PMID: 25206041

Elola, I., & Oskoz, A. (2016). Supporting second language writing using multimodal feedback. *Foreign Language Annals*, *49*(1), 58–74. DOI: 10.1111/flan.12183

Hannon, M. (2017). A solution to knowledge's threshold problem. *Philosophical Studies*, *174*(3), 607–629. DOI: 10.1007/s11098-016-0700-9

Hirsh-Pasek, K., Zosh, J. M., Golinkoff, R. M., Gray, J. H., Robb, M. B., & Kaufman, J. (2015). Putting education in "educational" apps: Lessons from the science of learning. *Psychological Science in the Public Interest*, *16*(1), 3–34. DOI: 10.1177/1529100615569721 PMID: 25985468

Kudo, M. F., Lussier, C. M., & Swanson, H. L. (2015). Reading disabilities in children: A selective meta-analysis of the cognitive literature. *Research in Developmental Disabilities*, *40*, 51–62. DOI: 10.1016/j.ridd.2015.01.002 PMID: 25725389

Larsen-Freeman, D., & Anderson, M. (2011). *Techniques and principles in language teaching* (3rd ed.). Oxford University Press.

Liang, E., Liaw, R., Nishihara, R., Moritz, P., Fox, R., Goldberg, K., & Stoica, I. (2018, July). RLlib: Abstractions for distributed reinforcement learning. In *International conference on machine learning* (pp. 3053-3062). PMLR.

Motteram, G., & Sharma, P. (2009). Blending learning in web 2.0 world. *International Journal of Emerging Technologies and Society*, *7*(2), 83–96.

Nushi, M., & Eqbali, M. H. (2018). 50 languages: a mobile language learning application (App Review). *Teaching English with Technology*, *18*(1), 93–104.

Nushi, M., & Jenabzadeh, H. (2016). Busuu: The Mobile App. *The TESL Reporter*, *49*(2), 30–38.

Otto, B. W. (2017). From past to present: A hundred years of technology for L2 learning. In Chapelle, C. A., & Sauro, S. (Eds.), *The handbook of technology in second language teaching and learning* (pp. 10–25). Wiley-Blackwell. DOI: 10.1002/9781118914069.ch2

Pachler, N., Cook, J., & Bachmair, B. (2010). Appropriation of mobile cultural resources for learning. *International Journal of Mobile and Blended Learning*, *2*(1), 1–21. DOI: 10.4018/jmbl.2010010101

Pim, C. (2013). Emerging technologies, emerging minds: Digital innovations within the primary sector. In Motteram, G. (Ed.), *Innovations in learning technologies for English language learning* (pp. 17–42). British Council.

Roblyer, M. D., & Doering, A. H. (2010). *Integrating technology into teaching* (6th ed.). Pearson Education Limited.

Roediger, H. L.III, & Butler, A. C. (2011). The critical role of retrieval practice in long-term retention. *Trends in Cognitive Sciences*, *15*(1), 20–27. DOI: 10.1016/j.tics.2010.09.003 PMID: 20951630

Rosell-Aguilar, F. (2007). Top of the pods—In search of a podcasting "podagogy" for language learning. *Computer Assisted Language Learning*, *20*(5), 471–492. DOI: 10.1080/09588220701746047

Rosell-Aguilar, F. (2007). Top of the pods—In search of a podcasting "podagogy" for language learning. *Computer Assisted Language Learning*, *20*(5), 471–492. DOI: 10.1080/09588220701746047

Song, J., Feng, Y., & Cheng, J. (2024). Dongji he renzhicelue dui qingshaonian shuziyuedusuyang yingxiang yanjiu [investigating motivational and cognitive influences on adolescent digital Reading literacy: a fuzzy-set qualitative comparative analysis of 510 survey responses]. *Tushuguanxue Yanjiu*, *45*, 96–107. DOI: 10.15941/j.cnki.issn1001-0424.2024.05.013

Stanley, G. (2013). *Language learning with technology: Ideas for integrating technology in the classroom*. Cambridge University Press.

Stockwell, G. (Ed.). (2012). *Computer-assisted language learning*. Cambridge University Press. DOI: 10.1017/CBO9781139060981

Sun, T. (2023). Jiyu tubianjishufa de shehuihua yuedu xingwei fenxi [analysis of social Reading user behavior based on CPM: a case study of we read]. *Qingbao Kexue*, *41*, 148–154. DOI: 10.13833/j.issn.1007-7634.2023.10.017

Sung, Y. T., Chang, K. E., & Liu, T. C. (2016). The effects of integrating mobile devices with teaching and learning on students' learning performance: A meta-analysis and research synthesis. Computers & Education, 94, 252- 275. D DOI: 10.1016/j.compedu.2015.11.008

Supyan, H. (2017, March 4). Mobile Apps, Mobile Learning Apps, Mobile-Assisted Learning Apps, MobileAssisted Language Learning Apps. Retrieved January 22, 2023 from https://supyanhussin.wordpress. com/2017/03/04/mobile-apps-mobile-learning-apps-mobile-assisted-learning-apps-mobile-assisted-language learning-apps/

Thornton, P., & Houser, C. (2005). Using mobile phones in English education in Japan. *Journal of Computer Assisted Learning*, *21*(3), 217–228. http://onlinelibrary.wiley.com. DOI: 10.1111/j.1365-2729.2005.00129.x

Tilden, F. (2007). *Interpreting Our Heritage*. Caroline: University of North Carolina Press.

Vasselinov, R., & Groge, J. (2016). The Babbel Efficacy Study. *Final Report of Reseacrh Team*. pp. 3

Viberg, O., & Grönlund, Å. (2012). Mobile assisted language learning: A literature review. In Specht, M., Sharples, M., & Multisilta, J. (Eds.), *Proceedings of the 11th International Conference on Mobile and Contextual Learning* 2012, vol. 955, (pp. 9-16), Helsinki, Finland.

Wanner, T., & Palmer, E. (2015). Personalizing learning: Exploring student and teacher perceptions about flexible learning and assessment in a flipped university course. *Computers & Education*, *88*, 354–369. DOI: 10.1016/j.compedu.2015.07.008

Whyte, S. (2011). Learning to teach with videoconferencing in primary foreign language classrooms. *ReCALL, 23*(3), 271–293. DOI: 10.1017/S0958344011000188

Wu, P. H., & Marek, M. (2016). Incorporating LINE smartphone affordances: Cross-cultural collaboration, willingness to communicate, and language learning. [IJCALLT]. *International Journal of Computer-Assisted Language Learning and Teaching*, *6*(2), 56–73. DOI: 10.4018/IJCALLT.2016040104

Xu, X., Daniel, Z., & Li, Y. (2024). 2023 niandu zhongguo shuziyuedu baogao fabu—shuziyuedu shichang guimo maishang xintaijie [the "2023 China digital Reading report" has been released: The scale of the digital Reading market has reached a new level]. *Guang Ming Ri Bao*. Advance online publication. DOI: 10.28273/n.cnki.ngmrb.2024.001843

Yang, C., Luo, L., Vadillo, M. A., Yu, R., & Shanks, D. R. (2021). Testing (quizzing) boosts classroom learning: A systematic and meta-analytic review. *Psychological Bulletin*, *147*(4), 399–435. DOI: 10.1037/bul0000309 PMID: 33683913

Zheng, Y., Ogata, H., & Yano, Y. (2005). A conceptual framework of computer-supported ubiquitous learning environment L. Li.

Chapter 5
Mobile-Assisted Pronunciation Learning in an EFL Context:
An Exploratory Study

Rizgar Qasim Mahmood
https://orcid.org/0000-0002-5987-8884
Wollongong University, Australia & Salahaddin University-Erbil, Iraq

ABSTRACT

Mobile-Assisted Pronunciation Learning (MAPL) provides digital tools with real-time feedback and exposure to native-like models. However, research remains limited in EFL contexts such as Iraqi Kurdistan. This chapter examines the engagement of Kurdish EFL learners with MAPL, focusing on their usage patterns, perceptions, motivations, and challenges, using surveys (n = 45) and interviews (n = 15). The data were analyzed descriptively and thematically. The study revealed that learners frequently utilized applications such as ChatGPT and Google Translate to practice vowel sounds, word stress, and connected speech. While learners valued MAPL's flexibility and effectiveness for segmental practice, they highlighted critical limitations, particularly in supporting suprasegmental features and providing complex feedback. Other limitations were also mentioned, such as technical shortcomings, distractions, and weak self-regulation. The study concludes with theoretical and pedagogical implications.

DOI: 10.4018/979-8-3373-3366-3.ch005

INTRODUCTION

Pronunciation is one of the essential components of mastering English as a Foreign Language (EFL) and successful communication (Levis, 2018; Nguyen et al., 2021). Research in applied linguistics emphasizes that pronunciation is fundamental to communicative competence, similar to vocabulary and grammar (Celce-Murcia et al., 2010; Thomson & Derwing, 2014). Achieving a native-like accent is not the primary goal for most learners; instead, the focus is on intelligibility and comprehensibility – being easy to understand in conversation (Levis, 2005; Munro & Derwing, 2015). Notably, a noticeable foreign accent does not necessarily impede understanding; a speaker can have a strong accent yet remain highly comprehensible to listeners (Tergujeff, 2021). Conversely, serious mispronunciations can hinder listener comprehension even if grammar and vocabulary are correct. Therefore, developing intelligible and comprehensible pronunciation is essential for EFL learners to communicate confidently and avoid misunderstandings (Jones, 2018). In the context of second language (L2) acquisition, pronunciation learning contributes to overall proficiency and can even benefit other skills. For example, empirical studies have shown that improving pronunciation can enhance L2 listening ability by sharpening learners' bottom-up processing of speech (Kissling, 2015). Despite its importance, pronunciation instruction has historically been somewhat neglected in language classrooms (Bajorek, 2017; Pardede, 2018).

Technological innovations have provided new avenues for addressing pronunciation learning in recent decades. In particular, the emergence of Mobile-Assisted Language Learning (MALL) has created fresh opportunities for practicing pronunciation outside the classroom. MALL refers to the use of smartphones and other portable devices to facilitate language learning, whether in formal or informal settings (Kumar & Goundar, 2023; Traxler & Crompton, 2015). It is essentially an extension of computer-assisted language learning to mobile platforms, allowing learners anytime, anywhere access to practice materials and interactive exercises (Amrate & Tsai, 2024; Chapelle, 2001; Chun & Jiang, 2022; Lan, 2022). With the rapid global proliferation of smartphones, MALL has become increasingly relevant to EFL contexts. Mobile devices are now ubiquitous among university students, including those in developing regions, and they offer multimedia capabilities, such as audio recording, playback, and speech recognition, which are well-suited for pronunciation practice. The relevance of MALL to pronunciation instruction lies in its capacity to provide learners with flexible, individualized practice and immediate feedback. For example, a student can use a pronunciation app to listen to model pronunciations, record their voice, and receive automated feedback on their production. This process can be done at the student's convenience, fitting practice into daily routines – a form of self-directed learning that aligns with sociocultural notions of learner autonomy and mediation through tools (Metruk, 2024). Although the benefits of technology integration and MALL have been emphasized in teaching and learning English pronunciation, research in the EFL context, especially the Kurdish EFL context, is scarce (Mahmood, 2024a, 2025). Therefore, this chapter aims to explore how EFL learners in an Iraqi Kurdistan university utilize and perceive mobile technology to enhance their English pronunciation and assess their perceptions of this practice.

Practically, the study addresses a recognized need in EFL education for effective pronunciation training. Pronunciation has often been underemphasized in curricula, particularly in Iraqi Kurdistan, where educational resources may be limited and traditional teaching methods prevail (Mahmood, 2024b). Mobile-assisted learning offers a cost-effective and accessible means to supplement formal instruction. The significance of this research lies in its potential to inform educators and curriculum designers about the benefits and challenges of integrating mobile tools into pronunciation teaching. If EFL learners are found to have positive attitudes and measurable improvements from using pronunciation apps, it would underscore the value of incorporating such tools into classroom practice or assignments. The study's findings could encourage teachers to guide students in using pronunciation apps and mobile resources by recommending high-quality apps or creating assignments that utilize mobile practice. Furthermore, this research is one of the first studies focusing on mobile-assisted pronunciation learning in the Iraqi Kurdistan context. It thus contributes region-specific insights to the growing global literature on MALL. By highlighting how Kurdish EFL learners leverage mobile technology for English pronunciation, the study can help identify contextually relevant factors (such as local students' access to technology, preferences for certain apps, or cultural attitudes towards autonomous learning) that might influence the success of MALL implementations. Ultimately, the study's significance lies in demonstrating how mobile-assisted pronunciation learning can enhance EFL education, potentially leading to more intelligible speakers, confident communicators, and autonomous and engaged language learners. It points towards pedagogical innovations that could improve learners' oral and intercultural communicative competence in an era where English pronunciation proficiency is increasingly important for academic and professional opportunities. Therefore, the following research questions are addressed:

1. To what extent do Kurdish EFL learners use mobile applications to improve pronunciation?
2. What are the perceptions of Kurdish EFL learners of the effectiveness of mobile applications on pronunciation learning?
3. What are the primary challenges that Kurdish EFL learners face when using mobile applications to improve their pronunciation?

LITERATURE REVIEW

Theoretical Framework

The theoretical framework guiding this study is built upon the interconnected pillars of Sociocultural Theory (Lantolf & Thorne, 2006; Vygotsky, 1978), Self-Regulated Learning (Zimmerman, 2002; Zimmerman & Risemberg, 1997), and Technology-Enhanced Language Learning (TELL) (Stockwell, 2010, 2013, 2022). These theoretical lenses collectively provide a robust and comprehensive framework for examining the multifaceted nature of mobile-assisted pronunciation learning (MAPL). They offer a lens through which to analyze how learners interact with mobile tools as mediational artifacts, actively regulate their pronunciation learning processes, and engage in technology-mediated pronunciation training. By integrating these theories, the study aims to capture both the cognitive and sociocultural dimensions of pronunciation learning within a mobile environment.

Sociocultural Theory, primarily developed by Vygotsky (1978), emphasizes the fundamental role of social interaction, cultural tools, and activity in developing cognitive functions. Key concepts include the Zone of Proximal Development (ZPD), mediation, and the social construction of knowledge. In the context of this study, Sociocultural Theory provides a framework for understanding how mobile devices and applications function as mediational tools that shape and support pronunciation development (Lantolf, 2000; Lantolf & Pavlenko, 1995; Poehner & Lantolf, 2024). These tools are not neutral; they actively influence how learners engage with pronunciation practice, receive feedback, and interact with others or the technology itself during the learning process. For example, a mobile app with speech recognition software mediates the learner's experience by providing feedback that might not be available from a teacher or peer (Metruk, 2024). The study examines how Kurdish EFL learners utilize mobile tools within their specific sociocultural context, acknowledging that cultural factors influence the adoption and use of technology for learning. Sociocultural Theory helps analyze MAPL as a socially mediated activity, even when learners use mobile devices independently. The design of the apps, the feedback they provide, and the opportunities for online interaction (if available) all contribute to a social context for learning.

Self-Regulated Learning Theory (SRLT), prominently associated with the work of Zimmerman (2002), focuses on the processes through which learners actively take control of and manage their learning. SRL involves metacognitive, motivational, and behavioral aspects, including goal setting, self-monitoring, strategy use, and self-evaluation. SRL theory is crucial for examining how learners independently use mobile applications to improve pronunciation. It provides a framework for investigating learners' strategies to monitor their pronunciation, identify areas for improvement, use mobile tools effectively, and evaluate their progress. This study focuses on understanding how Kurdish EFL learners self-regulate their pronunciation learning in a mobile-assisted environment, including their motivation, use of feedback, and persistence in overcoming challenges. SRL is inherently linked to MAPL, as mobile devices offer learners increased autonomy and control over their learning. Mobile apps can provide learners with tools to set goals, track their progress, and receive feedback; however, the learners' self-regulatory skills determine how effectively they utilize these tools. This study examines how learners utilize the affordances of mobile technology to facilitate their self-regulated pronunciation learning.

Technology-Enhanced Language Learning (TELL) is a broad field that explores the use of technology to enhance language learning and teaching (Stockwell, 2010, 2013, 2022). It encompasses a wide range of technologies, from computer-assisted language learning (CALL) to mobile-assisted language learning (MALL) (Chapelle, 2001). It focuses on how technology can create more engaging, effective, and efficient learning experiences. TELL provides a framework for understanding how mobile technology can be used to support pronunciation development. This includes using mobile apps to deliver pronunciation instruction, provide feedback, and facilitate practice. This study examines the affordances and limitations of mobile technology for pronunciation learning in the EFL context and its impact on learners' experiences and outcomes. TELL is directly connected to MAPL, as MAPL is a subfield within TELL. By integrating these three theoretical perspectives, the study offers a more comprehensive and nuanced understanding of the intricate interplay between technology, learner agency, and sociocultural factors in developing pronunciation skills in an EFL context.

The Importance of Pronunciation

Pronunciation is widely acknowledged as a cornerstone of L2 acquisition. It plays a pivotal role in shaping learners' ability to communicate effectively and ensuring they are readily understood by native speakers and other listeners (Celce-Murcia et al., 2010; Munro & Derwing, 2015). Accurate and intelligible pronunciation not only enhances learners' confidence and fluency in spoken communication but also facilitates successful interactions across a wide range of social, academic, and professional settings (Couper, 2021; Levis, 2005; Saito, 2012). Beyond its functional role in conveying meaning, pronunciation also carries significant social and identity-related implications. A learner's pronunciation can influence how they are perceived by others, potentially affecting their sense of belonging and their willingness to engage in communication (Derwing & Munro, 2022; Kang et al., 2018).

Therefore, developing effective pronunciation skills is crucial for learners to fully participate in and integrate into communities where the target language is used.

Conversely, persistent pronunciation difficulties can impede effective communication, leading to misunderstandings, communication breakdowns, and potentially negative perceptions of learners' overall language proficiency and communicative competence (Derwing & Munro, 2022; Levis, 2018). These challenges can lead to frustration, demotivation, and an unwillingness to communicate orally. Furthermore, in an increasingly globalized world, where English is often used as a lingua franca, clear and intelligible pronunciation is essential for effective communication across diverse linguistic backgrounds (Jenkins, 2000; Seidlhofer, 2017).

The significance of pronunciation instruction in EFL contexts has been extensively recognized and emphasized by language educators and researchers alike, resulting in the development and implementation of a diverse array of approaches and techniques aimed at improving learners' pronunciation skills and fostering greater communicative competence (Grant & Brinton, 2014; Pennington & Rogerson-Revell, 2019). These instructional strategies encompass explicit instruction on the intricacies of phonetics and phonology, providing learners with explicit knowledge of the sound system of the target language (Ehrlich & Avery, 2013; Murphy, 2017). Implicit learning, facilitated through exposure to authentic and naturalistic language input, also plays a crucial role, allowing learners to develop their pronunciation skills through observation and imitation (Fraser, 2000; Munro & Derwing, 2015). The provision of corrective feedback on pronunciation errors is another key component of effective pronunciation instruction, guiding learners towards more remarkable accuracy and fluency (Lyster & Ranta, 1997; Yates & Zielinski, 2009). This feedback can take various forms, including explicit correction, recasts, and clarification requests, and its effectiveness depends on factors such as the learner's level, the type of error, and the context in which it is given (Ellis, 2009; Sheen, 2010)Therefore, it is imperative to explore learners' perceptions of pronunciation learning and the available technological tools, such as mobile applications.

Mobile-Assisted Language Learning (MALL)

Mobile-Assisted Language Learning (MALL) has emerged as a prominent educational technology subfield, defined by mobile devices' use to support language learning processes. MALL is essentially the mobile extension of computer-assisted language learning (CALL), leveraging smartphones, tablets, and other handheld devices as platforms for language practice and instruction. A widely cited definition by Kukulska-Hulme and Shield (2008) describes MALL as formal or informal learning mediated via handheld devices, potentially available anytime, anywhere. It can even be helpful while performing regular jobs. Similarly, Trifanova et al.

(2004, p. 3) define mobile devices as "...any device that is small, autonomous and unobtrusive enough to accompany us in every moment". Unlike traditional classroom-bound resources, mobile devices accompany learners daily, enabling spontaneous practice and access to learning materials anytime. The personal and always-on nature of smartphones means learners can engage with language apps during commutes, while waiting, or whenever a free moment arises, thus blurring the boundary between classroom learning and real-world contexts. Researchers note that the portability of mobile devices allows for more personalized and contextually rich learning experiences, as learners can customize their use and take advantage of contextual learning opportunities (e.g., practicing vocabulary relevant to their location or needs) (Kukulska-Hulme, 2005; Kukulska-Hulme & Shield, 2008; Levy & Kennedy, 2005; Stockwell, 2007)

The historical development of MALL can be traced back to the early 2000s, with the advent of personal digital assistants (PDAs) and mobile phones being used experimentally for language learning. Early studies in mobile learning often focused on SMS (text messaging) for delivering learning content. For example, Thornton and Houser (2005) sent vocabulary lessons via SMS to Japanese university students and found notable gains in vocabulary retention compared to traditional methods, as well as positive student attitudes towards the convenience of learning via phone . Such pioneering projects demonstrated that even the limited technology of basic mobile phones could effectively supplement language instruction. In Thornton and Houser's (2005) study, students who received SMS vocabulary quizzes learned roughly twice as many new words as those who used only paper materials. As mobile technology advanced, so did MALL applications. By the 2010s, with the proliferation of internet-enabled smartphones and app stores, MALL research expanded rapidly. Scholars documented a survey of studies examining various language learning apps and mobile-based activities across different language skills. Vocabulary and reading were among the earliest areas extensively explored via mobile tools (due to the ease of delivering text-based content), but gradually, other skills like listening, writing, speaking, and pronunciation also garnered attention (Bensalem, 2018; Metruk, 2024).

MALL for Pronunciation Instruction and Learners' Perceptions

Applying mobile-assisted learning to pronunciation instruction is a relatively new but rapidly growing area of interest. Because pronunciation practice benefits from frequent, focused repetition and feedback – something difficult to achieve in limited classroom hours – mobile applications are particularly well-suited to supplement this aspect of language learning. Over the last decade, an increasing number of empirical studies have investigated the effectiveness of mobile apps and tools in teaching L2 pronunciation (Dai & Wu, 2023; Lan, 2022; Daniel Liakin et al., 2014). Overall, the findings from these studies suggest that mobile-assisted pronunciation training can lead to measurable improvements in learners' pronunciation skills, often comparable to or even exceeding gains from traditional instruction. For example, Arashnia and Shahrokhi (2016) experimented with Iranian EFL students. They found that those who received pronunciation instruction via mobile phone outperformed a control group regarding post-test pronunciation scores. Similarly, Amalia (2020) reported a significant increase in Indonesian secondary students' pronunciation ability after integrating a mobile-assisted learning program into their phonetics instruction. These studies align with earlier work by Kim and Kwon (2012, p. 31), who reviewed several English learning apps and concluded that well-designed mobile apps provide "ubiquitously accessible and flexible practices" for pronunciation and speaking, creating a personal, learner-centered environment for practice. In essence, mobile tools can provide intensive practice and individualized feedback that pronunciation learning requires, thereby accelerating improvement. Several controlled studies have compared mobile-assisted pronunciation practice with traditional methods. For instance, a study by Denis Liakin et al. (2014) examined the learning of a French vowel sound by English-speaking learners using different methods: one group practiced with a mobile speech recognition app, another group practiced with a teacher providing oral feedback, and a control group had no focused pronunciation practice.

The literature on mobile-assisted pronunciation instruction reveals a promising convergence of need and technology. Pronunciation, a challenging area for EFL learners and teachers, can be enhanced through judicious use of mobile apps that provide practice, modeling, and feedback. Empirical studies have documented improved intelligibility, more accurate production of difficult sounds, and greater learner engagement when mobile tools are employed (Daniel Liakin et al., 2014). Learners generally hold positive attitudes toward these tools, citing benefits such as convenience, increased practice time, and reduced anxiety. For example, although Dai and Wu (2023) found that the peer group outperformed the Collaborative-non-ASR Group (or the Co-non-ASR Group), both of which had positive perceptions of MAPL, especially when individual students wanted to engage in sustained, self-paced, and stress-free practice (p. 877). Similar results have been reported by previous studies as well (Bashori et al., 2020; Dillon & Wells, 2021; Neri et al., 2008). When implemented thoughtfully, mobile-assisted learning has the potential to make a significant contribution to pronunciation proficiency in EFL contexts. However, it is worth mentioning that a systematic review study found that not all studies on MAPL had positive effects (Stoughton & Kang, 2024). They further found that MAPT effectiveness was inconsistent, with some studies reporting significant effects and others minimal gains. Differences in participants, pronunciation targets, and platforms limit the ability to draw firm conclusions. The focus on segmental features likely had little influence on overall intelligibility (Stoughton & Kang, 2024). This sets the stage for the present study, which will explore how these global findings are applied in a specific context – that of Kurdish EFL learners – and how mobile-assisted pronunciation learning can be optimized to support learners in achieving clearer, more confident speech.

METHODOLOGY

Research Design

This study employed a mixed-methods research design, following an embedded approach that incorporated both quantitative and qualitative data collection (Creswell & Creswell, 2018). Specifically, a sequential embedded strategy was employed, wherein quantitative data were initially collected through an online survey questionnaire, followed by semi-structured interviews (Creswell & Plano Clark, 2017). In line with Greene et al.'s (1989) complementarity integration strategy, qualitative data were collected to further elaborate on and clarify the findings derived from the quantitative phase (Braun & Clarke, 2006). This methodological framework was intentionally selected to enable a thorough analysis and interpretation of Kurdish EFL learners' perceptions of mobile-assisted pronunciation learning.

Participants

The participants in this study were forty-five Kurdish learners of English as a Foreign Language (EFL), aged between 18 and 38 years ($M = 23.5$) from various universities. At the time of data collection for this study, they were pursuing their bachelor's degree in English language and linguistics. All participants were residents of the Kurdistan Region of Iraq and volunteered to participate after an open call was distributed across social media platforms, including Facebook, Telegram, and Instagram. The recruitment strategy was designed to be intentionally broad, requiring no specific academic background, to attract a wide and varied group of learners motivated by their personal interest in enhancing their English proficiency.

The cohort comprised 25 female participants (approximately 56%) and 20 male participants (approximately 44%), reflecting a balanced gender distribution compared to the initial outreach. Almost all participants identified Kurdish as their native language, with only a very small proportion (approximately 2%) not explicitly stating their linguistic background. Participants represented a broad age range; however, the majority (more than 75%) fell within the 18 to 24-year age range, characteristic of young adults in higher education or early stages of professional employment in the region.

Regarding English proficiency, participants' self-assessments revealed a diverse spectrum of levels. Approximately 36% of the learners reported attaining a B2 (upper-intermediate) level, while about 26% identified their proficiency as B1 (intermediate). A smaller segment, comprising 19% of the participants, described themselves as C1 (advanced) users of English. Additionally, three participants (approximately 7%) self-reported an A2 (pre-intermediate) level, and a small group (approximately 7%) were uncertain about their precise proficiency level.

Data Collection

Data were collected through an online survey questionnaire created using Google Forms. The questionnaire items were explicitly designed for this study and reviewed by two experts in the field—both professors specializing in second language acquisition and technology-assisted language learning—to ensure their reliability and content validity. The survey focused on participants' Mobile-Assisted Pronunciation Learning (MAPL) habits, including the frequency of mobile app use for pronunciation improvement, the types of applications employed, and the specific features they utilized, such as automated speech recognition, recordings of native speakers, International Phonetic Alphabet (IPA) symbols, and recorded instructional videos. Participants were also asked about the contexts in which they used these applications, whether they combined mobile learning with other pronunciation strategies, and their perceptions regarding the effectiveness and challenges of using mobile applications for pronunciation development.

Following the survey, semi-structured interviews were conducted with a subset of 15 participants to gain deeper insights into their experiences and perspectives (see Appendix). All data collection procedures, including the survey and interviews, were conducted online via Google Forms and Zoom.

Data Analysis

A mixed-methods approach was employed to collect and analyze the data, integrating both quantitative and qualitative strands to understand the participants' experiences and perceptions comprehensively. Data collection involved two instruments: an online survey questionnaire and semi-structured interviews.

The quantitative data, obtained from the survey questionnaire, were analyzed descriptively using the statistical software *Jamovi* (The Jamovi Project, 2021). Measures of central tendency, mean, and dispersion (standard deviation) were calculated to summarize participants' responses, according to the following recommendations for descriptive quantitative analysis in survey research (Creswell & Creswell, 2018).

The qualitative data, collected through semi-structured interviews, were transcribed verbatim to represent participants' narratives accurately. A thematic analysis approach, following the widely recognized procedure outlined by Braun and Clarke (2006) was adopted. This process consisted of six phases: familiarization with the data, generating initial codes, identifying themes, reviewing the themes, defining and naming the themes, and producing the final report. Thematic analysis was selected for its flexibility and suitability in identifying patterns of meaning across a dataset, particularly in studies that aim to capture participants' perspectives in depth.

Throughout the analysis, methodological rigor was maintained by adhering to guidelines for thematic analysis, including iterative coding and theme refinement, and by ensuring that both inductive and theoretically driven insights were considered (Nowell et al., 2017).

RESULTS

Findings: Survey Questionnaire

Figure 1 Which mobile applications or tools have you used for pronunciation learning?

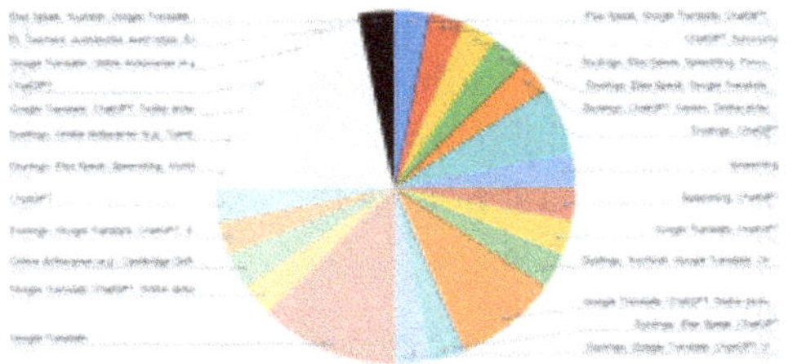

The first four items of the survey were designed to explore the participants' general habits of using Mobile-Assisted Pronunciation Learning. Participants (N = 45) were asked to indicate which mobile applications or tools they had used for pronunciation learning. Responses indicated that ChatGPT was the most frequently used tool, reported by 68.8% of participants. Google Translate was also widely utilized, with 62.5% of participants indicating its use. In addition, online dictionaries such as the Cambridge Dictionary and the Oxford Learner's Dictionary were reported by 53.1% of the sample.

Among dedicated language learning applications, Duolingo was selected by 34.4% of participants, while specialized pronunciation-focused tools, such as Elsa Speak (18.8%), Speechling (12.5%), and YouGlish (12.5%), were used less frequently. Other tools mentioned with lower frequencies (each by 3.1% of participants) included Forvo, Gemini, Claudia, various English Pronunciation IPA applications, and a few resources related to teachers' recordings and audiobook materials.

These findings indicate that participants tend to favor multifunctional AI-driven applications (e.g., ChatGPT and Google Translate) over specialized pronunciation training applications. The use of online dictionaries remains significant, highlighting that traditional pronunciation models provided by established dictionaries continue to play a substantial role.

Despite the availability of targeted mobile-assisted pronunciation learning tools, the relatively lower rates of their use suggest either limited awareness among learners or a preference for tools that integrate pronunciation within broader language learning functionalities.

Furthermore, they were asked to indicate which features of mobile applications they used most for pronunciation learning. Results revealed that the most commonly utilized feature was phonetic transcription and IPA symbols, selected by 53.1% of the respondents. This finding suggests a strong preference for visual phonological representations to support the development of pronunciation.

The second most frequently reported feature was video lessons (46.9%), indicating the importance of multimodal audiovisual input for learners. Both speech recognition and feedback, as well as audio recordings of native speakers, were cited equally by 28.1% of participants, reflecting learners' appreciation for immediate corrective feedback and authentic pronunciation models (See Figure 2).

Figure 2. What features of these mobile applications do you use the most?

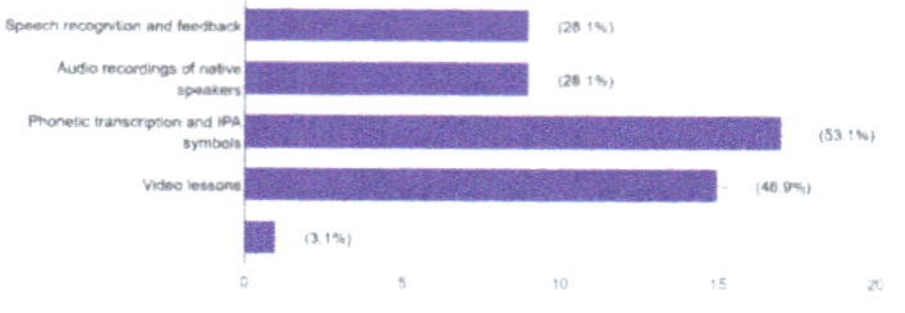

These results demonstrate that learners prioritize features that provide clear, structured phonological information (phonetic transcription) and authentic exemplars (native speaker recordings), followed closely by interactive feedback mechanisms and visual media resources, such as video lessons.

Additionally, the participants were asked to report the time they spend weekly using mobile applications for pronunciation learning. As shown in Figure 1, half of the participants (50%) reported spending less than 30 minutes per week on mobile-assisted pronunciation practice.

A substantial portion, 37.5%, indicated spending 30 minutes to 1 hour per week, suggesting that for many learners, mobile pronunciation practice constitutes a limited but regular part of their study habits. A smaller group, 9.4%, reported engaging with mobile tools for 1 to 2 hours weekly, while only 3.1% of participants practiced for more than 2 hours per week.

These findings suggest that, although most learners widely use mobile pronunciation tools, the overall time commitment remains relatively low. This may suggest that mobile applications are used as supplementary tools rather than primary methods for pronunciation development.

Figure 3. How much time do you spend using mobile tools for pronunciation practice per week?

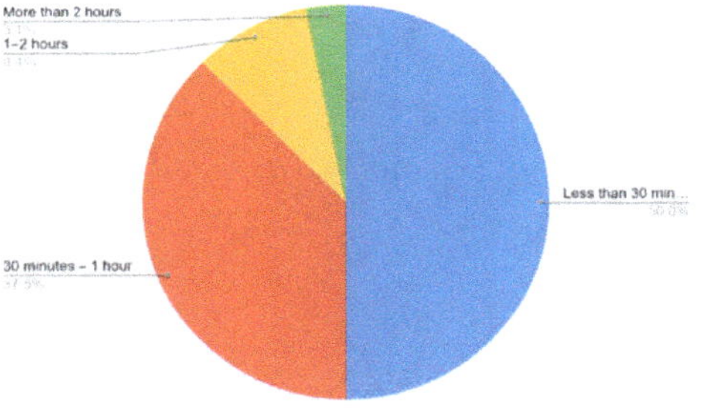

Furthermore, Participants were asked where they typically used mobile applications for pronunciation practice. The majority of respondents (50.0%) reported using mobile pronunciation tools at home. A significant proportion (21.9%) indicated using these tools in a classroom setting, while 15.6% reported usage during free time outside structured settings.

Smaller percentages of participants indicated other usage contexts: 6.3% reported using mobile tools during commuting, 3.1% reported using them whenever they encountered an unfamiliar word or pronunciation, and 3.1% indicated using them both at home and in the classroom. These findings emphasize that mobile-assisted pronunciation learning is primarily conducted in informal, self-directed environments, particularly at home.

Figure 4. Where do you typically use mobile pronunciation learning apps?

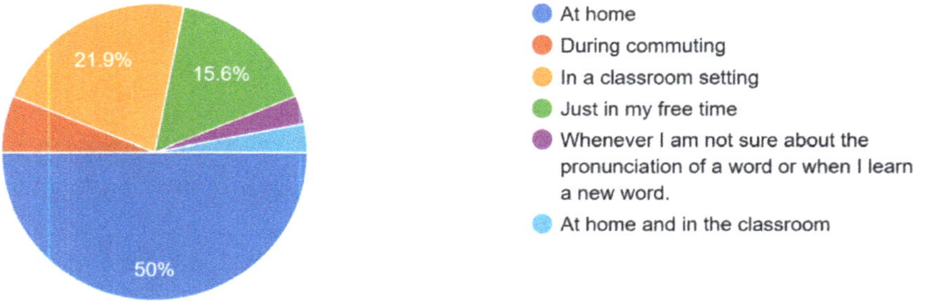

Another aspect of the survey was to explore the participants' perceptions and the challenges they face when using mobile applications to improve their pronunciation. For this reason, they were asked to rate the perceived effectiveness of mobile-assisted pronunciation learning.

Figure 5. How effective do you think mobile-assisted pronunciation learning is?

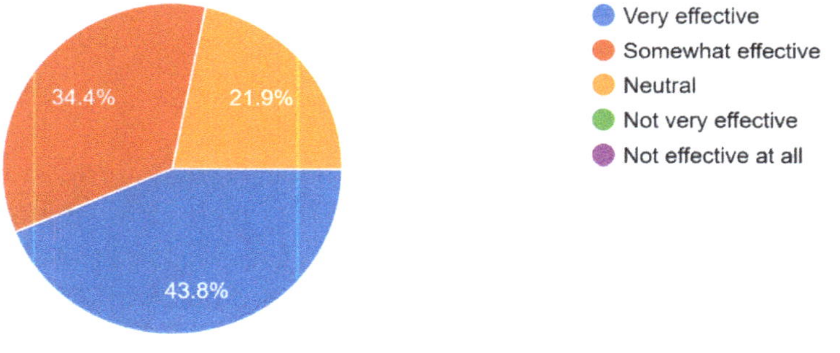

As illustrated in Figure 5, most respondents (43.8%) considered mobile-assisted pronunciation learning very effective. An additional 34.4% of participants rated it somewhat effective, indicating a generally positive perception among most users.

Meanwhile, 21.9% of participants reported a neutral stance toward its effectiveness. Notably, no participants selected the options "not very effective" or "not effective at all," suggesting that this sample had minimal negative perceptions regarding mobile-assisted pronunciation tools. Furthermore, these results suggest learners hold favourable perceptions of mobile-assisted pronunciation tools. A combined 78.2% (very effective + somewhat effective) of participants acknowledged at least some degree of effectiveness. The absence of negative evaluations (i.e., no selections of "not very effective" or "not effective at all") highlights high user satisfaction with the role of mobile technologies in pronunciation development.

Another aspect of the survey was to explore which aspects of pronunciation mobile applications can help achieve further improvement. Thus, the participants were asked to select which aspects of pronunciation they perceived as having improved the most through the use of mobile learning tools.

Figure 6. Which aspects of pronunciation have improved the most through mobile learning?

As shown in Figure 6, the most frequently reported improvement was in vowel sounds, selected by 62.5% of respondents. Similarly, word stress (59.4%) and connected speech (56.3%) were also widely acknowledged as areas of noticeable development.

Other aspects noted by learners included sentence intonation (46.9%) and consonant sounds (43.8%). A minority of respondents (9.4%) indicated that they had not observed any improvement in their pronunciation through mobile learning.

Additionally, this study examined the motivations behind the use of mobile applications among Kurdish EFL learners to enhance their pronunciation. Hence, they were asked to indicate their primary motivation for using mobile applications to enhance their pronunciation. As shown in Figure 7, the most frequently reported motivation was to improve fluency and accuracy, cited by 53.1% of respondents.

Figure 7. What is your primary motivation for using mobile tools for pronunciation learning?

Additionally, 28.1% of learners indicated their primary motivation was to communicate better in daily life or work contexts, highlighting a practical, communicative orientation. 18.8% of participants reported their goal was to achieve native-like pronunciation and accent. Notably, no participants selected exam preparation (e.g., IELTS, TOEFL) as their main motivation.

Research indicates that learners' attitudes and preferences on learning techniques and strategies are essential. Therefore, this study was specifically interested in exploring these aspects of learners while using mobile applications for pronunciation learning. In this regard, the participants were asked to express their feelings about using AI-based pronunciation feedback, such as automatic speech recognition or ChatGPT evaluations. As depicted in Figure 8, 40.6% of respondents rated AI-based feedback as very useful, while 34.4% considered it somewhat useful.

Figure 8. How do you feel about using AI-based pronunciation feedback, such as automatic speech recognition and scoring, or ChatGPT?

Approximately 21.9% of participants reported a neutral stance. Only 3.1% of respondents indicated that they found AI-based pronunciation feedback not useful at all, and no participants rated it as not very useful. The findings reveal an overall positive attitude toward AI-based pronunciation feedback among learners. A combined 75% of respondents viewed it as either very useful or somewhat helpful, suggesting general acceptance of automated feedback technologies in pronunciation learning.

Finally, this study investigated whether learners prefer using traditional instructions, using only their mobile device, or a combination of both when working on pronunciation improvement. As presented in Figure 9, the vast majority (87.5%) of participants preferred a combination of mobile applications and traditional classroom instruction.

Figure 9. Would you prefer learning pronunciation through mobile applications, traditional classroom instruction, or a combination of both?

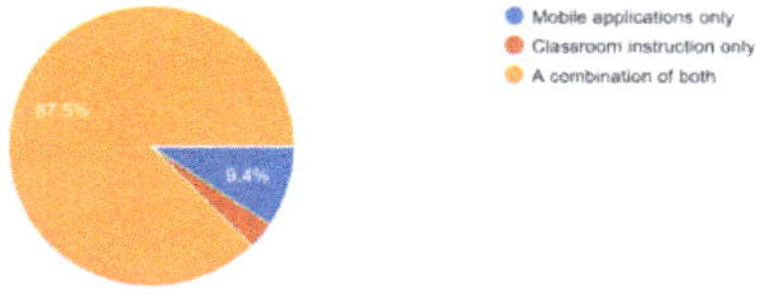

Only 9.4% of participants preferred using mobile applications, while 3.1% preferred classroom instruction only without integrating mobile tools. The results indicate a strong preference among learners for blended learning approaches, combining the benefits of mobile applications with the structure of classroom instruction and interpersonal interaction.

Interview Results

Perceived Benefits of Mobile-Assisted Pronunciation Learning

Participants overwhelmingly reported that mobile-assisted pronunciation learning positively impacted their pronunciation development. Several learners highlighted noticeable improvements in accuracy, fluency, and confidence. As one participant reflected, *"It has had a positive impact on my pronunciation, and I can clearly notice the improvement"* (Participant 2). Another elaborated that *"Mobile-assisted pronunciation learning has significantly improved my pronunciation by providing instant feedback, interactive exercises, and exposure to native speech"* (Participant 5).

Furthermore, the accessibility and motivational aspects of mobile tools were noted. For instance, one learner emphasized the value of audiobook-based learning, stating, *"My absolute favourite method is listening to audiobooks and reading the book with my eyes so that I hear the pronunciation and spelling"* (Participant 1). However, a few participants indicated that they used mobile tools only occasionally, describing them as supplementary resources rather than comprehensive solutions (*"I only searched for a couple of words when needed, not all the time,"* Participant 9).

Challenges Encountered in Mobile-Assisted Pronunciation Learning

Although generally positive, participants reported several challenges associated with mobile-assisted pronunciation learning. Technical issues, such as inaccurate feedback and limited contextual awareness, were frequently mentioned. One participant explained, *"The biggest challenges include inaccurate speech recognition, limited contextual feedback, and a lack of real-life conversation practice"* (Participant 15).

Other participants noted the absence of real-time clarification opportunities, with one remarking, *"It's not like a real-life teacher or friend that if you don't understand it, you ask them to try teaching in another way"* (Participant 10). Additionally, some participants cited cognitive and motivational challenges associated with mobile use, particularly the distractions caused by entertainment applications. For instance, one participant noted, *"The mental challenge arises from using mobile phones, particularly TikTok and Instagram, where short videos flood your mind with dopamine and joy"* (Participant 13).

Pronunciation Skills Improved

Participants frequently reported improvements in specific pronunciation skills, particularly in stress patterns, segmental features (including vowel and consonant sounds), and intonation. One participant shared, *"I successfully improved my intonation and word stress using a mobile app with speech analysis features"* (Participant 11). Another described improving difficult sounds: *"Using my mobile phone, I improved my pronunciation of difficult sounds, such as the difference between /θ/ and /ð/ (as in 'think' and 'this')"* (Participant 14).

However, while segmental improvements were commonly reported, challenges with connected speech and natural rhythm persisted, reflecting the partial effectiveness of mobile tools in addressing suprasegmental features.

Limitations of Mobile Tools Compared to Classroom Instruction

Participants strongly agreed that while valuable, mobile tools cannot fully replace classroom instruction. Learners emphasized the irreplaceable role of human teachers in providing emotional sensitivity, nuanced feedback, and motivational support. One participant noted, *"A machine, such as a mobile phone, cannot replace humans because we possess unique qualities that distinguish us from machines"* (Participant

12). Another reinforced this view by stating, *"Teachers provide real-time feedback, personalized guidance, and interactive learning"* (Participant 20).

Moreover, several participants suggested that mobile learning could lead to decreased discipline, with one participant explaining, "In the classroom, you are obliged to practice, but through applications, *you might be lazy and ignore it"* (Participant 6).

Teachers' Role in Supporting Mobile-Assisted Learning

Participants reported teachers as essential guides and facilitators in integrating mobile tools into pronunciation learning. For example, several participants recommend that teachers curate effective apps and monitor learner progress. One participant proposed, *"Teachers should guide students in selecting effective mobile tools, provide structured practice, and offer personalized feedback that apps may lack"* (Participant 15). Another added, *"Teachers can recommend useful mobile-assisted apps for students and integrate them into interactive speaking exercises"* (Participant 12). Overall, the responses underscored the importance of a blended learning approach, combining the benefits of mobile flexibility with structured classroom interaction.

Reliability of AI-Generated Feedback

Views on the reliability of AI-generated feedback were mixed. Some participants expressed confidence in its standardized accuracy, with one stating, *"The pronunciation feedback I receive comes from standardized pronunciations used globally. These applications provide accurate and consistent pronunciation"* (Participant 5). Others, however, expressed reservations about AI's ability to capture the nuances, accents, and intonation patterns of natural speech. One participant critically noted, *"AI-generated feedback in pronunciation apps is helpful but not always completely reliable; human feedback is still essential for accurate and meaningful pronunciation improvement"* (Participant 13). These mixed perceptions underscore that while AI is a valuable tool for self-directed practice, it cannot fully replace expert human feedback, particularly for refining advanced pronunciation.

Overall, the interview findings reveal that learners perceive mobile-assisted pronunciation learning as a beneficial supplement for improving specific aspects of pronunciation. However, the results also demonstrate a clear recognition of the limitations of mobile applications and a continued need for teacher involvement, real-time interaction, and balanced learning strategies to fully support pronunciation development.

DISCUSSION

This study critically examined the engagement of Kurdish EFL learners with Mobile-Assisted Pronunciation Learning (MAPL), guided by Sociocultural Theory (Lantolf & Thorne, 2006; Vygotsky, 1978), Self-Regulated Learning Theory (Zimmerman, 2002), and Technology-Enhanced Language Learning (TELL) perspectives (Stockwell, 2010, 2013, 2022). This section presents the discussion in accordance with the research questions, critically evaluating the findings.

The first research question aimed to investigate the extent to which Kurdish EFL learners utilize mobile applications to enhance their pronunciation. The results show that participants predominantly relied on tools such as ChatGPT, Google Translate, and online dictionaries for practicing vowel sounds, stress patterns, and connected speech, echoing trends reported by Amalia (2020) and Dai and Wu (2023). However, usage was often brief, sporadic, and motivated by immediate communicative needs rather than sustained learning goals. This limited engagement contrasts sharply with the more intensive, structured use of mobile tools envisioned by Stockwell (2013), suggesting that simply having access to technology does not guarantee profound or transformative learning outcomes.

Furthermore, this study aimed to explore the participants' perceptions of the effectiveness of MAPL, which reflects duality (RQ 2). On the one hand, learners valued the autonomy, flexibility, and immediate feedback provided by mobile applications, supporting claims by Kim and Kwon (2012) and Burston (2015) regarding the motivational advantages of MALLs. On the other hand, participants were acutely aware of the limitations of these tools. They noted that mobile apps often failed to capture the nuances of authentic speech, particularly suprasegmental features such as intonation, rhythm, and connected speech patterns—issues also highlighted by Stoughton and Kang (2024) and Derwing and Munro (2022). Furthermore, AI-generated feedback was critiqued for its inconsistency and insensitivity to accent variations, corroborating concerns raised by Pennington (2021) and Pennington (1999) about the current shortcomings of automated pronunciation evaluation systems. Thus, learners considered MAPL a practical but incomplete component of a broader, socially mediated learning ecology.

In addressing the challenges faced by Kurdish EFL learners (RQ 3), the study identified significant technical, pedagogical, and motivational obstacles. Technically, inaccuracies in speech recognition and limited adaptability of feedback hindered learners' ability to refine their pronunciation systematically. Pedagogically, the lack of dialogic interaction is essential for scaffolding in the Zone of Proximal Development (Poehner & Lantolf, 2024; Vygotsky, 1978). This meant learners were often left to navigate pronunciation complexities without sufficient guidance. This absence of contingent support starkly contrasts with the principles of Sociocultural Theory, which emphasize the importance of socially mediated learning. Motivationally, learners struggled with maintaining focus in environments saturated with distractions, such as social media platforms—a phenomenon that Dai and Wu (2023) and Mahmood (2025) also identified as a significant impediment to mobile learning success. These findings reinforce the argument that mobile learning environments must be carefully structured to support, rather than undermine, self-regulation and goal-directed learning.

Moreover, while participants demonstrated agency in selecting and using mobile tools, their reliance on general-purpose applications rather than specialized pronunciation apps such as Elsa Speak or Speechling suggests a gap in digital literacy. As noted by Bensalem (2018) and Lan (2022), effective mobile-assisted learning requires access to technology and informed decision-making about tool selection, usage strategies, and learning goals. Without explicit guidance from teachers, learners may default to less effective practices, thus limiting the potential benefits of MAPL.

The preference among participants for blended learning models—combining mobile practice with traditional classroom instruction—further highlights the limitations of mobile-only approaches. This preference aligns with Graham et al.'s (2013) advocacy for blended learning and Stockwell's (2022) assertion that technology is most effective when integrated into structured, socially interactive learning environments. Participants' recognition that "*a mobile phone cannot replace humans*" underscores the enduring value of teacher-mediated feedback, peer interaction, and contextually rich communicative practice.

In summary, while MAPL offers valuable supplementary opportunities for pronunciation development, its effectiveness is limited by technological constraints, challenges in learner self-regulation, and the lack of socially contingent support. To maximize the benefits of MAPL, future pedagogical interventions must focus on integrating mobile tools within broader, scaffolded, and socially interactive learning frameworks.

CONCLUSION

This study critically revealed that Kurdish EFL learners value mobile-assisted pronunciation tools for their flexibility, accessibility, and ability to enhance specific pronunciation features. Nevertheless, mobile applications were perceived as insufficient stand-alone solutions for comprehensive pronunciation development. Learners emphasized the need for human interaction, nuanced feedback, and structured support to complement mobile practice. Furthermore, it can be concluded that although using mobile applications for pronunciation learning has the potential to lead learners to pronunciation improvement, they sometimes face a lack of motivation and encounter challenges when trying to understand the provided feedback, especially regarding the connected speech aspects of L2 pronunciation. Therefore, although these findings reinforce the sociocultural, self-regulatory, and technological dimensions articulated in the theoretical framework, offering a contextualized contribution to MAPL research in EFL settings, those applications need further modifications to meet learners' needs.

THEORETICAL AND PEDAGOGICAL IMPLICATIONS

Theoretically, the study validates Sociocultural Theory's assertion that tools mediate learning but highlights that tool-mediated learning is most effective when coupled with social interaction. It extends Self-Regulated Learning Theory by illustrating that technological autonomy must be supported by strategies to combat motivational distractions. Finally, it nuances Technology-Enhanced Language Learning models by demonstrating that technological affordances must be critically matched with pedagogical interventions to maximize effectiveness.

Pedagogically, educators should integrate mobile pronunciation tools into curriculum design, but not as replacements for classroom interaction. Teachers should scaffold learners' mobile use through explicit strategy training, goal-setting support, and curated app recommendations. Blended learning models that synergize MAPL with peer feedback, teacher modeling, and corrective feedback will most likely enhance learners' pronunciation comprehensibility and engagement. Training learners to critically evaluate and effectively utilize AI-based pronunciation feedback should also be a pedagogical priority.

LIMITATIONS AND FUTURE RESEARCH

This study focused on a specific regional EFL context (Iraqi Kurdistan) with a limited sample size, which may constrain broader generalization. Self-reported measures of pronunciation improvement may also be subject to learner bias. Future studies should employ longitudinal designs, incorporate objective pre- and post-pronunciation measures, and explore the effectiveness of MAPL across varied sociolinguistic contexts. Further research should also investigate the design of AI-enhanced feedback systems that can effectively address suprasegmental and discourse-level pronunciation features. Moreover, future research could examine how self-regulation and digital literacy training influence the effectiveness of mobile-assisted pronunciation learning.

AI USAGE ACKNOWLEDGEMENT

I acknowledge the use of Grammarly in helping me to review my writing at the final stage of preparing my chapter. However, all the information has been checked and verified by the author, and he is responsible for the accuracy of the resources.

REFERENCES

Amalia, I. (2020). The application of mobile assisted language learning (MALL) in teaching pronunciation. *International Journal of Language Education and Cultural Review*, *6*(2), 194–203. DOI: 10.21009/IJLECR.062.20

Amrate, M., & Tsai, P. (2024). Computer-assisted pronunciation training: A systematic review. *ReCALL*. Advance online publication. DOI: 10.1017/S0958344024000181

Arashnia, M., & Shahrokhi, M. (2016). Mobile assisted language learning: English pronunciation among Iranian pre-intermediate EFL learners. *Journal of Applied Linguistics and Language Research*, *3*(4), 149–162. http://www.jallr.com/index.php/JALLR/article/view/433

Bajorek, J. (2017). L2 pronunciation in CALL: The unrealized potential of Rosetta Stone, Duolingo, Babbel, and Mango languages. *Issues and Trends in Educational Technology*, *5*(1), 60–87. DOI: 10.2458/azu_itet_v5i1_bajorek

Bashori, M., van Hout, R., Strik, H., & Cucchiarini, C. (2020). Web-based language learning and speaking anxiety. *Computer Assisted Language Learning*, ●●●, 1–32. DOI: 10.1080/09588221.2020.1770293

Bensalem, E. (2018). The impact of WhatsApp on EFL students' vocabulary learning. *Arab World English Journal*, *9*(1), 23–38. Advance online publication. DOI: 10.24093/awej/vol9no1.2

Braun, V., & Clarke, V. (2006). Using thematic analysis in psychology. *Qualitative Research in Psychology*, *3*(2), 77–101. DOI: 10.1191/1478088706qp063oa

Burston, J. (2015). Twenty years of MALL project implementation: A meta-analysis of learning outcomes. *ReCALL*, *27*(1), 4–20. DOI: 10.1017/S0958344014000159

Celce-Murcia, M., Brinton, D. M., & Goodwin, J. M. (2010). *Teaching pronunciation: A course book and reference guide*. Cambridge University Press.

Chapelle, C. A. (2001). *Computer applications in second language acquisition*. Cambridge University Press., DOI: 10.1017/CBO9781139524681

Chun, D. M., & Jiang, Y. (2022). Using technology to explore L2 pronunciation. In Levis, J., Derwing, T., & Sonsaat-Hegelheimer, S. (Eds.), *Second Language Pronunciation: Bridging the Gap Between Research and Teaching* (pp. 129–150). Wiley-Blackwell., DOI: 10.1002/9781394259663.ch7

Couper, G. (2021). Pronunciation teaching issues: Answering teachers' questions. *RELC Journal*, *52*(1), 128–143. DOI: 10.1177/0033688220964041

Creswell, J. W., & Creswell, J. D. (2018). Research design: Qualitative, quantitative, and mixed methods approaches. *Sage (Atlanta, Ga.).*

Creswell, J. W., & Plano Clark, V. L. (2017). Designing and conducting mixed methods research. *Sage (Atlanta, Ga.).*

Dai, Y., & Wu, Z. (2023). Mobile-assisted pronunciation learning with feedback from peers and/or automatic speech recognition: A mixed-methods study. *Computer Assisted Language Learning*, *36*(5-6), 861–884. DOI: 10.1080/09588221.2021.1952272

Derwing, T. M., & Munro, M. J. (2022). Pronunciation learning and teaching. In Derwing, T. M., Munro, M. J., & Thomson, R. I. (Eds.), *The Routledge handbook of second language acquisition and speaking* (pp. 147–159). Routledge. DOI: 10.4324/9781003022497-14

Dillon, T., & Wells, D. (2021). Student perceptions of mobile automated speech recognition for pronunciation study and testing. *English Teaching*, *76*(4), 101–122. DOI: 10.15858/engtea.76.4.202112.101

Ehrlich, S., & Avery, P. (2013). *Teaching American English pronunciation*. Oxford University Press.

Ellis, R. (2009). Corrective feedback and teacher development. *L2 Journal: An electronic refereed journal for foreign and second language educators, 1*(1). https://doi.org/DOI: 10.5070/L2.V1I1.9054

Fraser, H. (2000). *Coordinating improvements in pronunciation teaching for adult learners of English as a second language*. DETYA (ANTA Innovative project).

Graham, C. R., Woodfield, W., & Harrison, J. B. (2013). A framework for institutional adoption and implementation of blended learning in higher education. *The Internet and Higher Education*, *18*, 4–14. DOI: 10.1016/j.iheduc.2012.09.003

Grant, L., & Brinton, D. (2014). *Pronunciation myths: Applying second language research to classroom teaching*. University of Michigan Press. DOI: 10.3998/mpub.4584330

Greene, J. C., Caracelli, V. J., & Graham, W. F. (1989). Toward a conceptual framework for mixed-method evaluation designs. *Educational Evaluation and Policy Analysis*, *11*(3), 255–274. DOI: 10.3102/01623737011003255

Jenkins, J. (2000). *The phonology of English as an international language*. Oxford university press.

Jones, T. (2018). Pronunciation with other areas of language. In Kang, O., Thomson, R. I., & Murphy, J. M. (Eds.), *The Routledge Handbook of Contemporary English Pronunciation* (pp. 370–384). Routledge.

Kang, O., Thomson, R. I., & Murphy, J. (2018). *The Routledge handbook of contemporary English pronunciation*. Routledge.

Kim, H., & Kwon, Y. (2012). Exploring smartphone applications for effective mobile-assisted language learning. *Multimedia-Assisted Language Learning*, *15*(1), 31–57. DOI: 10.15702/mall.2012.15.1.31

Kissling, E. M. (2015). Phonetics instruction improves learners' perception of L2 sounds. *Language Teaching Research*, *19*(3), 254–275. DOI: 10.1177/1362168814541735

Kukulska-Hulme, A. (2005). Introduction. In Kukulska-Hulme, A., & Traxler, J. (Eds.), *Mobile Learning: A Handbook for Educators and Trainers* (pp. 1–7). Routledge.

Kukulska-Hulme, A., & Shield, L. (2008). An overview of mobile assisted language learning: From content delivery to supported collaboration and interaction. *ReCALL*, *20*(3), 271–289. DOI: 10.1017/S0958344008000335

Kumar, B. A., & Goundar, S. (2023). Developing mobile language learning applications: A systematic literature review. *Education and Information Technologies*, *28*(5), 5651–5671. DOI: 10.1007/s10639-022-11377-x

Lan, Y.-J. E. (2022). A comparative study of computer and mobile-assisted pronunciation training: The case of university students in Taiwan. *Education and Information Technologies*, *27*(2), 1559–1583. DOI: 10.1007/s10639-021-10647-4

Lantolf, J. P. (2000). *Sociocultural theory and second language learning* (Vol. 78). Oxford university press.

Lantolf, J. P., & Pavlenko, A. (1995). Sociocultural theory and second language acquisition. *Annual Review of Applied Linguistics*, *15*, 108–124. DOI: 10.1017/S0267190500002646

Lantolf, J. P., & Thorne, S. L. (2006). *Sociocultural theory and the genesis of second language development*. Oxford University Press.

Levis, J. M. (2005). Changing contexts and shifting paradigms in pronunciation teaching. *TESOL Quarterly*, *39*(3), 369–377. DOI: 10.2307/3588485

Levis, J. M. (2018). *Intelligibility, oral communication, and the teaching of pronunciation*. Cambridge University Press., DOI: 10.1017/9781108241564

Levy, M., & Kennedy, C. (2005). Learning Italian via mobile SMS. In Kukulska-Hulme, A., & Traxler, J. (Eds.), *Mobile Learning: A Handbook for Educators and Trainers* (pp. 76–83). Taylor & Francis.

Liakin, D., Cardoso, W., & Liakina, N. (2014). Learning L2 pronunciation with a mobile speech recognizer: French /y. *CALICO Journal*, *32*(1), 1–25. DOI: 10.1558/cj.v32i1.25962

Liakin, D., Cardoso, W., & Liakina, N. (2014). Learning L2 pronunciation with a mobile speech recognizer: French/y. *CALICO Journal*, *32*(1), 1–25. DOI: 10.1558/cj.v32i1.25962

Lyster, R., & Ranta, L. (1997). Corrective feedback and learner uptake: Negotiation of form in communicative classrooms. *Studies in Second Language Acquisition*, *19*(1), 37–66. DOI: 10.1017/S0272263197001034

Mahmood, R. Q. (2024a). ESL learners' perceptions of using High-Variability Phonetic Training (HVPT) through YouGlish to improve pronunciation skills. *Australian Journal of Applied Linguistics*, *7*(3), 1818. DOI: 10.29140/ajal.v7n3.1818

Mahmood, R. Q. (2024b). Still matters: Kurdish EFL learners' perceptions towards non-native English teachers' pronunciation skills. *Círculo de Lingüística Aplicada a la Comunicación*, *100*, 245–252. DOI: 10.5209/clac.81012

Mahmood, R. Q. (2025). Exploring educational technology in language education: Empirical evidence from teachers' perspectives. In Stevkovska, M., Klemenchich, M., & Kavaklı Ulutaş, N. (Eds.), *Reimagining Intelligent Computer-Assisted Language Education* (pp. 129–162). IGI Global., DOI: 10.4018/979-8-3693-4310-4.ch004

Metruk, R. (2024). Mobile-assisted language learning and pronunciation instruction: A systematic literature review. *Education and Information Technologies*, *29*(13), 16255–16282. DOI: 10.1007/s10639-024-12453-0

Munro, M. J., & Derwing, T. M. (2015). Intelligibility in research and practice. In Reed, M., & Levis, J. M. (Eds.), *The handbook of English pronunciation* (pp. 375–396). Wiley-Blackwell., DOI: 10.1002/9781118346952.ch21

Murphy, J. M. (2017). Teacher training in the teaching of pronunciation. In Kang, O., Thomson, R. I., & Murphy, J. M. (Eds.), *The Routledge handbook of contemporary English pronunciation* (pp. 298–319). Routledge. DOI: 10.4324/9781315145006-19

Neri, A., Mich, O., Gerosa, M., & Giuliani, D. (2008). The effectiveness of computer assisted pronunciation training for foreign language learning by children. *Computer Assisted Language Learning*, *21*(5), 393–408. DOI: 10.1080/09588220802447651

Nguyen, L. T., Hung, B. P., Duong, U. T. T., & Le, T. T. (2021). Teachers' and learners' beliefs about pronunciation instruction in tertiary English as a foreign language education. *Frontiers in Psychology*, *12*, 739842. DOI: 10.3389/fpsyg.2021.739842 PMID: 34489841

Nowell, L. S., Norris, J. M., White, D. E., & Moules, N. J. (2017). Thematic analysis: Striving to meet the trustworthiness criteria. *International Journal of Qualitative Methods*, *16*(1), 1609406917733847. DOI: 10.1177/1609406917733847

Pardede, P. (2018). Improving EFL students' English pronunciation by using the explicit teaching approach. *Journal of English Teaching*, *4*(3), 143–155. DOI: 10.33541/jet.v4i3.852

Pennington, M. C. (1999). Computer-aided pronunciation pedagogy: Promise, limitations, directions. *Computer Assisted Language Learning*, *12*(5), 427–440. DOI: 10.1076/call.12.5.427.5693

Pennington, M. C. (2021). Teaching pronunciation: The state of the art 2021. *RELC Journal*, *52*(1), 3–21. DOI: 10.1177/00336882211002283

Pennington, M. C., & Rogerson-Revell, P. (2019). *English pronunciation teaching and research* (Vol. 10). Palgrave Macmillan London. DOI: 10.1057/978-1-137-47677-7

Poehner, M. E., & Lantolf, J. P. (2024). *Sociocultural theory and second language developmental education*. Cambridge University Press., https://doi.org/DOI, DOI: 10.1017/9781009189422

Saito, K. (2012). Effects of instruction on L2 pronunciation development: A synthesis of 15 quasi-experimental intervention studies. *TESOL Quarterly*, *46*(4), 842–854. DOI: 10.1002/tesq.67

Seidlhofer, B. (2017). Standard English and the dynamics of ELF variation. In J. Jenkins, W. Baker, & M. Dewey (Eds.), *The Routledge handbook of English as a lingua franca* (pp. 85-100). Routledge. DOI: 10.4324/9781315717173-8

Sheen, Y. (2010). Differential effects of oral and written corrective feedback in the ESL classroom. *Studies in Second Language Acquisition*, *32*(2), 203–234. DOI: 10.1017/S0272263109990507

Stockwell, G. (2007). Vocabulary on the move: Investigating an intelligent mobile phone-based vocabulary tutor. *Computer Assisted Language Learning*, *20*(4), 365–383. DOI: 10.1080/09588220701745817

Stockwell, G. (2010). Using mobile phones for vocabulary activities: Examining the effect of the platform. *Language Learning & Technology*, *14*(2), 95–110. http://llt.msu.edu/vol14num2/stockwell.pdf. DOI: 10.64152/10125/44216

Stockwell, G. (2013). Tracking learner usage of mobile phones for language learning outside of the classroom. *CALICO Journal*, *30*, 118–136. https://www.jstor.org/stable/calicojournal.30.118. DOI: 10.1558/cj.v30i0.118-136

Stockwell, G. (2022). *Mobile assisted language learning: Concepts, contexts and challenges*. Cambridge University Press. DOI: 10.1017/9781108652087

Stoughton, A. M., & Kang, O. (2024). A systematic review of empirical mobile-assisted pronunciation studies through a perception–production lens. *Languages (Basel, Switzerland)*, *9*(7), 251. https://www.mdpi.com/2226-471X/9/7/251. DOI: 10.3390/languages9070251

Tergujeff, E. (2021). Second language comprehensibility and accentedness across oral proficiency levels: A comparison of two L1s. *System*, *100*, 102567. https://doi.org/https://doi.org/10.1016/j.system.2021.102567. DOI: 10.1016/j.system.2021.102567

Thomson, R. I., & Derwing, T. M. (2014). The effectiveness of L2 pronunciation instruction: A narrative review. *Applied Linguistics*, *36*(3), 326–344. DOI: 10.1093/applin/amu076

Thornton, P., & Houser, C. (2005). Using mobile phones in English education in Japan. *Journal of Computer Assisted Learning*, *21*(3), 217–228. DOI: 10.1111/j.1365-2729.2005.00129.x

Traxler, J. M., & Crompton, H. (2015). Mobile learning. In *Encyclopedia of Mobile Phone Behavior* (pp. 506–518). IGI Global. DOI: 10.4018/978-1-4666-8239-9.ch042

Trifanova, A., Knapp, J., Ronchetti, M., & Gamper, J. (2004). *Mobile ELDIT: Challenges in the transitions from an e-learning to an m-learning system*. http://eprints.biblio.unitn.it/archive/00000532/01/paper4911.pdf

Vygotsky, L. S. (1978). *Mind in society: The development of higher psychological processes*. Harvard University Press.

Yates, L., & Zielinski, B. (2009). *Give it a go: Teaching pronunciation to adults*. Adult Migrant English Program (AMEP) Research Centre, Macquarie University.

Zimmerman, B. J. (2002). Becoming a self-regulated learner: An overview. *Theory into Practice*, *41*(2), 64–70. DOI: 10.1207/s15430421tip4102_2

Zimmerman, B. J., & Risemberg, R. (1997). Becoming a self-regulated writer: A social cognitive perspective. *Contemporary Educational Psychology*, *22*(1), 73–101. DOI: 10.1006/ceps.1997.0919

ADDITIONAL READING

2. Bui, H. P., Kumar, R., & Kamila, N. K. (Eds.). (2024). *Innovations and applications of technology in language education*. CRC Press., DOI: 10.1201/9781003473916

1. Bui, H. P., & Namaziandost, E. (2024). *Innovations in technologies for language teaching and learning*. Springer Nature Switzerland., DOI: 10.1007/978-3-031-63447-5

3. Keengwe, J. (Ed.). (2018). *Handbook of research on mobile technology, constructivism, and meaningful learning*. Information Science Reference., DOI: 10.4018/978-1-5225-3949-0

4. Sadeghi, K. (Ed.). (2025). *Routledge handbook of technological advances in researching language learning*. Routledge., DOI: 10.4324/9781003459088

5. Zhang, F. (Ed.). (2012). *Computer-enhanced and mobile-assisted language learning: Emerging issues and trends*. IGI Global., DOI: 10.4018/978-1-61350-065-1

KEY TERMS AND DEFINITIONS

Autonomous Learners: Learners who take independent responsibility for managing their learning.

Corrective Feedback (CF): Responses that address learner errors to guide more accurate language use.

Learner Perceptions: Students' attitudes and evaluations of their learning experiences.

Mobile-Assisted Pronunciation Learning (MAPL): Using mobile devices for flexible, autonomous pronunciation practice.

Self-Regulation: Learners' ability to plan, monitor, and control their own learning.

Technology-Assisted Pronunciation Training (TAPT): Using digital technologies to improve learners' pronunciation through interactive feedback.

APPENDIX

Semi-Structured Interview Procedure

1) Describe your overall experience with mobile-assisted pronunciation learning. How has it impacted your pronunciation development?

What are your biggest challenges while using mobile applications for pronunciation learning? How do you overcome them?

2) Can you share an example of a specific pronunciation skill or feature you successfully improved using a mobile tool?
3) Do you think mobile-assisted pronunciation learning can fully replace traditional classroom instruction and the role of teachers? Why or why not?
4) What role should teachers play in supporting students who use mobile-assisted pronunciation learning tools?

Do you find the AI-generated feedback in pronunciation apps reliable? Why or why not?

Chapter 6
The Role of Gamification in Enhancing Mobile Language Learning Experiences

Mustafa Kayyali
https://orcid.org/0000-0003-3300-262X
Maaref University of Applied Sciences, Syria

ABSTRACT

As mobile technology reshapes the landscape of language education, gamification has emerged as a powerful catalyst for transforming learner engagement. This chapter explores how game elements—such as points, levels, challenges, and rewards—can enhance motivation, increase persistence, and deepen cognitive engagement in mobile language learning environments. Drawing from both pedagogical theory and practical implementations, it examines how well-designed gamified experiences can personalize the learning journey, create emotional resonance, and foster a sense of agency and progress. Yet the chapter also offers a critical lens, considering the limitations of over-simplified or poorly contextualized gamification strategies. It underscores the importance of culturally responsive design and learner diversity in shaping meaningful experiences. Ultimately, this chapter argues for a balanced, intentional use of gamification—not as a superficial layer, but as an integrated framework that supports authentic, human-centered language learning through mobile technologies.

DOI: 10.4018/979-8-3373-3366-3.ch006

INTRODUCTION

The act of learning a language has always been more than the mastery of vocabulary and grammar—it is a deeply human endeavor marked by identity, curiosity, emotion, and the desire to connect. For generations, language learning was tethered to classrooms, textbooks, and scheduled sessions, often reduced to repetition and memorization. But with the rapid expansion of mobile technologies, particularly smartphones and tablets, new frontiers have emerged—ones that are reshaping not only where learning happens, but *how* it is experienced. Within this evolving ecosystem, gamification has become a strikingly influential concept, offering a bridge between the worlds of education and entertainment, rigor and reward, engagement and achievement.

Gamification, at its core, refers to the integration of game elements—points, challenges, levels, leaderboards, narratives, and rewards—into non-game contexts (Christopoulos & Mystakidis, 2023). When thoughtfully applied, these elements do more than entertain; they tap into intrinsic psychological motivators like autonomy, competence, and relatedness. In mobile language learning, gamification serves not just as a motivational boost, but as a design philosophy that places the learner at the center of an adaptive, responsive, and emotionally resonant journey. Whether through digital flashcard battles, narrative quests that require linguistic choices, or collaborative missions among global learners, gamification is recasting the learner not as a passive recipient of content but as an active participant in a transformative experience. Yet, despite its popularity, gamification is often misunderstood or oversimplified. It is not a one-size-fits-all solution, nor a magical wrapper to make dull content suddenly engaging. In the context of language education, especially on mobile platforms where attention spans are fleeting and distractions abound, gamification must be handled with nuance. Poorly designed gamified systems can feel manipulative, shallow, or even alienating—especially when they ignore cultural, linguistic, or accessibility considerations (Hadi Mogavi et al, 2022). The challenge, then, lies in designing gamified environments that not only captivate but also educate—systems that are rooted in sound pedagogical principles, responsive to learners' needs, and mindful of the broader socio-cultural context.

This chapter embarks on an exploration of gamification's role in enhancing mobile language learning experiences. It begins by unpacking the theoretical underpinnings of gamification and its relevance to motivation in language acquisition. It then analyzes the ways in which mobile platforms afford unique opportunities for personalized, immersive, and on-the-go learning through game mechanics (Kayyali, 2025 c). Drawing on examples from popular applications, academic studies, and real-world classroom integrations, the discussion will move beyond surface-level praise and critique to examine what truly makes gamified mobile learning effective—or ineffective. It will also raise critical questions around ethics, inclusivity, and learner diversity, reminding us that technology, no matter how advanced, must always be in service of the human being at its heart. This chapter is a call to educators, developers, and researchers to approach gamification not as a trend, but as a transformative tool—one with the power to reimagine language learning for the mobile generation. In doing so, it hopes to illuminate a path toward more joyful, meaningful, and impactful learning experiences, driven not by gimmicks, but by intentional design rooted in empathy, insight, and innovation.

RETHINKING LANGUAGE LEARNING: THE EMERGENCE OF GAMIFICATION IN MOBILE CONTEXTS

There was a time when language learning was a solitary, rigid pursuit—one that often took place behind desks, with heavy textbooks, flashcards, and rote memorization (Wu, 2024). Lessons were linear, assessments were fixed, and the learner's journey was rarely personalized. Success depended not only on cognitive ability, but on discipline, repetition, and often, endurance. In many ways, the process was more mechanical than meaningful. But with the digital revolution—and more specifically, the explosion of mobile technology—language learning began to move, quite literally, into the learner's hands. And with it came a subtle but seismic shift: a reimagining of not just where learning happens, but how it feels, how it is experienced, and who drives it. At the heart of this transformation lies gamification—a quiet force that is reshaping education, not by replacing learning with play, but by recognizing that the two are not as distant as we once thought. To understand the emergence of gamification in mobile contexts, we must begin by reconsidering what we mean by "learning." For decades, education systems have equated learning with information transfer—content in, performance out. But real learning, especially language learning, is rarely that straightforward. It is messy, emotional, contextual (Wang et al, 2024). It depends on rhythm, feedback, curiosity, and connection. It is influenced by how safe we feel, how motivated we are, and whether we believe we belong in the world we are trying to understand. In short, learning is human. And gamification, when done right, speaks directly to that humanity.

Gamification is not simply the use of games, nor is it about turning education into entertainment (Luo, 2022). It is the thoughtful integration of game-based mechanics—such as rewards, progress indicators, social challenges, quests, surprise elements, and narrative arcs—into environments that are not inherently games. In the context of mobile language learning, these mechanics are no longer confined to software add-ons or gimmicky apps. They have become embedded in the very architecture of learning platforms, subtly guiding how users interact, persist, and grow. Mobile devices have created a uniquely fertile ground for gamification to take root. Unlike desktop computers or classroom tools, smartphones are intimate, personal, and always within reach (Manyeredzi & Mpofu, 2022). They accompany the learner through the rhythms of everyday life—on the bus, in cafés, during lunch breaks, and late at night when motivation flickers between fatigue and ambition. This portability allows gamification to operate in micro-moments: five-minute vocabulary challenges, spontaneous pronunciation games, or daily streaks that nudge learners toward consistency. In these small windows of time, gamification doesn't just fill the space—it transforms it into opportunity.

Consider, for instance, the psychological effect of completing a level or unlocking a badge after a short lesson. For many learners, especially those who have historically struggled with traditional methods, these moments are not trivial—they are validating. They offer a sense of accomplishment, no matter how small, and signal that progress is being made. Over time, this builds a sense of agency, of ownership over the learning process. Gamification taps into our natural desire to advance, to master, to be seen. It encourages persistence in ways that static content alone cannot. But what truly makes gamification powerful in mobile language learning is its emotional dimension. Unlike conventional curricula, gamified systems often introduce narrative and choice—two elements that humanize the learning experience (Toledo Palomino et al, 2023). A learner may play the role of a traveler navigating a foreign city, solving language puzzles to advance their journey. Or they might be part of a global community competing in challenges, sharing tips, or celebrating streaks together. These experiences aren't just instructional; they're immersive. They evoke curiosity, imagination, and a sense of play that reconnects learning with joy—a feeling that too many adult learners have lost along the way.

Yet, gamification is not a magic wand. Its rise in mobile contexts has also raised important questions about ethics, quality, and inclusivity (Tomé Klock et al, 2024). For one, there is a danger in reducing learning to points and leaderboards, especially if the game mechanics overshadow the pedagogical intent. When learners begin to chase rewards rather than meaning, or feel discouraged by perceived failures within the system, the entire purpose is lost. Badly designed gamification can feel manipulative, even infantilizing, particularly to adult learners who expect autonomy and depth. Moreover, if developers overlook cultural sensitivity or linguistic diversity, gamified experiences can alienate those they aim to engage. In rethinking language learning through gamification, then, it is essential to move beyond surface-level mechanics. What's needed is intentionality—a deep awareness of how and why game elements are used, and how they align with learning goals. Gamification should support linguistic skills: listening, speaking, reading, writing, and the nuanced competencies in between, such as pragmatics, intonation, or cultural literacy. It should adapt to different learner profiles—novices and advanced users, introverts and extroverts, children and adults—without assuming one path fits all.

Some of the most successful mobile language apps—like Duolingo, Memrise, and LingQ—have understood this balance, at least in part (Silva, 2024). They use gamification not merely to keep users addicted, but to help them feel competent and connected. They leverage progress tracking to offer feedback loops, not surveillance. They use rewards to foster consistency, not dependency. Still, the field is evolving, and not all attempts at gamification strike this balance. This is why educators, linguists, and designers must work together—not just to implement gamification, but to *reimagine it* as a form of pedagogical storytelling, capable of making learning feel both rigorous and alive. Another important shift that gamification enables is the decentralization of authority in the learning process. Traditionally, teachers controlled the pace, content, and assessment. But gamified mobile platforms allow learners to self-direct, to choose their own challenges, to fail safely and try again. This builds not only language skills but confidence—a trait that is just as vital for real-world communication. When learners feel in control, they are more likely to take risks, speak up, and experiment with language in authentic contexts. In this way, gamification empowers learners not just to learn a language, but to *use it*—to live it.

What emerges, then, is a more holistic view of learning—one that is not just about knowledge acquisition, but about identity formation. Gamified mobile learning becomes a space where users can try on new linguistic selves, where they can practice being someone else—someone who orders confidently in a café in Rome, or who jokes in Japanese with a friend across the world (Hadi Mogavi et al, 2022). These experiences are not trivial; they shape how learners see themselves, how they imagine their future, and how they relate to the cultures embedded within the languages they study. This rethinking of language learning through gamification is not merely a pedagogical trend—it is a philosophical shift. It challenges the idea that learning must be serious to be effective, that it must be slow to be meaningful, or that it must be standardized to be legitimate. Instead, it affirms that learning can be joyful, fast-paced, and diverse—qualities that mirror the very nature of language itself. After all, language is not a fixed system; it is a living, breathing, playful act (Heidari-Shahreza, 2024). And perhaps it is no coincidence that the word "play" exists in so many languages not just as entertainment, but as experimentation, improvisation, and creation. As mobile technologies continue to evolve—with artificial intelligence, augmented reality, and adaptive systems—gamification will undoubtedly take on new forms. But its essence will remain the same: to make learning feel less like a task and more like a journey. A journey that is not only educational but emotional, not only efficient but transformative. To rethink language learning in this way is to believe in the possibility that, through play, we can become more open, more connected, and more fluent—not just in words, but in worlds.

MOTIVATIONAL DYNAMICS: HOW GAME ELEMENTS RESHAPE LEARNER ENGAGEMENT

Motivation has long stood at the heart of education. No amount of resources, strategies, or technologies can fully compensate for its absence. In the context of language learning—where the path is often long, the progress incremental, and the rewards delayed—motivation becomes not just a desirable trait but a lifeline (Dörnyei & Henry, 2022). It is what keeps a learner returning to unfamiliar words, challenging grammar, and moments of confusion that can feel like failure. But motivation is fragile. It ebbs and flows, often influenced by the emotional, cognitive, and even physical state of the learner. That is why the entrance of gamification into language learning—especially in mobile settings—has opened up a new conversation: not about replacing intrinsic motivation, but about sustaining and amplifying it through thoughtfully designed experiences. It's worth remembering that the act of learning a language is fundamentally social and psychological. It involves risk, exposure, and vulnerability. When learners begin their journey, they're not just grappling with unfamiliar syntax—they're stepping into a space where they may feel inadequate, misunderstood, or even judged. The traditional classroom, despite its structure, doesn't always create the safest or most energizing environment to persist through these difficulties (Kayyali, 2025 d). Mobile gamified platforms, by contrast, offer a space that is both private and playful—two qualities that can radically shift how learners engage (Burden, 2025).

To understand the motivational pull of gamified elements, we need to look beyond the surface mechanics and examine what they *mean* to the learner. A point is never just a point. A badge is not merely a graphic. A streak is more than a number. These features, when implemented thoughtfully, become symbols of progress, validation, consistency, and personal investment. They transform abstract effort into tangible outcomes, creating a feedback loop that sustains engagement not through pressure, but through recognition. At the center of this dynamic lies the psychology of achievement and progression. Humans are wired to find satisfaction in advancement. The concept of "leveling up" is deeply ingrained in how we conceptualize growth—not just in games, but in life. By dividing language learning into levels, stages, or quests, gamified systems provide learners with manageable goals and the satisfying sense of forward motion (Nechifor, 2024). These small wins matter. They shift the focus from what the learner has yet to master to what they *have already accomplished.* Over time, this accumulation of wins builds confidence—a quality far more predictive of long-term success than raw ability. And then there's the matter of autonomy—the sense of control over one's own learning. In traditional settings, learners are often passive recipients of a preordained curriculum. But in mobile gamified environments, learners choose when to engage, what to focus on, how much to practice, and even the style of learning they prefer. This freedom, paradoxically, does not lead to disengagement but to deeper commitment. When learners feel that their choices matter, their motivation becomes self-directed. They are not doing something *for* the teacher, the grade, or the institution—they are doing it for themselves. This is perhaps gamification's most underestimated strength: its quiet invitation to take ownership.

Social features add another layer of motivational richness. Leaderboards, peer challenges, team quests—these are not just additions to the gamified model; they are bridges to community. Language, after all, is social. We learn to speak not to pass tests, but to communicate. When learners see others struggling, achieving, sharing tips, or celebrating streaks, a sense of belonging emerges (Qarooni, 2023). Suddenly, learning is no longer a solitary act. It becomes part of a shared endeavor, a collective rhythm. Even for introverted or self-paced learners, the simple knowledge that others are walking a similar path can reinforce their own momentum. And yet, not all game elements are equal, nor are they always beneficial. Motivation is not just about quantity—how many points or how often learners return—but about quality: what kind of experience the learner is having. Poorly designed gamification can backfire. If rewards are handed out too easily, they lose their meaning. If challenges are too hard, they create frustration. If leaderboards create constant comparison, they can diminish self-esteem, especially for those who progress more slowly. This is why balance is crucial. The goal is not to addict, but to engage; not to manipulate, but to invite. What makes game elements powerful is not their novelty but their alignment with deeper psychological needs. According to self-determination theory—a well-regarded framework in motivational psychology—three core needs drive human motivation: autonomy, competence, and relatedness. Game elements, when used wisely, can nurture all three. Autonomy is supported through flexible pathways and user choice. Competence is reinforced through progressive challenges and instant feedback (Morgan, 2024). Relatedness is enhanced through community features, messaging, and collaborative missions. In this sense, gamification isn't about adding glitter to learning. It's about designing systems that respect and respond to the emotional realities of being a learner.

Mobile technology amplifies these dynamics by adding ubiquity and intimacy. Learners are no longer confined to scheduled sessions. They can engage while waiting in line, commuting, or lying in bed. Motivation becomes embedded in everyday life. A five-minute lesson is not an interruption—it's a ritual. Over time, these rituals accumulate into habits, and habits are the scaffolding of achievement. Gamified elements like daily goals, progress streaks, and push notifications gently reinforce these habits without feeling oppressive (Petkevičius, 2024). They act as soft nudges rather than hard commands—suggesting rather than forcing engagement. But there's another layer to all of this—something deeper and more personal:

emotion. Learning, especially language learning, is profoundly emotional. Success brings joy. Failure can trigger shame. Uncertainty, frustration, curiosity, pride—all of these play a role in shaping motivation. Game elements have the unique ability to modulate these emotions. They can soften the sting of failure through retry mechanics. They can increase anticipation through surprise rewards. They can celebrate small victories in ways that traditional classrooms often overlook. In doing so, they transform emotion from a barrier into a resource.

Moreover, game elements can help reframe failure. In many traditional learning settings, failure is stigmatized. It signals inadequacy, invites judgment, and discourages further effort. But in gamified contexts, failure is expected. It is part of the process. Learners are encouraged to try again, to adapt strategies, to approach problems from different angles (Adipat et al, 2021). This shift—from failure as defeat to failure as feedback—is perhaps one of gamification's greatest contributions. It nurtures resilience, and resilience is the quiet engine of sustained motivation. At the same time, it's important to recognize that motivation is not purely individual. It is shaped by culture, identity, and context. A gamified system that works brilliantly in one country may fall flat in another. Game elements must therefore be designed with cultural sensitivity. What motivates one learner may feel trivial or even inappropriate to another. Developers and educators must listen—to feedback, to user data, to the diverse voices of learners from around the world. Only then can gamification truly fulfill its potential as a motivational tool.

The motivational power of gamification does not lie in its mechanics alone. It lies in the *meaning* those mechanics carry for the learner (Villegas et al, 2021). A badge that marks the completion of a difficult module is not just a virtual sticker—it's a symbol of effort, persistence, and growth. A streak is not just a number—it is a story of commitment. A challenge completed is not just a task done—it's a personal victory. When gamification is approached with this level of respect and intentionality, it becomes something far more than a design strategy. It becomes an act of empathy. The true test of gamification's value is not whether learners stay longer on an app, but whether they come away changed. More confident. More curious. More willing to keep learning, even when it's hard. If game elements can help learners find joy in the journey, take pride in their progress, and feel seen in their struggle—then they have done more than motivate. They have inspired.

CULTURAL SENSITIVITY AND INCLUSIVITY IN GAMIFIED MOBILE LEARNING ENVIRONMENTS

In the world of language learning, culture is not an accessory—it is the essence. Language lives through culture. It is shaped by it, steeped in it, and inseparable from the values, stories, gestures, and worldviews of the people who speak it. To learn a new language is to step into unfamiliar ways of thinking, into new rhythms of politeness, into different ways of telling a story or saying no or showing respect (Gigerenzer, 2022). When we gamify language learning, we must therefore take great care—because we are not just building games. We are shaping encounters between cultures. We are designing experiences that may become someone's first intimate exposure to a world beyond their own. Cultural sensitivity in gamified mobile learning environments is not a box to tick; it is a foundational mindset (Asrifan et al, 2025). It requires an understanding that learners come from a vast array of cultural, religious, linguistic, and social backgrounds—and that the design choices we make will either open the door to a sense of belonging or quietly close it. A seemingly harmless cartoon, an idiom used out of context, a character dressed in a particular way—these small elements carry weight. They can reflect respect, or they can reinforce stereotypes. They can invite learners in, or they can alienate them entirely. When learners feel that their identities are not represented—or worse, misrepresented—they are less likely to engage, persist, or trust the learning experience.

At the heart of inclusivity is representation—and not just token gestures, but authentic, respectful inclusion. Who gets to be the hero in the story? Who is cast as the teacher or the shopkeeper or the friend in dialogue exercises? Are there characters wearing religious garments, speaking with regional accents, using diverse names and gestures? Is disability visible? Are non-Western cultures treated as rich and modern, not as exotic backdrops? These questions go to the core of what inclusivity means in design. It is not about diversity for diversity's sake. It is about reflecting the complexity of the world that learners actually live in—and the world that the language opens up. Mobile platforms have a unique responsibility in this regard. Because they are global by default, they are used in homes, villages, cities, and refugee camps in every corner of the world (Kayyali, 2025 a). The same app might be downloaded by a high school student in Paris, a grandmother in Manila, or a Syrian child learning English in Istanbul. The gamified experience must therefore be both globally accessible and locally meaningful. This is no small task. But it begins with listening—listening to the learners themselves, involving them in testing, gathering stories, and adapting content to reflect local contexts without diluting linguistic depth. Take idioms, for example. They are deeply cultural. Teaching them without explanation, or without context, can lead to confusion or misinterpretation. A phrase like "kick the bucket" or "spill the beans" may be engaging for a learner from one background and completely baffling to another. In a gamified setting, idioms might be introduced in a playful way—perhaps as part of a mystery game or a storytelling challenge. But care must be taken to explain them, to show alternatives, and to respect that not all learners will find them relatable or even appropriate. Cultural sensitivity is not about avoiding complexity—it is about approaching it with humility and care.

Inclusivity also means language variety. Many gamified platforms default to a single "standard" version of a language—Parisian French, for example, or American English. (Langin, 2022) But learners may be preparing to move to Senegal, to Canada, to India. Variants matter. Accents, vocabulary choices, even tone and formality levels vary widely. A truly inclusive gamified environment might allow learners to choose their target variant—or even better, expose them to multiple forms, helping them build flexibility and awareness. Just as important, the platform should not treat one variant as superior to another. All are valid. All are worthy of respect. Accessibility must also be addressed—not just in the technical sense, but in the social and emotional sense as well. A gamified app that assumes every learner has fast internet, a private device, or a quiet space is excluding millions. Design must account for intermittent connectivity, low-bandwidth modes, text-to-speech options, offline storage. For learners with visual or hearing impairments, there must be alternatives for audio, readable fonts, high-contrast visuals, and simple navigation. And for learners with cognitive differences, such as ADHD or dyslexia, features like clear instructions, the ability to slow down or repeat content, and calm, uncluttered design are essential (Dunne, 2024). But inclusivity is not just about software. It is also about tone—the emotional language of the app. How are mistakes treated? Are learners mocked for errors, or gently guided? Are jokes inclusive, or do they rely on sarcasm, stereotypes, or cultural references that exclude? Do avatars reflect a range of skin tones, body types, clothing styles, and names? Does the user feel they are being spoken to with respect? These are the subtler aspects of design, often overlooked, but they are the ones that shape the emotional experience of learning. A respectful tone builds trust. Trust builds confidence. Confidence builds fluency.

The reality is that language learning is often a deeply personal journey. For immigrants, it may be tied to survival or dignity. For travelers, it may be a bridge to discovery. For children of diaspora, it may be a path back to roots. Every learner brings a story. The best gamified platforms honor those stories. They do not reduce learners to data points or game scores. They see them as people, with histories, hopes, and identities that matter. Gamified environments must also be sensitive to pace and pressure. Some learners thrive in competitive environments; others do not. A leaderboard can be exciting for one learner and anxiety-inducing for another. A countdown timer can feel like a challenge—or a punishment (Adkins-Jablonsky et al, 2021). Design must allow for flexibility. Perhaps users can opt in or out of competitive features. Perhaps rewards can be structured to value consistency, effort, or reflection—not just speed. Inclusivity means designing for different temperaments, learning styles, and emotional needs. Inclusivity also stretches to gender and identity. Language is full of gendered constructions, and learners come with a wide range of identities. Platforms must offer non-binary options for avatars, inclusive pronoun usage in example sentences, and storylines that do not rely on outdated gender roles.

A romance storyline, for instance, can include queer characters without making it a spectacle. A workplace simulation can include a woman as a CEO. These choices may seem small, but they communicate something powerful: that the language world learners are stepping into is one where they, too, belong.

A particularly sensitive issue is that of religious and cultural values (Pentaris & Christodoulou, 2021). What is considered a harmless joke in one culture may be offensive in another. What seems like a fun party scene to some may feel uncomfortable or inappropriate to others. Gamified content must be created with an awareness of these lines—not to sanitize or censor learning, but to ensure respect. This doesn't mean removing cultural elements; it means providing context, offering choice, and avoiding assumptions. A food game, for instance, might allow learners to select from different cultural menus—not just burgers and hot dogs, but biryani, falafel, sushi, or injera. These options are not just fun—they're affirmations. Importantly, cultural sensitivity is not static. It evolves. New social conversations, shifting norms, emerging voices—all require continual reflection and adaptation (Mills et al, 2023). This is why gamified platforms must build in feedback loops. Users should be able to flag content, suggest improvements, request new features, or share their experiences. Designers must treat this feedback not as criticism, but as a gift. It is a window into the lived reality of the learners they serve.

The question, ultimately, is not whether gamified mobile language learning environments *can* be inclusive. It is whether we choose to make them so. The tools are there. The research is there. What remains is the will—the ethical commitment—to create experiences that welcome rather than exclude, that reflect rather than erase, that invite rather than impose. When gamification is designed with cultural sensitivity and inclusivity at its core, it becomes something far more than a method of engagement. It becomes an act of hospitality. It says to the learner, *You matter. Your world matters. And you are not alone in this journey.* That message, more than any reward or badge, is what makes a learning experience unforgettable. It is what makes language not just something we learn, but something we live.

MEASURING IMPACT: OUTCOMES, LIMITATIONS, AND FUTURE PATHWAYS

At the intersection of education and innovation lies a perennial challenge: how do we measure what matters? In the case of gamified mobile language learning, this question is both urgent and complicated. While we've seen the rise of sleek interfaces, immersive challenges, and impressive user metrics, the real test of impact lies deeper. It's not in the number of downloads or hours spent on an app, but in the quiet, steady transformations that take place in a learner's mind, their confidence, their habits, and ultimately, their ability to live through another language. To measure the true impact of gamification in language learning is to grapple with the paradox of modern education (Kayyali, 2025 b). On one hand, technology offers us tools of unprecedented precision: real-time analytics, adaptive algorithms, data dashboards, and behavioral heatmaps. On the other hand, language learning is profoundly human—filled with ambiguity, emotion, cultural nuance, and deeply personal goals (Hossain, 2024). A learner might gain vocabulary but still fear speaking. They might complete levels but struggle with real-world comprehension. They might improve fluency in an app but fail to feel belonging in a new linguistic community. These realities complicate the notion of "progress," especially when progress is measured primarily through gamified metrics.

Yet, some outcomes are indeed observable, and some trends have begun to emerge. Learners who engage regularly with well-designed gamified language apps tend to demonstrate increased consistency in study routines. Gamification's power lies not just in making learning more entertaining, but in making it habitual (Torresan & Hinterhuber, 2023). Features like daily streaks, achievement badges, and leveling systems tap into a learner's sense of rhythm and routine. Over time, this consistency compounds. Just as a plant grows through daily light rather than sudden downpours, language mastery often emerges not through cramming but through repetition, review, and return. Motivation, too, is visibly affected. Many learners who may have found traditional approaches dry or intimidating report feeling more energized and empowered through gamified systems. The presence of clear, immediate feedback loops—whether in the form of in-game rewards, animations, or encouraging prompts—helps reframe failure not as an endpoint but as a part of the process. The stakes are lower. The tone is lighter. Learners feel safe enough to experiment, to make mistakes, and to try again. That safety fosters resilience, and resilience fosters growth.

There is also evidence that gamification supports incremental mastery. By breaking down large learning goals into smaller, trackable milestones, learners are less likely to feel overwhelmed. Whether it's mastering ten verbs in one day, or completing a short listening challenge on a bus ride, the accessibility and segmentation of content allow learning to slip into the edges of daily life. This modularity is more than a convenience—it's a pedagogical advantage. It helps learners internalize content at a sustainable pace, in contexts that feel manageable. But for all the promise gamification holds, its limitations are real—and they must be confronted with honesty. One of the most persistent issues is the gap between game success and real-world fluency. A learner may be adept at navigating a gamified app—collecting points, acing quizzes, reaching high scores—yet struggle to carry out a simple conversation in the target language (Fune, 2024). This is not necessarily a failure of the learner, but of the design itself. If gamified environments prioritize recognition over use, memorization over expression, they risk creating the illusion of competence without the substance. In this sense, gamification can become a seductive distraction—rewarding shallow engagement while neglecting the deep, often uncomfortable work of developing communicative confidence.

Another limitation is the one-size-fits-all design that many platforms adopt. While gamification thrives on uniformity for the sake of simplicity, learners are not uniform (Vintaloro, 2021). They bring different prior knowledge, different cognitive preferences, different emotional triggers. A competitive feature like a leaderboard may spur one learner on and cause another to withdraw. A progress bar may be encouraging to some, anxiety-inducing to others. The assumption that all learners will respond positively to the same stimuli ignores the diversity of human motivation. It also raises ethical questions about how engagement is engineered and for whom. There are also concerns around overjustification—a well-documented phenomenon in educational psychology, where external rewards can diminish intrinsic motivation. When learners begin to associate success with badges, coins, or level-ups, they may lose touch with the deeper reasons they came to learn in the first place. Language, which should be about connection, curiosity, and exploration, becomes a means to a badge rather than a bridge between people (Soye, 2024). When the rewards are removed or the novelty wears off, what remains? Does the learner continue out of genuine interest, or do they stop altogether?

Furthermore, gamification does not always accommodate cultural and contextual variation. A reward system designed for learners in the United States may not resonate with learners in East Africa or the Middle East. Idioms, jokes, character designs, and even assumptions about learning pace can reflect implicit biases about who the "ideal" learner is. Without localization, gamification can inadvertently replicate the very barriers it seeks to break down. Despite these limitations, gamification's future remains bright—but it must be reimagined with care. The next generation of gamified learning environments will need to move beyond surface-level metrics and adopt holistic models of assessment (Jiménez-Valverde et al, 2025). These might include self-reflection prompts, voice-recorded responses that track oral fluency over time, journaling features that allow learners to narrate their own learning stories, and peer-reviewed dialogues that simulate authentic communication. These kinds of tools emphasize not only *what* the learner knows, but *how* they are evolving as speakers, listeners, and cultural participants. Long-term impact must also be studied through qualitative lenses. Numbers are useful, but they don't tell the whole story. What do learners feel when they return to a gamified app each day? How do they describe their changing sense of identity as they grow more confident in a new language? What barriers remain invisible to data but are very present in the learner's lived experience? These are the questions that academic researchers, developers, and educators must ask—and keep asking.

There is also room for greater ethical transparency in how gamification is deployed. Too often, game elements are used to keep users on platforms for as long as possible, echoing techniques from the world of social media and digital entertainment. But education must hold itself to a higher standard. Engagement should not be engineered merely for app usage—it should be cultivated for meaningful learning. Learners deserve to know why they are being prompted, rewarded, nudged. They deserve experiences that respect their time, their dignity, and their humanity. Looking ahead, the future of gamification in language learning lies not in more features, but in *better questions*. Instead of asking, "How can we make this addictive?" we might ask, "How can we make this empowering?" Instead of designing for maximum retention, we can design for maximum transformation (Thurairasu, 2022). Can learners walk away from the app and use what they've learned to make a new friend, to navigate a city, to tell a story from their life in another language? That is the true measure of impact—not just acquisition, but application.

The pathway forward also involves collaboration. Designers need to work closely with educators, linguists, psychologists, and, most importantly, learners themselves. Participatory design—where users help shape the very systems they'll engage with—can uncover insights no algorithm could generate. Language is human, and its learning must be humanized as well. Perhaps the most profound shift we need is a redefinition of success. Success in gamified language learning is not necessarily fluency in a short time, or even mastering all the vocabulary in a course (Waluyo & Bucol, 2021). It may be the learner who logs in after a long break because they felt encouraged rather than ashamed. It may be the learner who, for the first time, says hello in a language they once feared. It may be the smile that comes from recognizing a phrase in a song, or the courage to ask for directions abroad. These moments are quiet, but they are powerful. They are the real outcomes—the ones no badge can capture, but that change lives nonetheless. Measuring impact in gamified mobile learning is not just a technical task. It is an ethical, emotional, and educational responsibility. It asks us to go beyond what is easy to count and reach toward what truly counts. And in doing so, it reminds us that behind every app tap, every level-up, every correct answer, there is a person—a learner—hoping not just to learn a language, but to find their voice within it.

CONCLUSION

As we stand at the crossroads of education, technology, and play, it is clear that gamification has moved beyond novelty. In mobile language learning, it has become a defining feature of how learners engage with content, interact with platforms, and shape their own trajectories. Yet its true value lies not in flashy interfaces or endless points and badges, but in its capacity to speak to something much older, much deeper—our inherent drive to grow, to be challenged, to belong, and to find meaning in the learning process. This chapter has shown that gamification, when carefully designed and thoughtfully implemented, can unlock powerful dimensions of learner motivation and agency. Through progress mechanics, narrative immersion, immediate feedback, and collaborative dynamics, mobile gamified environments can foster not just knowledge acquisition, but confidence, persistence, and enjoyment. It has also emphasized that the effectiveness of gamification does not depend on the number of game elements used, but on how well those elements align with pedagogical goals and learner identities. A well-crafted challenge, a culturally relevant reward system, or a simple narrative layer can do more for learning than the most elaborate point systems if it resonates with the learner's context and needs.

Yet, gamification is not without its tensions. It can unintentionally promote extrinsic over intrinsic motivation, or create hierarchies among learners. It can obscure the complexity of language learning behind a veneer of progress. It can, when carelessly implemented, alienate rather than engage. These risks remind us that gamification is not inherently good or bad—it is a tool, and like all tools, its impact depends on the hands that wield it. Educators, designers, and policymakers must therefore approach gamification with critical awareness, cultural sensitivity, and a commitment to ethical design. Looking forward, the next phase of gamification in mobile language education should embrace personalization, accessibility, and cross-cultural adaptability. It must go beyond uniform mechanics and static modules to offer flexible experiences that reflect the diversity of learners' linguistic backgrounds, motivations, and emotional landscapes. It must also engage with real-world data and feedback loops to evolve continuously based on how learners are responding—not just to the content, but to the *experience* of learning itself. In essence, gamification should not be seen as a superficial layer of engagement, but as an opportunity to reframe learning as a lived, dynamic, and human-centered journey. Language, after all, is not just a system of rules or words—it is a way of being in the world. And when mobile technologies can make that journey feel like an adventure—personal, empowering, and joyful—then we have not merely added games to education. We have made learning feel alive again.

REFERENCES

Adipat, S., Laksana, K., Busayanon, K., Asawasowan, A., & Adipat, B. (2021). Engaging students in the learning process with game-based learning: The fundamental concepts. *International Journal of Technology in Education, 4*(3), 542–552. DOI: 10.46328/ijte.169

Adkins-Jablonsky, S. J., Shaffer, J. F., Morris, J. J., England, B., & Raut, S. (2021). A tale of two institutions: Analyzing the impact of gamified student response systems on student anxiety in two different introductory biology courses. *CBE Life Sciences Education, 20*(2), ar19. DOI: 10.1187/cbe.20-08-0187 PMID: 33797283

Asrifan, A., Kaddas, B., Mulyadi, M., Pratiwi, W. R., Supriadi, S., & Jabu, B. (2025). Cross-Cultural Gamification: Understanding Design and User Experience in Different Cultural Contexts. In Enhancing Engagement With Gamification: Education, Business, and Healthcare Perspectives (pp. 1-28). IGI Global Scientific Publishing.

Burden, P. R. (2025). *Classroom management: Creating a successful K-12 learning community*. John Wiley & Sons.

Christopoulos, A., & Mystakidis, S. (2023). Gamification in education. *Encyclopedia, 3*(4), 1223–1243. DOI: 10.3390/encyclopedia3040089

Dörnyei, Z., & Henry, A. (2022). Accounting for long-term motivation and sustained motivated learning: Motivational currents, self-concordant vision, and persistence in language learning. In *Advances in motivation science* (Vol. 9, pp. 89–134). Elsevier.

Dunne, C. (2024). Design strategies and dyslexia: Improving the accessibility of course material for third-level students with dyslexia (Doctoral dissertation, Institute of Art, Design+ Technology).

Fune, A. (2024). *A Study of Quizizz's User Experience (UX)*. Measuring the Impact of Gamification and Usability on Language Acquisition.

Gigerenzer, G. (2022). *How to stay smart in a smart world: Why human intelligence still beats algorithms*. MIT Press.

Hadi Mogavi, R., Guo, B., Zhang, Y., Haq, E. U., Hui, P., & Ma, X. (2022, June). When gamification spoils your learning: A qualitative case study of gamification misuse in a language-learning app. In Proceedings of the Ninth ACM Conference on Learning@ Scale (pp. 175-188). DOI: 10.1145/3491140.3528274

Heidari-Shahreza, M. A. (2024). When language is play: Looking into linguistic creativity through language play. Creativity Journal: Theories. *Research & Applications*, *11*(1), 18–35.

Hossain, K. I. (2024). Reviewing the role of culture in English language learning: Challenges and opportunities for educators. *Social Sciences & Humanities Open*, *9*, 100781. DOI: 10.1016/j.ssaho.2023.100781

Jiménez-Valverde, G., Fabre-Mitjans, N., Heras-Paniagua, C., & Guimerà-Ballesta, G. (2025). Tailoring Gamification in a Science Course to Enhance Intrinsic Motivation in Preservice Primary Teachers. *Education Sciences*, *15*(3), 300. DOI: 10.3390/educsci15030300

Kayyali, M. (2025 a). E-Learning Platforms and Their Influence on Lifelong Learning. In Impact of Digitalization on Education and Social Sustainability (pp. 155-176). IGI Global.

Kayyali, M. (2025 b). Transforming Education: The Future of Mastery-Based Learning and Its Impact. In Cultivating Flourishing Practices and Environments by Embracing Positive Education (pp. 119-148). IGI Global Scientific Publishing.

Kayyali, M. (2025 c). Transforming Education: The Future of Mastery-Based Learning and Its Impact. In Cultivating Flourishing Practices and Environments by Embracing Positive Education (pp. 119-148). IGI Global Scientific Publishing.

Kayyali, M. (2025 d). Classroom Management: Techniques for Creating a Positive, Friendly Learning Environment. In Exploration of K-12 Teaching and Learning for Teacher Educators (pp. 53-78). IGI Global Scientific Publishing.

Langin, F. (2022). *Teaching french as a second language to primary students online: implementing a gamified routine system*. McGill University.

Luo, Z. (2022). Gamification for educational purposes: What are the factors contributing to varied effectiveness? *Education and Information Technologies*, *27*(1), 891–915. DOI: 10.1007/s10639-021-10642-9

Manyeredzi, T., & Mpofu, V. (2022). Smartphones as digital instructional interface devices: The teacher's perspective. *Research in Learning Technology*, *30*, 30. DOI: 10.25304/rlt.v30.2639

Mills, A., Bali, M., & Eaton, L. (2023). How do we respond to generative AI in education? Open educational practices give us a framework for an ongoing process. *Journal of Applied Learning and Teaching*, *6*(1), 16–30.

Morgan, J. D. (2024). Developing a capacity for lifelong learning: self-regulation and autonomous learning competencies within the European Framework (Doctoral dissertation).

Nechifor, A. (2024). Gamified Romanian for Specific Purposes: A Journey to Engaged Language Acquisition. Analele Universităţii Ovidius din Constanţa. *Seria Filologie*, *35*(2), 348–371.

Pentaris, P., & Christodoulou, P. (2021). Qualities of culturally and religiously sensitive practice: A cross-sectional study. *Journal of Palliative Care*, •••, 08258597211050742. PMID: 34841969

Petkevičius, M. (2024). The impact of gamification elements on the intention to use health related applications (Doctoral dissertation, Vilniaus universitetas.).

Qarooni, N. (2023). Process: The Journey is the Learning. In Nourishing Caregiver Collaborations (pp. 43-69). Routledge.

Silva, C. V. C. (2024). Communicative language skills: a comparative analysis of duolingo and memrise applications for language learning.

Soye, E. (2024). Funny Language? Curiosity, Contact, and Humour. In Peer Relationships at School (pp. 71-88). Bristol University Press. DOI: 10.2307/jj.12348141.9

Thurairasu, V. (2022). Gamification-based learning as the future of language learning: An overview. *European Journal of Humanities and Social Sciences*, *2*(6), 62–69. DOI: 10.24018/ejsocial.2022.2.6.353

Toledo Palomino, P., Nacke, L., & Isotani, S. (2023, October). Gamification of virtual learning environments: a narrative and user experience approach. In *Proceedings of the XXII Brazilian Symposium on Human Factors in Computing Systems* (pp. 1-10). DOI: 10.1145/3638067.3638103

Tomé Klock, A. C., Palomino, P. T., Rodrigues, L. A. L., Toda, A. M., Simanke, S., Spors, V., ... & Hamari, J. (2024). Gamification towards and alongside equity, diversity and inclusion: Looking back to move forward. new media & society, 14614448241254028.

Torresan, S., & Hinterhuber, A. (2023). Continuous learning at work: The power of gamification. *Management Decision*, *61*(13), 386–412. DOI: 10.1108/MD-12-2020-1669

Villegas, E., Fonseca, D., Peña, E., Bonet, P., & Fernández-Guinea, S. (2021). Qualitative assessment of effective gamification design processes using motivators to identify game mechanics. *Sensors (Basel), 21*(7), 2556. DOI: 10.3390/s21072556 PMID: 33917409

Vintaloro, G. (2021). Teaching English in Italy: A Tale of Two Systems.

Waluyo, B., & Bucol, J. L. (2021). The impact of gamified vocabulary learning using Quizlet on low-proficiency students. *Computer Assisted Language Learning, 22*(1), 158–179.

Wang, P., Ganushchak, L., Welie, C., & van Steensel, R. (2024). The dynamic nature of emotions in language learning context: Theory, method, and analysis. *Educational Psychology Review, 36*(4), 105. DOI: 10.1007/s10648-024-09946-2

Wu, R. (2024). Teaching Strategies for Vocabulary Acquisition in Junior High School English Language Learning. *Frontiers in Educational Research, 7*(9).

Chapter 7
Digital Gamification in Times of Crisis:
Perspectives of Pre-Service Elementary School Teachers on Language Teaching

Tetiana Holovatenko
https://orcid.org/0000-0002-7545-3253
V.N. Karazin Kharkiv National University, Ukraine

ABSTRACT

This chapter investigates pre-service elementary school teachers' perspectives on digital gamification for English language teaching in the challenging context of wartime Ukraine. Quantitative data was gathered through surveying, and qualitative data was obtained through the analysis of artefacts (language teaching activities). Findings reveal strong positive attitudes toward gamification's effectiveness, particularly for lowering learners' Affective Filter and supporting trauma-informed pedagogy. The analysis showed that Ukrainian students accept the uncertainty of wartime teaching and gamify the activities using low-stress game mechanics and tend to use digital tools, which can be easily used in both online and offline settings. This study provides suggestions for teacher training programs, educational institutions and curriculum developers on the implementation of digital gamification in language teaching, with a critical focus on overcoming pedagogical barriers and developing trauma-informed education.

DOI: 10.4018/979-8-3373-3366-3.ch007

INTRODUCTION

The extensive digitalization of education, facilitated by the COVID-19 pandemic and the shift to online, blended, and hybrid learning, requires systemic changes in approaches to language teaching to retain learners' motivation. Researchers have claimed both positive (Bailey et al., 2021; Fauzi & Anindiati, 2021; Panagiotidis et al., 2023) and negative impacts (Hamad et al., 2021) of digitalization on sustaining learners' motivation. Despite the documented effects of online instruction on students' motivation, recent scholarly research has demonstrated that motivation is a significant predictor of learning outcomes for language learners (Teng et al., 2021). Moreover, scholarly research also shows the correlation between teachers' perceptions of the efficacy of student-centred teaching using digital technologies and students' engagement and motivation to learning (De Souza et al., 2021). Scholars claim that among the top reasons why students lose motivation are materials, lack of communication between students and teachers, and lack of commitment to finish the course (Meşe & Sevilen, 2021). As a result, instruction in digital learning environments heavily relies on the instructors' perceived efficacy and is attributed to the resources, methods, and approaches used to organize collaboration between students and teachers. Another crucial dimension of the digital learning environment, particularly within the Ukrainian context, is the profound socio-emotional impact of the ongoing full-scale russian aggression. Since February 2022, Ukrainian pre-service teachers and their future elementary students operate under the constant threat of air raids, displacement, and power outages, demanding a systemic shift to trauma-informed pedagogy. This exceptionally stressful wartime context impacts students' executive functioning and their capacity to learn and process information.

Another thing to consider when teaching in a digital environment is the challenging environment students often find themselves in. Either they are learning online due to lockdown, or because of security reasons, the stressful setting impacts both their ability to learn and process information and executive functioning (Dombo & Sabatino, 2019; Forbes & Maki, 2020; Klaus, 2022; Venet, 2021). According to Krashen's Affective Filter hypothesis, language classes are particularly challenging for learners for a number of reasons. The scholar states that requiring students to speak a language different from their native one makes them feel anxious and worried about how they are perceived by their peers (Krashen, 1995). As a result, they hesitate to speak in order to make fewer mistakes.

One way to reduce the impact of stressful situations on students and lower their Affective Filter is to introduce gamification activities, which are suitable for both brick-and-mortar settings and online language teaching. The gamification of language teaching classrooms has been well-documented (Inayati & Waloyo, 2022; Peterson & Jabbari, 2023; Samson & Karthiga, 2020; Szabó & Kopinska, 2023;

Zhou & Ao, 2020). Gamification of the learning process has also been proven to be an effective tool to engage students and promote collaborative learning (Zhao & McClure, 2022; Wulantari et al., 2023).

Gamification is particularly important in Ukrainian EFL classrooms. The reality of Ukrainian education under russian military aggression poses unprecedented challenges on educators. For instance, during the 2022-2023 academic year, school hours were interrupted by over 16,000 air raid alerts, with typical alerts lasting approximately one hour (Air-alarms.in.ua, 2023). By early 2025, Russian attacks had damaged more than 4,000 educational facilities across Ukraine, with nearly 400 rendered completely uninhabitable (SaveSchools.in.ua, 2025). In regions experiencing active combat, classroom instruction frequently occurs in underground bomb shelters, basement spaces, or adapted metro infrastructure. Recent clinical research conducted with 1,291 Ukrainian schoolchildren between ages 6 and 16 identified that more than one-third demonstrated clinically significant distress indicators or post-traumatic stress disorder symptomatology attributable to war-related trauma exposure (Redlener et al., 2025). Hence, Ukrainian educators need to foster a sense of safety, stability, and control, which aligns directly with the goal of gamification: to lower the Affective Filter and create situations of success. Therefore, digital gamification is not only an innovation but also an essential tool for building resilience and supporting mental well-being within the language classroom during martial law.

However, scholars state that the introduction of gaming into the educational process is closely tied to the perspectives teachers have on the gamification of the learning process (Lamb, 2017; Zhou & Ao, 2020). Despite the proven effectiveness of gamification in language learning, a critical gap remains in understanding how pre-service elementary school language teachers, operating within a challenging and uncertain environment, especially in Ukraine, perceive and practically apply digital gamification for trauma-informed language teaching. This study addresses this gap, focusing specifically on their perspectives, perceived challenges, and the digital tools and mechanics they have chosen.

Hence, this paper is guided by four research questions:

RQ1. What are the opinions of pre-service elementary school teachers on using games for language teaching in challenging times?

RQ2. What challenges do pre-service elementary school teachers face in using games for language teaching in challenging times?

RQ3. What are pre-service elementary school teachers' perceptions of the effectiveness of the gamification of the language teaching in challenging times?

RQ4. What digital tools and gamification techniques do pre-service teachers incorporate to facilitate gamification of language teaching in elementary classrooms in challenging times?

To address the outlined research questions, it is first necessary to establish clear conceptual and terminological frameworks. This requires delineating the concepts of "gamification" and "game-based activities" and selecting an adequate model of game mechanics for subsequent analysis. Consequently, this chapter begins with the Theoretical Background section, which provides the foundation for the empirical investigation. The paper then considers the methodology underpinning the suggested study, followed by the analysis of the research results. The paper continues with a discussion of the results. The chapter concludes by outlining the study's implications for the field of teacher training.

THEORETICAL BACKGROUND

Trauma-Informed Language Teaching

Trauma-informed pedagogy is an educational approach based on the recognition that all people might have experienced traumatic events or stress and these adverse experiences change how students learn and behave in educational settings. This approach requires educators to understand how trauma affects the brain and body, particularly the ways chronic stress impairs executive functions such as working memory, attention regulation, and impulse control (Forbes & Maki, 2020; Venet, 2021). For language learners specifically, these neurobiological impacts create significant challenges.

Van der Kolk's (2000) research on trauma's neurobiological effects provides an important insight for educators: when triggered by stress or trauma, critical brain functions related to learning are compromised. Broca's area, responsible for speech production and articulation, becomes inactive during traumatic stress responses. The limbic system activates while the neocortex, which is responsible for reasoning and language processing, is suppressed. For language learners navigating traumatic experiences, this creates a double burden: the inherent stress of language learning combines with the cognitive impairment caused by trauma.

Creating trauma-informed language classrooms requires specific strategies that address both the heightened Affective Filter and compromised cognitive functioning. These include establishing predictable routines to provide stability, offering choices to restore agency, incorporating movement and regulation activities, and designing instruction that accommodates reduced attention spans (Forbes & Maki, 2020; Venet, 2021). For language teachers, this means adapting communicative activities to reduce performance anxiety, scaffolding language production tasks extensively, and accepting that fluency may temporarily matter more than accuracy when students need to express urgent emotional or practical needs.

Language classrooms can either heighten or help heal trauma symptoms. When structured according to trauma-informed principles (safety, trustworthiness, peer support, collaboration, empowerment, and cultural responsiveness) language instruction can provide not just linguistic development but also emotional regulation and social connection (Holovatenko, 2023). This dual function becomes essential in crisis contexts where language learning must continue despite destabilized conditions.

Gamification of the Educational Process

Because of the challenges mentioned above, teaching methods need to support students' thinking, feelings, and social needs at the same time. Using gamification can help with this and supports ideas from trauma-informed teaching. The use of games has gained the interest of both researchers and practitioners. For instance, empirical evidence exists on the effectiveness of using video games and computer games (Tobias et al., 2014), simulation games (Goi, 2018), serious games (Yu, 2019), and other game types in teaching and learning. However, these studies interchange the use of two concepts to designate the use of games: game-based activities and gamification. Brown and Lee define games as "any activity that formalizes a technique in units that can be scored in some way" (Brown, 2015). Game-based activities are defined as "Learning by playing games which are developed to reach predefined learning outcomes" (Şendurur & Sendurur, 2022). Gamification in its turn is defined as the use of game-play mechanics for non-game contexts (Deterding et al., 2011). Scholars include education and teaching into non-gaming contexts. In this paper, the author will use this definition of gamification by Deteding et al. (2011) and attempt to identify the preferred game mechanics of pre-service elementary school language teachers.

There are various approaches and techniques of the gamification. They are grounded in psychology and take into account the dynamics in the group when playing games. For instance, an overview of research into gamification of language teaching in digital environments showed that the most common techniques are progress bars, medals, challenges, rewards, leaderboards, levels, badges, appreciation, and storytelling (Dehghanzadeh et al., 2021). It is worth mentioning that even though these game mechanics are grounded in the individual characteristics of players, they are diverse in its nature.

A more comprehensive approach towards structuring game mechanics is outlined in the Octalysis Framework for Gamification and Behavioral Design by Yu-kai Chou. The Octalysis Framework comprises eight domains, referred to as "drives": Epic Meaning and Calling, Development and Accomplishment, Empowerment of Creativity and Feedback, Ownership and Possession, Social Influence and Relatedness, Scarcity and Impatience, Unpredictability and Curiosity, and Loss and

Avoidance (Chou, 2021). All the mechanics previously mentioned are distributed among these domains. The Epic Meaning and Calling domain is attributed to the motivation of the person to be the chosen one or the one meant for a specific thing. Development and accomplishment are the domains revolving around skills development and overcoming challenges. Empowerment of Creativity and Feedback is closely related to finding creative solutions to various situations. The ownership and Possession domain is grounded in the motivation of people to own something and make it better. Social Influence and Relatedness is focused on the innate social characteristics of people, such as mentorship, acceptance, competition, and envy. Scarcity and Impatience are grounded in humans' desire for things they cannot have right now. Unpredictability and Curiosity are grounded in people's innate curiosity about what will happen next. Loss and Avoidance are based on the human's wish to avoid the negative consequences of some events (Chou, 2021).

Scholars have empirically proven the effectiveness of the Octalysis Framework (Fortunato, Moreira, & Simões, 2023; Medina-Gómez, Cely-Betancourt, & Abreus-González, 2023; Nadif & Bidari, 2023; Sulispera & Recard, 2020). At the same time, Sulispera raises concerns about the key role classroom management plays in implementing game mechanics (Sulispera & Recard, 2020).

In the author's opinion, the Octalysis framework presents a comprehensive way to gamify any type of activity in various educational settings. However, it is important to introduce it into the scope of teaching pre-service language teachers, along with feasible ways to manage classrooms. At the same time, as the learning process takes place in blended mode, it is important for pre-service teachers to be able to apply this Framework both in online and offline modes. Next, the author reviews the scholarly research on gamification in digital settings.

The active development of ICT has significantly influenced approaches to organizing games in classrooms. Current research on the gamification of language teaching focuses on integrating digital technologies into language instruction to enhance students' motivation (Boller & Kapp, 2017; Hanson-Smith, 2016; Mast, 2017; Szabó & Kopinska, 2023). Scholars argue that the gamification of language teaching offers students instant feedback and continuous assessment, thereby developing their autonomy (Szabó & Kopinska, 2023). Systemic review of scholarly research by Dehganzadeh and Dehganzadeh (2020) has shown that online platforms with in-build solutions for gamifying the educational process are preferred in language teaching. Among the most commonly used environments for gamification are ClassDojo, Duolingo, Brainscap, Feelsbot, Kahoot, Jeopardy, etc (Dehghanzadeh et al., 2021).

In the author's opinion, these websites use a diverse range of game mechanics. For instance, ClassDojo uses timers, points, badges, and leaderboards. Duolingo uses various game mechanics based on the Octalysis framework. Brainscape uses points and badges as key game mechanics. Arce and Valdivia have proven that language learners feel more motivated to learn a foreign language when obtaining medals, progress bars, and ranking gamification mechanics are used (Arce & Valdivia, 2020). However, even though the game mechanics are entirely automated, Demirbilek states that teachers experience various challenges with the technical settings of games (Demirbilek et al., 2022). As a result, it is important to consider the perceptions of teachers on the gamification of language teaching and on choosing the game mechanics.

Pre-service Elementary School Language Teachers' Views on the Gamification of the Educational Process

A significant strand of research focuses on the opinions of pre-service elementary school language teachers regarding the gamification of language teaching. The ability to implement gamification techniques is contingent upon the development of teachers' competencies. It is important for pre-service elementary school language teachers to have the knowledge about gamification, the ability to gamify the language teaching process both in online and offline settings, and have a positive attitude towards gamification of the educational process.

Teacher perceptions of the gamification of language teaching in higher education were experimentally measured by Zhou and Ao, who have also empirically tested the Goal, Access, Feedback, Challenge, and Collaboration model of gamification (Zhou & Ao, 2020). Scholars found out that most teachers positively percept the model. However, in contrast to pre-service teachers, the data showed that instructors with 6-10 years of teaching experience tend to disagree with the model. Zhou and Ao attribute this result to taking into account a broad perspective of the language teaching field and employing diverse approaches of teaching languages, which are not limited to gamification techniques only. The study of perceptions of pre-service teachers was also carried out by Mee et al. Scholars have empirically proven that games develop learners' soft skills, motivate them, and entertain. However, respondents were concerned about their level of preparedness to deliver gamified activities (Mee et al., 2020).

Gómez-Carrasco et al. investigated the opinions of pre-service elementary school students regarding the training program, which employed gamification and flipped classroom strategies (Gómez-Carrasco et al., 2020). Scholars found that students have a very positive opinion of the training program where gamification is used, and perceive their learning in a positive light (p. 19). Aldemir et al. studied students' opinions of gamification elements (challenges, narrative, leaderboards, rewards, badges, teams, points, and a win-state) in the teacher training course (Aldemir et al., 2018). Participants of the study found challenges to be a reengaging feature, which is perceived as a reinforcement feature to read the content and a self-assessment strategy (Aldemir et al., 2018). However, scholars suggest that challenges should increase in difficulty to retain participants' interest. The narrative was positively perceived by participants because it provided a comprehensive framework for all gamification mechanics and a background to the story and tasks. Leaderboards in the educational process have created a sense of competition among participants, yet they have been criticized by some for the same reason (p. 249). Participants had a positive opinion about the rewards, both digital and tangible, and badges, which increased their confidence and served as a means of self-assessment (p. 249). Teams turned to be a powerful tool for socializing between participants and a way of relationship-building (p.250). Participants viewed points as a means of continuous feedback during the game (p. 250). There was an overall positive perception of gamification mechanics implemented in the teacher training course.

Based on previous analysis of empirical studies, it is evident that perceptions of the gamification of the educational process have a number of nuances. While it has the potential to engage students and raise the level of their motivation, teachers' perceptions of its effectiveness vary based on teacher experience, student proficiency level, and specific game mechanics employed. It requires in-service teachers to choose approaches to gamification based on the dynamic interplay between the teaching content, learning environment, and individual characteristics of learners. It also highlights the importance of educating pre-service teachers about various feasible approaches towards gamification of language teaching and the choice of game mechanics.

Having established the theoretical framework, including the definition of gamification and the Octalysis Framework, and identified the research questions, the subsequent section details the methodology used to empirically investigate the pre-service teachers' perspectives and practices.

METHODOLOGY

This paper is a mixed-method study. The author aims to develop a comprehensive understanding of pre-service teachers' perspectives on gamification in language teaching in a challenging context. The quantitative methodology includes surveying pre-service elementary school language teachers. The paper is also grounded in qualitative methodology through the analysis of pre-service elementary school teachers' artifacts in the form of lesson plans to teach English as a foreign language using gamification. Combining both quantitative and qualitative methods in the paper allows the author not only to identify pre-service teachers' opinions but also to delve deeper into the choice of game mechanics and the way students tend to employ them in language teaching.

This study is limited to the perspectives and practices of Year 2 and 3 pre-service elementary school teachers enrolled in a specific minor at a Ukrainian university. The research focuses exclusively on their views regarding the gamification of English as a Foreign Language (EFL) teaching in elementary classrooms, specifically within the context of adapting to blended learning in wartime environments. It does not assess the long-term impact of the intervention or the actual in-class performance of the participants.

Data collection

The quantitative data in this paper were obtained through an adapted survey developed by Mee et al. (2020). Participants were asked to share their opinion on using games to teach English in elementary schools in a wartime context using a Likert Scale with 23 statements. The survey was based on the first three research questions and consisted of 4 blocks: background information, views on using games in language teaching, challenges in using games in language teaching, and pre-service teachers' perception of the effectiveness of gamification. Participants could choose their answer on the scale from 1 (Totally Disagree) to 5 (Totally Agree). Cronbach α of the pretest survey is .783, which shows an acceptable internal consistency of the tool.

The intervention was organized in the form of a module on language teaching for Year 3 undergraduates. The one-credit module was introduced within the course of Teaching Foreign Language in Elementary Schools (12 credits). The outline of the module is provided in Table 1.

Table 1. Outline of the study intervention

Module 4. Gamification in Teaching Foreign Languages to Young Learners				
Topics	**Lectures**	**Practical classes**	**Office hours**	**Self-study work**
Topic 1. The Concept and Structure of Gamification in Teaching Foreign Languages to Elementary School Students	2			
Topic 2. Using Gamification to Form Phonetics, Grammar, and Vocabulary Foreign Language Skills of Elementary School Stu- dents		4		6
Topic 3. Using Gamification to Form Listening, Reading, Speaking and Writing Foreign Language Skills of Elementary School Students		8		6
Topic 4. Gamification of the Assessment of the Level of Foreign Language Skill Formation of Elementary School Students		2		
Test			2	
Total 30h	2	14	2	12

The study was carried out in 2023, when Ukraine was amidst full-scale war and its territories and critical infrastructure was shelled by russian missiles causing rolling blackouts. This setting required the instructor to adapt the material and present it through the lens of using gamification to create trauma-informed language classrooms and provide for possible adaptations of the designed activities to be suitable both for online and offline language teaching to model the possible teaching environment they will teach in.

The posttest was administered for 8 participants, who had successfully completed the intervention module in the face-to-face mode. The quantitative data were analyzed using the descriptive statistics data analysis tool in Excel. The interpretation of the Likert scale is provided in Table 2. The Cronbach α of the posttest is .743, which shows an acceptable internal consistency of the tool.

Table 2. Interpreting Likert scale ratings

Interval	Adjectival Rating
1.00 – 1.80	Strongly Disagree (SD)
1.81 – 2.60	Disagree (D)
2.61 – 3.40	Neutral (N)
3.41 – 4.20	Agree (A)
4.21 – 5.00	Strongly Agree (SA)

In order to get in-depth qualitative data from the participants on the fourth research question, in addition to the survey, the students' artifacts were analyzed. Activities, designed by students were uploaded in NVivo, and coded to identify digital tools students used to bring gamification into language teaching, as well as the gamification techniques used for that.

Participants

The pretest survey was administered on the non-probability sample of 27 preservice elementary school teachers at one of the Ukrainian universities. It aimed to identify respondents' existing opinions. All students are Year 2 and 3 undergraduate elementary school students, enrolled in the Teaching English as a Foreign Language minor. 100% of participants are women in their twenties, 48.1% of whom have no teaching experience, 37% have 1-3 years of teaching experience, and 14.8% have more than 3 years of relevant teaching experience. The posttest was administered for 8 participants, who had successfully completed the intervention module in the face-to-face mode. The qualitative data (artifacts *n=37*) were gathered from all participants enrolled in the intervention module, reflecting work submitted by both the pretest and posttest groups, and those who continued in the module in both online and offline modes.

RESULTS

The results section presents findings from both quantitative survey data and qualitative analysis of pre-service elementary school language teachers' artifacts. The analysis addresses all four research questions, with particular attention to how the challenging wartime context in Ukraine shaped participants' perspectives and practical choices regarding digital gamification in language teaching.

Q1. What are the opinions of pre-service elementary school teachers on using games for language teaching in challenging times?

The interpretation of responses aiming to answer this research question is shown in Table 3. In the table the results are provided for pretest (T1) and the posttest (T2). The ratings in all the five items as well as the overall average fall under the interval for "Agree" based on the Likert scale questionnaire procedure of both in pretest and in posttest. However, the analysis of the mean of Likert items using the interpreting ratings shows that the opinion of participants has changed from "Neutral' to 'Disagree' for the statement 'Playing games requires too much engagement time' (M=2.89 to 2.5). This shift is particularly significant in the crisis context, as it suggests that participants recognized gamification as a time-efficient strategy rather than a divertion from essential curriculum coverage during unstable learning conditions. This perception change might be attributed to the way the instructor highlighted possible low-prep and no-prep games for language teachers, as well as classroom time management strategies that are essential when lessons may be interrupted by air raids or power outages.

The majority of the respondents strongly agree that playing games stimulates curiosity (M=4.85; 4.88 and SD=.456 and .354), helps development of social skills (M=4.67; 5 and SD=.620 and .000), and provides for cooperative learning environment (M=4.59; 4.38 and SD=.694 and 1.408). The agreement regarding social skills development is particularly meaningful in the Ukrainian wartime context, where students experience displacement, separation from peers, and social fragmentation. Creating opportunities for social connection through gamified language learning addresses not only linguistic but also socio-emotional needs of learners operating in isolation or under threat. The standard deviation (SD) implies some variability in responses, though this decreased slightly in the posttest for most items. At the same time, there is a need to dwell more on statement 5 'It is easy for me to use games in the classroom', as the opinion of participants remained neutral (M=3.19;

3.25). This persistent uncertainty about implementation ease, despite completing the intervention module, reflects the realistic challenges of teaching in an unpredictable environment where technological infrastructure may fail, students may need to evacuate mid-lesson, and rapid adaptation between online and offline formats is required. The neutral stance suggests that while participants recognize the value of gamification, they remain uncertain about their capacity to implement it effectively under such demanding circumstances.

Table 3. Descriptive statistics for opinions of pre-service elementary school teachers on using games for language teaching

No	Statement	Mean T1	Mean T2	Median T1	Median T2	Mode T1	Mode T2	SD T1	SD T2	Interpretation T1	Interpretation T2
1	Playing games requires too much engagement time	2.89	2.5	3	2.5	3	2; 3	1.086	.926	N	D
2	Playing game stimulates curiosity in learning the content	4.85	4.88	5	5	5	5	.456	.354	SA	SA
3	When games are played with a group, this helps development of social skills	4.67	5	5	5	5	5	.620	.000	SA	SA
4	Games provide cooperative learning environment	4.59	4.38	5	5	5	5	.694	1.408	SA	SA
5	It is easy for me to use games in the classroom	3.19	3.25	3	3	3	3	1.001	.707	N	N
	Average	4.04	4.0	4.2	4.1	4.2	4.2	.772	.679	A	A

Key to Table 3: **T1:** Pretest; **T2:** Posttest; **SA:** Strongly Agree; **A:** Agree; **N:** Neutral; **D:** Disagree

The analysis of the overall descriptive statistics suggests that participants have positive opinions on using games in language teaching. Participants do not view games in language teaching as a diversion of lesson time, and they strongly agree on their positive potential. Notably, the perception of games requiring "too much engagement time" shifted from Neutral (N) in the pretest to Disagree (D) in the posttest, suggesting the intervention successfully addressed this concern by demonstrating practical, time-efficient gamification strategies suitable for crisis environments. Hence, further analysis of the game design is needed through a qualitative methodology. Overall, participants perceive games as an asset in language teaching that can address both pedagogical and socio-emotional needs in challenging times.

Q2. What challenges do pre-service elementary school teachers face in using games for language teaching in challenging times?

The interpretation of responses to this research question is presented in Table 4. Participants were asked to express their opinion on nine statements regarding the challenges teachers face in using gamification in language teaching. On average, participants have shown a 'Neutral' opinion with no change after the posttest (M=3.06; 3.03).

Table 4. Descriptive statistics for challenges pre-service elementary school teachers face in using games for language teaching

No	Statement	Mean T1	Mean T2	Median T1	Median T2	Mode T1	Mode T2	SD T1	SD T2	Interpretation T1	Interpretation T2
1	Students get side-tracked and do not focus on learning when games are used	2.74	1.88	3	2	3	2	1.095	.641	N	D
2	It is difficult to evaluate students' performance/learning when they are playing games	2.78	2.88	3	3	2; 3	3	1.281	.641	N	N
3	Language teachers do not have enough skills on implementing gamification techniques	3.37	3.37	3	3	3	3	.839	1.414	N	N
4	There is lack of availability of games that match the lesson topic	2.93	2.5	3	2	3	2; 4;	.874	1.309	N	D
5	Some games are too complicated for language students	3.37	2.88	3	3	3	3	.884	1.356	N	N
6	The technological support available in the classroom is inadequate	3.41	3.75	3	4	3	5	.971	1.282	A	A
7	There is lack of supporting materials for how to best use games for learning English	3.33	3.88	3	4	3	3; 4;	3.26	.835	N	A
8	There is lack of teachers training in using gamification technology	3.26	3.38	3	4	3	5	.903	1.768	N	N
9	There is limited or no available time to play games due to the need to meet all of the required curriculum standards	3.3	3.75	4	4	4	4	1.103	1.035	N	A
	Average	4.04	4.0	4.2	4.1	4.2	4.2	.772	.679	A	A

Key to Table 4: **T1:** Pretest; **T2:** Posttest; **A:** Agree; **N:** Neutral; **D:** Disagree

However, the average in this case is not a representative indicator of the opinions of participants, as responses showed considerable variation across different types of challenges. Participants demonstrated a significant positive change in their opinions for the statement "Students get side-tracked and do not focus on learning when games are used" (M=2.74 to 1.88; M=3 to 2; SD=1.095 to .641), which means participants are more likely to disagree with the statement after the intervention. This shift is important in the crisis context: participants learned that well-designed gamification actually increases focus rather than creating distraction, which is essential when students' attention is already compromised by stress, trauma, and environmental disruptions such as air raid alerts. The decreased standard deviation indicates growing consensus among participants on this point.

A similar shift in perception is demonstrated with the statement "There is lack of availability of games that match the lesson topic" (M=2.93 to 2.5; Median=3 to 2; SD=.874 to 1.309). However, the standard deviation after the intervention is slightly higher, showing some variability in answers. This change suggests that the intervention successfully demonstrated how to create or adapt gamification activities to match specific learning objectives. The increased variability may reflect different levels of confidence among participants in their ability to design context-appropriate games under challenging conditions.

However, there were three areas where participants' concerns either remained stable or increased. First, the persistent neutral stance on assessment difficulty (M=2.78; 2.88) indicates that participants still struggle to envision how to evaluate learning outcomes when students are engaged in gameplay, particularly when assessment must occur rapidly or be adapted for sudden format changes between online and offline delivery. Second, the neutral perception that language teachers lack sufficient skills (M=3.37; 3.37) suggests that despite completing the intervention, participants recognize the complexity of gamification and the need for ongoing professional development—a realistic assessment given the demanding crisis context.

Most significantly, there were slightly negative changes in opinion for two statements directly reflecting the crisis environment. Agreement with "The technological support available in the classroom is inadequate" increased (M=3.41 to 3.75; Median=3 to 4), directly reflecting the Ukrainian reality of rolling blackouts, damaged infrastructure, and unreliable internet connectivity. This increased awareness may reflect participants' more sophisticated understanding after the intervention of what technological support would be ideal for gamification, making them more conscious of the gap between ideal and available resources. The mode shift to 5 (Strongly Agree) for some participants indicates that those teaching in particularly affected regions or institutions felt this constraint acutely.

Similarly, opinion shifted from "Neutral" to "Agree" for the statement "There is lack of supporting materials for how to best use games for learning English" (M=3.33 to 3.88; Median=3 to 4). Participants in the intervention were required to create their language teaching activities using gamification from scratch, which may have heightened their awareness of the limited ready-made resources available for their specific context. This finding is particularly relevant for wartime conditions where teachers need readily adaptable materials that function across platforms and can be implemented with minimal preparation time during emergencies.

The statement "There is limited or no available time to play games due to the need to meet all of the required curriculum standards" also shifted from "Neutral" to "Agree" (M=3.3 to 3.75; Median=4; Mode=4). This increasing concern about time constraints appears paradoxical given the decreased perception that games waste time (RQ1). However, this may reflect participants' growing understanding of the curriculum pressures they will face, particularly in a crisis context where instructional time is frequently interrupted by security concerns, evacuations, and infrastructure failures. The consistent mode of 4 suggests this is a widespread concern rather than an outlier perception.

In summary, the analysis of the descriptive statistics of the Likert scale indicates mixed and nuanced views of participants on the challenges in gamifying language teaching in elementary schools in challenging times. While participants demonstrated decreased concerns about student distraction and game availability—suggesting the intervention built their confidence in managing gamified classrooms—they simultaneously developed more realistic and increased awareness of systemic challenges. Specifically, they recognize ongoing concerns about inadequate technological infrastructure, insufficient supporting materials, lack of training opportunities, and time pressure to meet curriculum standards—all challenges amplified by the wartime context. Even though participants agree on the value of games in language teaching, they remain realistic about implementation barriers, particularly regarding assessing students' performance, developing sufficient skills to gamify language teaching effectively, choosing games of appropriate complexity for students, and obtaining adequate training in implementing gamification for language teaching under demanding conditions.

RQ3. What are pre-service elementary school teachers' perceptions of the effectiveness of the gamification of the language teaching in challenging times?

Table 5 provides the analysis of pre-service elementary school teachers' perceptions of the effectiveness of language teaching using gamification. The overall average shifted from "Agree" to "Strongly Agree" (M=4.04 to 4.40; Median=4.22 to 4.56), indicating that the intervention significantly strengthened participants' confidence in gamification's effectiveness. This substantial positive shift is particularly meaningful in a crisis context where educators seek evidence-based strategies that can demonstrably improve learning outcomes despite challenging conditions.

Table 5. Pre-service elementary school teachers' perceptions of the effectiveness of the gamification of the language teaching

No	Statement	Mean T1	Mean T2	Median T1	Median T2	Mode T1	Mode T2	SD T1	SD T2	Interpretation T1	Interpretation T2
1	Game-based learning contribute to the development of creative thinking skills	4.74	4.88	5	5	5	5	.447	.354	SA	SA
2	Game-based learning contribute to the development of critical thinking skills	4.44	4.5	5	5	5	5	.801	.756	SA	SA
3	Game-based learning contribute to the development of problem-solving skills	4.26	4.5	5	5	5	5	.984	1.069	SA	SA
4	Implementation of game-based learning in primary school can decrease teacher's anxiety level towards teaching English	3.85	4.5	4	5	5	5	1.134	.756	A	SA

No	Statement	Mean T1	Mean T2	Median T1	Median T2	Mode T1	Mode T2	SD T1	SD T2	Interpret-ation T1	Interpret-ation T2
5	I believe that I can use games I will prepare in this module effectively in my language teaching	3.93	4.63	4	5	3	5	.874	.518	A	SA
6	I believe that using games can lower learners' anxiety during English classes	4.52	4.75	5	5	5	5	.643	.463	SA	SA
7	I believe that games are not very effective in language learning	1.89	2.13	1	1	1	1	1.476	1.808	D	D
8	I believe that using games are highly motivating and entertaining way of teaching language especially for weak learners	4.7	5	5	5	5	5	.609	.0	SA	SA
9	I believe that while playing a game, learners are learning language subconsciously	4.07	4.75	4	5	5	5	1.107	.463	A	SA
	Average	4.04	4.40	4.22	4.56	4.33	4.56	.90	.69	A	SA

Key to Table 5: **SA:** Strongly Agree, **A:** Agree, **D:** Disagree, **T1:** Pretest, **T2:** Posttest

The majority of respondents after the intervention changed their opinion to "Strongly Agree" that gamification contributes to the development of creative thinking skills of students (M=4.74 to 4.88), their critical thinking (M=4.44 to 4.5), problem-solving skills (M=4.26 to 4.5). These cognitive skill developments are important for students navigating the uncertainties and complex challenges of wartime environments, where adaptability and creative problem-solving are survival skills. The consistently high scores and mode of 5 for all three items indicate strong consensus among participants.

Three items showed particularly notable improvements that directly address the trauma-informed dimensions of teaching in crisis. First, agreement that "Implementation of game-based learning in primary school can decrease teacher's anxiety level towards teaching English" increased substantially from "Agree" to "Strongly Agree" (M=3.85 to 4.5; Median=4 to 5; Mode=5; SD=1.134 to .756). This finding is very important for Ukraine, where future teachers may have to teach English – even though it is often their hardest subject – while also dealing with their own trauma and the stress of uncertain teaching situations. The decreased standard deviation indicates growing consensus that gamification provides psychological support not only for students but also for teachers themselves. Gamification's structured, playful approach may reduce instructional anxiety by providing clear frameworks and positive emotional engagement that counterbalances the pervasive stress of the crisis environment.

Second, participants' confidence in their own implementation capacity increased: "I believe that I can use games I will prepare in this module effectively in my language teaching" shifted from "Agree" to "Strongly Agree" (M=3.93 to 4.63; Median=4 to 5; Mode=3 to 5; SD=.874 to .518). The shift in mode from 3 to 5 is particularly striking, indicating that participants moved from neutral uncertainty to strong confidence. The decreased standard deviation shows this confidence-building was widespread rather than limited to high-performing students. This self-efficacy development is essential for successful implementation, particularly when teachers must adapt strategies independently during crises without ready access to mentoring or support.

Third, agreement with "I believe that using games can lower learners' anxiety during English classes" strengthened from an already high level (M=4.52 to 4.75; Median=5; Mode=5; SD=.643 to .463). This finding directly addresses Krashen's Affective Filter hypothesis, which the theoretical framework identified as particularly relevant for language learning in stressful contexts. Strong agreement (Mode=5, SD=.463) indicates that participants recognized gamification as a practical tool for trauma-informed pedagogy—specifically, as a mechanism to create psychological safety, reduce language learning anxiety, and lower the Affective Filter that is elevated by the chronic stress of wartime conditions.

Importantly, participants' perception of implicit language learning through gameplay also strengthened significantly: "I believe that while playing a game, learners are learning language subconsciously" moved from "Agree" to "Strongly Agree" (M=4.07 to 4.75; Median=4 to 5; Mode=5; SD=1.107 to .463). This understanding particularly relevant for trauma-affected learners whose executive functioning and explicit learning capacity may be compromised by stress. The recognition that games facilitate implicit, subconscious language acquisition suggests participants understand how gamification can enable learning to continue even when students' cognitive abilities are compromised by trauma.

The statement "I believe that using games are highly motivating and entertaining way of teaching language especially for weak learners" maintained perfect strong agreement in the posttest (M=4.7 to 5.0; Mode=5; SD=.609 to .000). The zero standard deviation in the posttest indicates strong agreement. This is critical in the crisis context where many students experience learning regression, gaps in instruction, and reduced motivation due to displacement, trauma, or irregular school attendance.

After the intervention, most participants chose extreme answers like "Strongly Agree" or "Strongly Disagree" (with the most common answers being 1 or 5), instead of choosing middle options. The standard deviation stayed below 2 for all statements. This shows that the intervention helped participants form clear and confident opinions about how effective gamification is, instead of staying unsure or neutral. The lower standard deviations also suggest that participants agreed more with each other and felt more confident in their views.

Overall, the quantitative findings demonstrate that the intervention successfully strengthened participants' perceptions of gamification's effectiveness across multiple dimensions—cognitive skill development, motivation, anxiety reduction, and implicit learning. Most significantly for the chapter's focus on challenging times, participants developed strong confidence in gamification's capacity to address the socio-emotional and psychological needs of both teachers and learners operating under crisis conditions. Strong agreement on anxiety reduction, teacher confidence, and effectiveness for struggling learners indicates that participants recognized gamification as an important tool for trauma-informed, resilient pedagogy in wartime contexts.

RQ 4. What digital tools and gamification techniques do pre-service teachers incorporate to facilitate gamification of language teaching in elementary classrooms in challenging times?

The author performed a qualitative analysis of artifacts (*n=37*) contributed by students while studying the module on gamification. The author developed three codes to process the data: format adaptability, digital tools used, and gamification mechanics employed. This analysis provides insight into how pre-service elementary school language teachers translate theoretical knowledge about gamification into practical teaching strategies suitable for the uncertain Ukrainian educational context.

Format Adaptability: Universal Design for Crisis Contexts

The first code analyzed whether activities were analyzed to determine whether students designed them for offline, online usage, or for blended learning. The analysis results show that 100% of artifacts were designed to be used in blended learning. This universal adaptability reflects participants' acute awareness that they cannot predict from one day to the next whether they will teach online (due to air raids, power outages, etc.), offline (when infrastructure and security permit), or in hybrid configurations (with some students physically present and others joining remotely due to displacement or safety concerns).

Some of the students' ideas provided for playing games on students' own devices (BYOD, Bring Your Own Device), recognizing that personal smartphones may be the most reliable technology available when institutional infrastructure fails. Others designed games to be shared on the screen of an interactive board for in-person instruction or printed out as physical materials for low-tech implementation when electricity is unavailable. This strategic flexibility demonstrates that participants get ready for a dynamic, unpredictable teaching conditions.

The 100% adoption of blended-ready design represents a significant shift from traditional teacher education, where pre-service elementary school language teachers typically design for a single, stable instructional mode. In the Ukrainian crisis context, this adaptability is essential. The finding suggests that the intervention successfully prepared participants to think flexibly and maintain instructional continuity across disruptions—a core competency for trauma-informed teaching in unstable environments.

Digital Tools: Preference for Simplicity and Accessibility

The second code comprised digital tools used by students to design their gamified activities. The author has created a broad code category "Digital tool" and added sub-codes during the analysis process. The results are presented in Table 6.

Table 6. Digital tools used by pre-service elementary school language teachers to design their activities

Resource	Number of Occurrences	Percentage
Canva	9	21.45%
Genial.ly	1	2.43%
Google Jamboard	6	14.63%
LearningApps	2	4.87%
Microsoft Word	21	51.2%
Other resources with ready solutions	2	4.87%
Total	41	100%

The most common resource used is Microsoft Word, (51.2%, n=21), a finding that initially may seem surprising given the availability of specialized gamification platforms. The author can explain it by the ease of integration of the Microsoft Word with other programs, the ability to share it online and print it out if necessary for brick-and-mortar classes or to copy-paste the text in other programs, including videoconferencing. This preference reflects sophisticated strategic thinking of participants adapted to crisis conditions.

The second most popular digital tool is Canva (21.45%, n=9), which was used to design gameboards and make them interactive by sharing access links. Canva's popularity reflects its balance of visual appeal and accessibility: it offers attractive, professional-looking designs that can engage students while remaining relatively simple to use. Importantly, Canva allows both online sharing (when connectivity permits) and offline use through downloaded copies. The visual nature of Canva-designed materials also addresses the needs of language learners operating under stress, for whom visual scaffolding can reduce cognitive load and support comprehension when executive functioning is compromised.

Google Jamboard (14.63%, n=6) was used similarly to design interactive, collaborative gameboards. The result of the analysis of artifacts were based on ready solutions with in-build gamification mechanics. The overall tendency is to use simple digital tools that can be easily adapted for blended learning (both online and offline) in the uncertain Ukrainian context. Jamboard's synchronous collaboration features make it valuable for maintaining social connection and cooperative learning when students are physically dispersed due to displacement or security concerns. However, its lower usage compared to Word and Canva may reflect concerns about its dependence on stable internet connectivity and Google account access. Notably, Google discontinued Jamboard in October 2024, after this study's data collection but before publication. This discontinuation validates participants' instinct for caution regarding specialized platforms and reinforces the wisdom of the majority's preference for more stable, established tools like Microsoft Word.

Gamification Mechanics: Strategic Alignment with Trauma-Informed Pedagogy

The third code was developed to analyze dispositions of pre-service elementary school language teachers towards the use of various types of gamification techniques structured by the Octalysis Framework. The analysis reveals patterns in mechanics selection that reflect both pedagogical sophistication and trauma-informed sensitivity (Table 7).

Table 7. Pre-service elementary school language teachers' use of gamification mechanics structured according the Octalysis Framework

Gamification Mechanic (Octalysis Drive)	Frequency of Usage
Accomplishment	29
Social influence	28
Ownership	18
Scarcity	15
Unpredictability	12
Avoidance	4
Empowerment	2
Epic meaning	0
Total	108

The most common game mechanic used by pre-service elementary school teachers in language teaching was Accomplishment *(n=29)*. It was utilized through progress bars, challenges, points, badges, and leaderboards. The author explains this choice of mechanics to the fact, that a majority of elementary school students have an extrinsic motivation, which can be supported by this type of positive reinforcement. In a context where students experience repeated disruptions, and loss of control over their lives, Accomplishment mechanics provide structured opportunities for success. Each earned point, completed challenge, or achieved badge represents a tangible accomplishment in an environment where a lot feels uncontrollable. These mechanics build self-efficacy and create success situations.

Pre-service elementary school language teachers used Social Influence *(n=28)* almost as frequently as the previous mechanic. Frequent use of teamwork and competition can be attributed to the specifics of language teaching classes, where communication and collaboration are required according to the Communication Approach to language teaching. However, in the wartime Ukrainian context, these social mechanics serve dual purposes: they develop language proficiency while simultaneously rebuilding social connections and combating the isolation which deepens trauma. Competition, when carefully structured, can provide healthy ways to exhibit agency and control.

Ownership (*n=18*) was the third most-used mechanic, often paired with Accomplishment through systems where students exchange points for tangible prizes, accumulate virtual currencies, or customize their learning avatars or materials. In a crisis context where students may have lost physical possessions, homes, or stability, creating opportunities to own, build, and improve virtual or symbolic assets provides psychological compensation. These mechanics also develop planning and resource management skills while giving students decision-making power and control, which is important for students who lack control and power in their lives.

Scarcity (*n=15*) was implemented primarily through delayed prizes, countdown timers, and limited-availability rewards. Scarcity mechanics leverage humans' increased desire for things perceived as rare or time-limited. The relatively lower frequency of its usage by participants may be due to caution about creating urgency or pressure for students already operating under chronic stress.

Unpredictability (*n=12*) was implemented through surprise elements and random rewards. The moderate use of this mechanic suggests participants saw value in creating positive surprises. However, while pleasant surprises can be engaging, unpredictability might trigger anxiety in students already coping with environmental instability.

Avoidance (*n=4*) and Empowerment (*n=2*) were minimally present, and Epic Meaning was completely absent (*n=0*). Avoidance mechanics (e.g. punishment, failure, or countdowns) were largely excluded, likely because Ukrainian students already face real-life threats such as air raids, loss, and displacement. Adding simulated danger would be unhelpful and potentially harmful. Empowerment mechanics, which involve complex tasks, were also rare. It might be due to demand of cognitive effort and stress tolerance that trauma-affected students may lack. Instead, participants opted for smaller, manageable challenges that build confidence without causing overload. Epic Meaning mechanics, which cast students as heroes in grand narratives, were entirely absent. This could reflect both the assignment's focus on short tasks. Participants may also have lacked examples of using this type of mechanic effectively.

Looking across all three aspects of RQ4, a clear pattern emerges. Participants created practical, resilient gamification strategies tailored to their challenging teaching environments. Rather than copying examples or relying on complex platforms, they preferred to design flexible, meaningful activities. Participants demonstrated the mindset of adaptive educators teaching in challenging circumstances. However, more research into this result is needed.

The artifact analysis has several limitations that should be acknowledged. First, they represent designed activities rather than implemented lessons, so actual effectiveness in real classrooms remains unknown. Second, the analysis focused on observable features of artifacts, but could not fully refer to participants' pedagogical reasoning. Third, coding game mechanics using the Octalysis Framework involved subjective judgment. As a result, frequency counts should be seen as approximate, showing general trends rather than exact measures. Finally, the sample came from pre-service elementary school language teachers in one Ukrainian university's minor program, which limits how broadly the findings can be applied. Despite these limitations, the artifact analysis offers useful insight into how pre-service elementary school language teachers apply gamification in crisis settings. Common patterns appeared across multiple participants, indicating shared understandings as a result of the intervention rather than isolated personal choices.

SOLUTIONS AND RECOMMENDATIONS

The findings of this study demonstrate the potential of digital gamification for language teaching in challenging times and highlights barriers which should be addressed to support its effective implementation. Based on the mixed-methods analysis, this section offers targeted recommendations for teacher training programs, educational institutions, and curriculum developers.

Teacher training programs should systematically integrate gamification pedagogy throughout methods courses rather than isolating it in a single module. The intervention demonstrated that focused training strengthens confidence (RQ3). However, concerns about assessment, skill development, and time management (RQ2) indicate need for an integrated approach to teaching pedagogics, teaching methods and gamification in language classrooms. Programs should provide supervised practicum where pre-service elementary school language teachers implement gamified activities with actual students in different teaching modes (online, offline, and hybrid formats). The artifacts analyzed (RQ4) represent designed but not implemented activities; pre-service teachers need opportunities to test designs, observe student responses, and refine approaches based on evidence. Teacher training programs should explicitly teach how to embed assessments within game structures, and how to interpret game results into the curriculum outcomes. Providing rubrics and example assessments would address this issue. Programs should prioritize simple gamification tools over specialized platforms. The preference for simple tools (Table 6) reveals that effective gamification does not require advanced technology. Finally, programs must explicitly connect gamification to trauma-informed teaching principles. While participants strategically avoided Avoidance and Empowerment mechanics (Table 7), demonstrating emerging trauma-sensitivity, this understanding should be explicitly taught. Programs should provide frameworks for analyzing game mechanics through a trauma-informed lens and strategies for creating psychologically safe gamified environments.

Educational institutions should provide ongoing professional development in gamification, recognizing that expertise develops with mentoring and practice. Institutions should establish professional learning communities where teachers share strategies, troubleshoot challenges, and collaboratively develop resources. Creating institutional repositories of teacher-created gamified activities with implementation notes and adaptation suggestions would address the lack of supporting materials (Table 4). Schools should prioritize resilient tech solutions: offline content libraries, printed resource collections, and low-tech teaching methods that function during infrastructure failures. This addresses the increased concern about inadequate technological support (Table 4, statement 6).

The lack of supporting materials (Table 4, statement 7) represents an opportunity for curriculum developers to create resources specifically designed for crisis-responsive education. These should be open-access or low-cost, modular and adaptable, and available in multiple formats (Word documents, printable PDFs, editable slides). Resource collections should include: ready-to-use activity templates, assessment rubrics and protocols, and adaptation guides for shifting between formats. Resources should align with evidence-based frameworks like Octalysis, explicitly labeling which game mechanics each activity employs and explaining pedagogical rationale. Resources must reflect trauma-informed design principles, with guidance about appropriate mechanics for different contexts, warnings about elements that may trigger anxiety, and alternatives for students who may not respond well.

FUTURE RESEARCH DIRECTIONS

This study has some limitations that should be discussed. Among them is a limitation of the sample. The pretest was administered for 27 participants and the posttest after the intervention was administered for 8 participants. Hence, administering the study for a larger sample can become a prospect for further research. Moreover, the pretest was administered to Year 2 and 3 pre-service elementary school language teachers, but the intervention and the post—test were administered only to Year 3 students. As a result, some variables between these two groups in terms of their knowledge, skills, and attitudes were not taken into account. The limitations connected to the surveying tool also need to be considered. Data collection depends on participants' understanding of statements and their sincerity in answers. To mitigate the effect of this limitation, a qualitative method was employed. At the same time, the opinions and perceptions of participants are limited to their previous lived experiences and the context of the study. It is a generally accepted fact, that the use of Likert scale can become a subject of such distortions as avoidance of extreme answers and an attempt to portray participants in a more favorable light.

CONCLUSION

The objective of the paper is to identify pre-service elementary school language teacher's perspectives on the use of digital activities in language teaching. The mixed-method study is grounded in the theory of gamification of the educational process. From the quantitative data obtained, it is possible to conclude that participants have a positive attitude towards gamification. Participants perceive games as a valuable asset in language teaching. Among the main challenges, participants have indicated the lack of supporting materials and lack of time to fulfill the curriculum requirements. Participants strongly agree with statements on the effectiveness of gamification. An additional insight into the practices pre-service elementary school language teachers acquired was obtained through the analysis of qualitative data. Participants have demonstrated their understanding of gamification principles through the lesson planning. The most commonly used digital tools were Microsoft Word, Canva, Google Jamboard. Participants tended to use simple digital tools. This tendency might be explained by their wish to develop gamification activities that can be easily adapted to both offline and online learning in uncertain Ukrainian conditions. Among the most frequently used gamification mechanics students employed was Accomplishment, Social influence, and Ownership. These findings contribute to the field of language teacher education and emphasize the importance of preparing pre-service elementary school language teachers to utilize gamification strategies in

ACKNOWLEDGMENT

This research received no specific grant from any funding agency in the public, commercial, or not-for-profit sectors. This study was made possible with the informational support of libraries of the University of Minnesota Twin Cities, where the author was a Visiting Scholar in 2022-2023 academic year. The author also wishes to acknowledge the participation of pre-service elementary school language teachers in the study. The author also stretches her acknowledgment to her family, who freed substantial amount of her time for academic writing.

REFERENCES

AirAlarms.in.ua. (2023). Statistics of air alarms in Ukraine, February 24, 2022–December 26, 2023. https://air-alarms.in.ua/?from=2022-02-24&to=2023-12-26#statistic

Aldemir, T., Celik, B., & Kaplan, G. (2018). A qualitative investigation of student perceptions of game elements in a gamified course. *Computers in Human Behavior*, *78*, 235–254. DOI: 10.1016/j.chb.2017.10.001

Arce, N. H., & Valdivia, A. C. (2020). Adapting competitiveness and gamification to a digital platform for foreign language learning. [iJET]. *International Journal of Emerging Technologies in Learning*, *15*(20), 194–209. DOI: 10.3991/ijet.v15i20.16135

Bailey, D., Almusharraf, N., & Hatcher, R. (2021). Finding satisfaction: Intrinsic motivation for synchronous and asynchronous communication in the online language learning context. *Education and Information Technologies*, *26*(3), 2563–2583. DOI: 10.1007/s10639-020-10369-z PMID: 33169066

Boller, S., & Kapp, K. (2017). *Play to learn: Everything you need to know about designing effective learning games*. ATD Press.

Brown, H. D. (2015). *Teaching by principles: An interactive approach to language pedagogy* (4th ed., eText). Pearson Education.

Chou, Y.-K. (2021). *The Octalysis framework for gamification & behavioral design*. https://yukaichou.com/gamification-examples/octalysis-complete-gamification-framework/

De Souza, R., Parveen, R., Chupradit, S., Velasco, L. G., Arcinas, M., Tabuena, A. C., Pentang, J., & Ventayen, R. J. M. (2021). Language teachers' pedagogical orientations in integrating technology in the online classroom: Its effect on students' motivation and engagement. *Turkish Journal of Computer and Mathematics Education, 12*. Advance online publication. DOI: 10.2139/ssrn.3844678

Dehganzadeh, H., & Dehganzadeh, H. (2020). Investigating effects of digital gamification-based language learning: A systematic review. *Two Quarterly Journal of English Language Teaching and Learning University of Tabriz, 12*, 53–93. DOI: 10.22034/elt.2020.10676

Dehghanzadeh, H., Fardanesh, H., Hatami, J., Talaee, E., & Noroozi, O. (2021). Using gamification to support learning English as a second language: A systematic review. *Computer Assisted Language Learning, 34*(6), 934–957. DOI: 10.1080/09588221.2019.1648298

Demirbilek, M., Talan, T., & Alzouebi, K. (2022). An examination of the factors and challenges to adopting gamification in English foreign language teaching. *International Journal of Technology in Education, 5*(4), 654–668. DOI: 10.46328/ijte.358

Deterding, S., Khaled, R., Nacke, L., & Dixon, D. (2011). *Gamification: Toward a definition*. In *CHI 2011, Gamification Workshop Proceedings* (pp. 1–9). Vancouver.

Dombo, E., & Sabatino, C. A. (2019). Trauma and its sequelae in children and adolescents. In *Creating trauma-informed schools: A guide to school social workers and educators*. Oxford University Press, Incorporated. DOI: 10.1093/oso/9780190873806.003.0002

Fauzi, M. F., & Anindiati, I. (2021). Improving the motivation of students in Arabic language learning through learning management system. *Izdihar: Journal of Arabic Language Teaching, Linguistics, and Literature, 4*(3), 257–274. DOI: 10.22219/jiz.v4i3.19110

Forbes, H., & Maki, D. (2020). *Classroom180: A framework for creating, sustaining, and assessing the trauma-informed classroom*. Beyond Consequences Institute.

Fortunato, M., Moreira, A., & Simões, A. R. (2023). 'All hands-on deck': Octalysis framework in primary English language learning. *Sisyphus—. Journal of Education, 10*, 225–243. DOI: 10.25749/sis.26985

Goi, C. L. (2018). The use of business simulation games in teaching and learning. *Journal of Education for Business, 94*(5), 342–349. DOI: 10.1080/08832323.2018.1536028

Gómez-Carrasco, C. J., Monteagudo-Fernández, J., Moreno-Vera, J. R., & Sainz-Gómez, M. (2020). Evaluation of a gamification and flipped-classroom program used in teacher training: Perception of learning and outcome. *PLoS One*, *15*(7), e0236083. DOI: 10.1371/journal.pone.0236083 PMID: 32673373

Hamad, D., Dafaallah, A. S., & Alhaj, A. A. (2021). The challenges of online English language teaching from EFL instructors' point of view in KKU, Tehama campus during COVID-19 period. *Journal of Language Teaching and Research*, *12*(4), 543–556. DOI: 10.17507/jltr.1204.04

Hanson-Smith, E. (2016). Games, gaming, and gamification: Some aspects of motivation. *TESOL Journal*, *7*(2), 227–232. DOI: 10.1002/tesj.233

Holovatenko, T. (2023). Implementation of the principles of a trauma-informed environment in English lessons in elementary grades: a meta-analysis of foreign practices. *Pedagogical Education: Theory and Practice. Psychology. Pedagogy*, *39*(1), 80–86. DOI: 10.28925/2311-2409.2023.3912

Inayati, N., & Waloyo, A. A. (2022). The influence of Quizziz-online gamification on learning engagement and outcomes in online English language teaching. *Journal of English as a Foreign Language*, *12*(2), 249–271. DOI: 10.23971/jefl.v12i2.3546

Klaus, D. (2022). *Implementing a trauma-informed compassionate classroom to increase student engagement and academic achievement*. ProQuest LLC.

Krashen, S. D. (1995). *Principles and practice in second language acquisition*. Phoenix ELT incorporating Prentice Hall Macmillan.

Lamb, M. (2017). The motivational dimension of language teaching. *Language Teaching*, *50*(3), 301–346. DOI: 10.1017/S0261444817000088

Mast, E. (2017). *The gamification of an online English composition course: An action research exploration of student motivation and engagement*. ProQuest Dissertations & Theses.

Medina-Gómez, D. C., Cely-Betancourt, B. L., & Abreus-González, A. (2023). Octalysis framework: A favourable environment to increase successful practices in English as a foreign language (EFL) classrooms. *Revista Metropolitana de Ciencias Aplicadas*, *6*, 70–78. DOI: 10.61707/p9156r22

Mee, M., Shahdan, R., Ismail, T., Ghani, M., Pek, K., Von, L., & Rao, R. (2020). Role of gamification in classroom teaching: Pre-service teachers' view. *International Journal of Evaluation and Research in Education*, *9*(3), 1–7. DOI: 10.11591/ijere.v9i3.20622

Meşe, E., & Sevilen, S. (2021). Factors influencing EFL students' motivation in online learning: A qualitative case study. *Journal of Educational Technology and Online Learning*, *4*, 11–22. https://dergipark.org.tr/en/pub/jetol/issue/60134/817680

Nadif, B., & Bidari, S. (2023). Advanced tools in the EFL classroom: Friend or foe for novice teachers. *The Journal of Quality in Education*, *13*(21), 12–20. DOI: 10.37870/joqie.v13i21.336

Panagiotidis, P., Krystalli, P., & Arvanitis, P. (2023). Technology as a motivational factor in foreign language learning. *European Journal of Education*, *6*(1), 69–84. DOI: 10.2478/ejed-2023-0007

Peterson, M., & Jabbari, N. (2023). *Digital games in language learning: Case studies and applications* (New directions in computer assisted language learning). Routledge.

Redlener, I., Dudashvili, K., Viontsek, O., Grant, R., & Roca, E. (2025). Intervention for Psychological Trauma in Children Impacted by War in Ukraine. *JAMA Network Open*, *8*(3), e253057. DOI: 10.1001/jamanetworkopen.2025.3057 PMID: 40085087

Samson, S., & Karthiga, D. S. V. (2020). Gamification as a tool in English language teaching. *International Journal of Early Childhood Special Education*, *12*(2), 99–102. DOI: 10.9756/INT-JECSE/V12I2.201061

SaveSchools.in.ua. (2025). Education under attack: Damage to Ukrainian educational institutions. https://saveschools.in.ua/en/

Şendurur, P., & Sendurur, E. (2022). Students as gamers: Design, code, and play. In Lane, C. (Ed.), *Handbook of research on acquiring 21st century literacy skills through game-based learning* (pp. 868–887). IGI Global., DOI: 10.4018/978-1-7998-7271-9.ch044

Sulispera, T., & Recard, M. (2020). Octalysis gamification framework for enhancing students' engagement in language learning. *Jurnal Dialektika Program Studi Pendidikan Bahasa Inggris, 8*, 103–128. https://scispace.com/pdf/octalysis-gamification-framework-for-enhancing-students-1wmk7scims.pdf

Szabó, F., & Kopinska, M. (2023). Gamification in foreign language teaching: A conceptual introduction. *Hungarian Educational Research Journal*, *13*(3), 1–11. DOI: 10.1556/063.2023.00202

Teng, M. F., Wang, C., & Wu, J. G. (2021). Metacognitive Strategies, Language Learning Motivation, Self-Efficacy Belief, and English Achievement During Remote Learning: A Structural Equation Modelling Approach. *RELC Journal*, 54(3), 648-666. (Original work published 2023).DOI: 10.1177/00336882211040268

Tobias, S., Fletcher, J. D., & Wind, A. P. (2014). Game-Based Learning. In Spector, J., Merrill, M., Elen, J., & Bishop, M. (Eds.), *Handbook of Research on Educational Communications and Technology*. Springer., DOI: 10.1007/978-1-4614-3185-5_38

Venet, S. (2021). *Equity-centered trauma-informed education* (Equity and Social Justice in Education). WW Norton & Company.

Wulantari, N. P., Rachman, A., Sari, M. N., Uktolseja, L. J., & Rofi'i, A. (2023). The role of gamification in English language teaching: A literature review. *Journal of Education*, *6*, 2847–2856. DOI: 10.31004/joe.v6i1.3328

Yu, Z. (2019). A meta-analysis of use of serious games in education over a decade. *International Journal of Computer Games Technology, 2019. International Journal of Computer Games Technology*, *4797032*, 1–8. DOI: 10.1155/2019/4797032

Zhao, X., & McClure, C. D. (2022). Gather.Town: A Gamification Tool to Promote Engagement and Establish Online Learning Communities for Language Learners. *RELC Journal, 55*(1), 240-245. https://doi.org/ (Original work published 2024)DOI: 10.1177/00336882221097216

Zhou, J., & Ao, L. (2020). Digital gamification in language teaching. *Dialog on Language Instruction*, *30*, 25–40.

ADDITIONAL READING

Holovatenko, T. (2023). *Pre-service teachers' perceptions on implementing the trauma-informed approach in educational institutions*. In *Proceedings of the 4th International Conference on History, Theory and Methodology of Learning (ICHTML)* (pp. 130–137). SciTePress. https://doi.org/DOI: 10.5220/0012648000003737

Holovatenko, T. (2023). Enhancing elementary classroom learning experiences with mobile learning: implications for practice. In Parks, M. (Ed.), *Impactful Classroom Experiences in Elementary Schools: Practices and Policies* (pp. 25–49). IGI Global., DOI: 10.4018/978-1-6684-8551-4.ch002

Holovatenko, T. (2025). Rethinking Competences for Initial Language Teacher Training Through Trauma-Informed Lens. In Rahmat, M., & Osborne, A. (Eds.), *Pedagogies of Compassion and Care in Education* (pp. 337–370). IGI Global Scientific Publishing., DOI: 10.4018/979-8-3693-3613-7.ch015

Hooper, A., Hyder, M., Colclough, T., & Mann, D. (2023). Re- Envisioning Learning through a Trauma- Informed Lens: Empowering Students in their Personal and Academic Growth. *The Journal of Scholarship of Teaching and Learning*, *23*(3). Advance online publication. DOI: 10.14434/josotl.v23i3.35234

Ponomarenko, T., Proshkin, V., Shynkar, T., & Holovatenko, T. (2025). Use of Media in Preschool Education: A Case Study from Ukraine. In Manditereza, B. (Ed.), *Digital Pedagogy in Early Childhood Language Development* (pp. 93–114). IGI Global Scientific Publishing., DOI: 10.4018/979-8-3693-7437-5.ch005

Reinhardt, J. (2019). *Gameful second and foreign Language teaching and learning : theory, research, and practice. Palgrave Macmillan. Forbes, H., & Maki, D. (2020). Classroom180: A Framework for Creating, Sustaining, and Assessing the Trauma- Informed Classroom*. Beyond Consequences Institute.

Taqdir, T. (2024). Global Trends and Impacts of Gamification in Language Learning: A Bibliometric Analysis. *Lingua Cultura*, *18*(2), 205–215. DOI: 10.21512/lc.v18i2.12338

KEY TERMS AND DEFINITIONS

Digital Gamification: The application of game-design elements and game mechanics to non-game contexts such as education, using digital tools and platforms to increase learner engagement, motivation, and achievement in instructional settings.

Octalysis Framework: A comprehensive gamification design framework developed by Yu-kai Chou comprising eight core drives that motivate human behavior: Epic Meaning and Calling, Development and Accomplishment, Empowerment of Creativity and Feedback, Ownership and Possession, Social Influence and Relatedness, Scarcity and Impatience, Unpredictability and Curiosity, and Loss and Avoidance.

Pre-service Elementary School Language Teacher Training: Formal training of student teachers at ISCED 6 and/or ISCED 7 level of education to become language teachers in elementary schools.

Trauma- Informed Pedagogy: A pedagogic approach grounded in the recognition of the prevalence of traumatic events and their influence on learners, aimed at assisting instructors in establishing an inclusive environment that caters to the diverse needs of learners with various traumatic backgrounds.

Chapter 8
Immersive Language Learning:
Exploring the Role of AR and VR in Education

Laylo Rakhimova
Urgench State University, Uzbekistan

Zebo Saidjanova
Mamun University, Uzbekistan

Yunusjon Khakkulov
Termez University of Economics and Service, Uzbekistan

Munir Ahmad
https://orcid.org/0000-0003-4836-6151
Survey of Pakistan, Pakistan

ABSTRACT

This chapter examines the transformative potential of Augmented Reality (AR) and Virtual Reality (VR) in immersive language learning. By enabling experiential, context-rich, and culturally embedded interactions, AR/VR redefine traditional pedagogy and foster greater engagement, retention, and communicative competence. Grounded in experiential learning and situated cognition, these technologies support diverse learning styles and personalized pathways. Despite challenges in cost, accessibility, and teacher readiness, emerging innovations such as mobile-based VR, browser-native platforms, and AI-driven tutors are expanding access. The chapter envisions a future where immersive tools democratize language education, making it more inclusive, global, and learner-centered.

DOI: 10.4018/979-8-3373-3366-3.ch008

INTRODUCTION

Rote learning, drilling from a book, grammar teaching in the classroom, and similar methodologies have been dominant in the classroom. These mechanisms may be able to develop basic knowledge, but they do not usually develop communication skills of fluency in real life, confidence in conversation, and cultural competence. Most of the learners have difficulties in applying the language skills they have learned in the classroom to real-life settings, particularly in spontaneous language, pronunciation of the language in a natural environment, or the cultural aspects of communication. Such breaks in theoretical knowledge and the real-life use of language may negatively affect the achievement of long-term memory and even motivation among learners (Ghosheh Wahbeh et al., 2021; Illés & Akcan, 2017; Shadiev et al., 2018).

As a response to such shortcomings, immersive learning has emerged as a potential learning paradigm. In the method of immersive learning, the learners are put in a very interactive and real-life situation to get exposed to the concept and apply knowledge in a practical context. In this context, Augmented Reality (AR) and Virtual Reality (VR) have become popular and are now recognized as powerful tools that may create the same immersive exposure that classical classrooms cannot emulate. AR is a digital overlay added onto the physical world and may include things like visual overlays, text prompts, or auditory cues (Arena et al., 2022; Kamińska et al., 2023; Rauschnabel et al., 2022; Vasarainen et al., 2021), whereas VR immerses the learners within an entirely virtual, usually life-like environment (Hamad & Jia, 2022; Wu et al., 2023). Both technologies make it possible to engage in contextual, experiential learning, so that learners are able to play with and engage with language in ways that it is used in authentic situations, and that it has meaning (Al-Ansi et al., 2023; Chen et al., 2022; Ikeda et al., 2008; Kovalchuk et al., 2020).

The chapter examines the application of AR and VR to revolutionize language education and provide a dynamic, personified, and culture-integrated learning experience. The positioning of learners in online markets, in the city, cultural festivals, or history discussions, means that these technologies have enabled language to become learned not as an idea or knowledge, but as an action too. This hands-on experience can help increase the size of working vocabulary, raise conversational proficiency, and develop cultural intelligence in some ways; traditional training is unable to do so. Synthesizing pedagogical analysis and exploring technological opportunities, as well as examining case examples of their application in practice, the chapter introduces AR and VR as something more than a tool: the catalyst of the change on the path of language learning progress.

FOUNDATIONS OF AR AND VR IN EDUCATION

Two immersive technologies are changing the face of education, VR and AR, which offer fresh means of stimulating learners beyond the constraints of traditional teaching methods. Albeit via radically different technologies, they are both great opportunities to change the very nature of knowledge delivery, exposure, and recollection, especially within language learning. The similarity of approach lies in there, in a way, converging in educational practice being driven by the capacity to provide learners with an opportunity to be brought into a media-rich interactive experience where abstract becomes concrete and language is a lived experience (Al-Ansi et al., 2023; Haoming & Wei, 2024; Tan et al., 2022; Zhao et al., 2023).

Functionally, AR displays digital content on a real-life visual world, which could be text, images, audio, video, or 3D animation, via a smartphone, tablet, or AR glasses. In contrast to a completely immersive VR, AR supplements the real world, not replacing it, and only covers it with superimposed layers of information and activities that the learner can use in real-time. In such integration of real and virtual, learners can be grounded in a situational context that they are used to and at the same time be subjected to language-relevant experiences, enriched input. As a case in point, learners learning a foreign language would simply aim their devices at ordinary items in their classrooms or their environment and observe that labels, translations, or pronunciation tables appear on the screens. Higher-end applications include an AR-enhanced textbook that may be made to jump alive with 3D avatars reading dialogues, or a museum tour that offers a cultural context and a vocabulary supplementation in pop-up notes. Dispositions within these types of AR usages help situated learning, meaning language isn't observed as memory blocks of words but, rather, used as labels, directions, or conversations made native into the actual environment. The learner is both cognitively and physically in his or her immediate surroundings, thus creating good links between the target language and its real-life associations (Buchner & Kerres, 2023; Chang et al., 2022; Garzón, 2021).

VR, in its turn, provides a fully immersive experience, moving users into the simulated environments (either modeling the real ones or being exclusively invented to teach). With the help of VR headsets, such as Oculus Rift or HTC Vive, a person can hear and see a new world and use multiple senses to immerse themselves in it and explore the space around them. The implications of this complete immersion in terms of language learning are extremely strong. In such a lesson, delivered in a VR setting, students can be put into a virtual restaurant in Tokyo, with all the background sounds that can be heard in such a place, NPC (non-player character) interactions, and culturally appropriate visual triggers. They may be encouraged to talk to a virtual waiter, order something out of a Japanese-based menu, and react to social norms, all three of those practices involving the use of the related words and phrases. During this time, the learning/experience divide is lost. VR has the potential to recreate situations of high stakes or deep emotional appeal, such as job interviews, crossing a border, or working on a team in a new workplace, that cannot be generated in a classroom at all, much less safely. Through such simulations, it is possible to have profound, embodied practice that not only reiterates language use but also adds cultural fluency and interpersonal communication skills (Marougkas et al., 2023; Pirker & Dengel, 2021; Rojas-Sánchez et al., 2023; Rosendahl & Wagner, 2024).

Notably, both AR and VR promote active and contextual learning in which learners are not passive consumers of information but valued participants in an endo-languo state. AR supplements the here and now with contextual prompts, multiple simultaneous significances, and therefore it is the ideal means of reinforcing vocabulary, pronunciation, and understanding in the moment of the action. VR, in contrast, allows experiencing storyline-focused or activity-based situations that need repetitive use of language under the conditions of complete immersion. Nonetheless, both technologies promote language interactions in socially significant, emotionally relevant, and experienced contexts, which contribute to memory, motivation, and competence required to communicate in a social context.

Multimodal Learning

Irrespective of the technological differences between AR and VR, these approaches are pedagogically intersecting in their support of multimodal learning, which is one of the core properties of successful language acquisition. Multimodal learning is a process involving the use of a variety of senses, such as the visual, auditory, and kinesthetic senses, which enable the learner to drink and digest information using a more colorful fabric of experience. In language learning, this is particularly important because it is only through engagement with language that is situated, embodied, with language that becomes contextual, situated interaction that understanding and competence in language gradually occur, not as the result of drilling vocabulary or grammar. Through their ability to provide a layer upon layer of sensory-rich exposure to language, both AR and VR can help learners internalize linguistic forms more easily on more natural, deeper levels, and help them approach more closely the natural process of language acquisition in natural contexts (Martin et al., 2022; Moesl et al., 2023; Xu et al., 2023; Yuan et al., 2025).

The physical world may often be the foundation of this multimodality in the situation with AR. For example, AR apps can overlay vocabulary labels on real-world items or give immediate audio feedback when a student says a word. These types of interaction enable the learners to establish a close and significant association between language and objects that they see, touch, or handle. The matching of words to what they touch, a sense experience, like when shown a chair and saying to them "la silla" in Spanish, makes the memory stick and develop an intuitive perception. AR can also react to the input of learners instantly, allow personal time customization, and interchange interactive. Students who move about the streets with the help of AR apps that translate what the street signs or restaurant menu say, or information placards, are not merely reading text; they are encountering language through visual and social contexts with which it is meant to be used.

By contrast, VR consists of a more engaging and typically fanciful multimodal experience. VR enables learners to be in fully simulated environments, therefore, living in linguistic contexts that are spatially and socially rich. A Mandarin student can become a participant of a busy virtual station in Beijing, where he would have to read the directions, request directions, and respond to instructions all in the target language. This kind of learning appeals to both the ears and the eyes, but also demands the involvement of the senses of kinesthetics. The user has to look around, approach a kiosk, or touch the right set of words either by gestures or by speaking. Such actions that are related to real-life situations stimulate memory pathways, which enhance retention and fluency. In addition, the ability to reproduce cultural space in VR (and vice versa) by including gestures, rules of etiquette, non-verbal communication, and other requirements provides learners a unique socio-linguistic experience that would hardly be possible inside a regular classroom.

AR and VR are complementary to each other. AR augments the real-world learning experience by introducing language learning directly to the immediate environment of the learner, providing direct reinforcement, and in real-time context walking. VR, in its turn, constructs artificial and yet realistic environments, in which an excess of interaction and immersion in the story can help the learners to master communicative skills and cultural familiarity (Anil & Batdi, 2023). Pedagogical synergy is epitomized in these modalities so that learners are no longer learning with regard to language, but rather they are living in it, absorbing it, and expressing it. These types of learning spaces can also be said to be in line with constructivist and experiential learning theories, which have focused on the effectiveness of learning through doing and on the educational utility of connected senses. Table 1 shows the spectrum of multimodal engagement in language learning across AR and VR environments, showing how each technology supports visual, auditory, and kinesthetic modalities through distinct mechanisms.

Table 1. Multimodal engagement spectrum in AR and VR language learning

Modality	Augmented Reality (AR)	Virtual Reality (VR)
Visual	Overlays on real-world objects, translations, and icons	Immersive virtual scenes, spatial cues, and simulated signage
Auditory	Pronunciation cues, language prompts, and voice-activated input	Ambient sounds, interactive conversations, and spoken dialogue
Kinesthetic	Tapping, scanning, and moving around real space	Navigating virtual spaces, performing actions in simulations

Experiential Learning and Situated Cognition

The effectiveness of pedagogical use of AR and VR in language learning can be explained by the two influential theories of education: the theory of experiential learning and situated cognition. These models are contrary to the conventional didactic ways of teaching because they focus on teaching being lively, participatory, and embedded in some context. In each of these theories, environments are encouraged in which learners are not recipients of information but constructors of their knowledge, physically, socially, and cognitively, through meaningful engagements with people, objects, tasks, and environments that have a series of goals. Such technologies, such as AR and VR, by their very nature, are perfectly tailored to achieving this goal and allow the creation of language acquisition experiences that would react to how language is naturally learned and engaged with (Cakmakci et al., 2025; Lebert & Vilarroya, 2024; Lui & Goel, 2022).

The experiential learning theory, perhaps best postulated by David Kolb (Wijnen-Meijer et al., 2022), suggests that knowledge can be built up cyclically, building on a concrete experience, reflecting on that concrete experience, conceptualizing after relating it to abstractions, and experimenting to put those abstractions to the test. These phases are dynamic and develop when the learners are exposed to immersive environments, in the case of AR and VR. For example, a learner participating in a foreign language simulation based on VR can be placed in a virtual supermarket in which he or she would have to navigate through the market, communicate with clerks, and inquire about different items by using the target language. This is the concrete experience. The learner can make reflections about their performance, know the communication gaps, and provide feedback after the interaction, and reflective observation can be facilitated. Later, the learning process can be followed by the module introducing grammar rules or specific vocabulary to achieve deeper comprehension, which is a form of abstract conceptualization. The last possibility is the learner using this honed knowledge in another VR scenario or a real-world application, then active experimentation helps in solidifying the learning cycle and extending it. The immersive behavior of AR and VR brings down the gap between theory and practice by facilitating quick and significant connectedness amidst these states. It has a positive effect in making encoding of memories stronger and language retention higher than through rote approaches because learners associate linguistic information with gastronomic and affective states.

To supplement this is situated cognition theory, which asserts that learning is always embedded in both social and physical contexts in which the actual learning takes place. Language, thus looked at, is not only a collection of symbols or rules decontextualized but a means of interaction with the social world. This type of contextual immersion is possible with the help of AR and VR environments, placing learners in an environment of realistic linguistic and cultural situations. Take the case of a student having a VR simulation of a local cafe in Paris: the learner can learn how to order food there in French, but that is not all. He or she has also learnt cultural norms of politeness, turn-taking in conversations, and the spaces of cafes in the Francophone cultures. This very stratified learning of vocabulary, pragmatics, and social indicators coupled in the real-life setting is not easy, not to mention impossible, to recreate in traditional classrooms. VR and AR provide the ability to experience the language not only as a cognitive game, but as lived communication and dramatically strengthen the bond between linguistic forms and their use in real-life communication situations. Table 2 shows the comparison of the pedagogical affordances of Augmented and Virtual Reality in immersive language learning, highlighting their complementary strengths and educational applications.

Table 2. Comparative pedagogical affordances of AR vs. VR in language learning

Feature/Aspect	Augmented Reality (AR)	Virtual Reality (VR)
Environment	Real world with digital overlays	Fully simulated virtual world
Contextual Use	Enhances physical surroundings (e.g., object labeling)	Recreates authentic scenarios (e.g., virtual cafés)
Learning Style Support	Visual, auditory, context-sensitive	Immersive, experiential, emotionally engaging
Typical Applications	Cultural tours, vocabulary reinforcement	Role-play, conversation simulations, storytelling
Degree of Immersion	Partial (augments reality)	Full (immerses the user in a new environment)
Device Requirement	Smartphone, tablet, AR glasses	VR headset (e.g., Oculus, HTC Vive)

The strength of these theoretical underpinnings is the fact that they all focus on authentic and meaningful interaction. AR and VR do not just offer contents that need to be memorized--rather, they immerse the learners in stories, activities, and cultural environments where they have to use language to take part. They get their motivation, cognitive, and social advantages of experiential and situated learning through this. This leads to increased engagement of the learners, greater confidence in communicating, and more permanent and deeper knowledge retention. Figure 1 presents a conceptual model illustrating how AR and VR support experiential learning through active, reflective, and contextual engagement, while simultaneously embedding knowledge within socially and culturally relevant scenarios as posited by situated cognition.

Figure 1. Experiential and situated learning model in immersive language environments

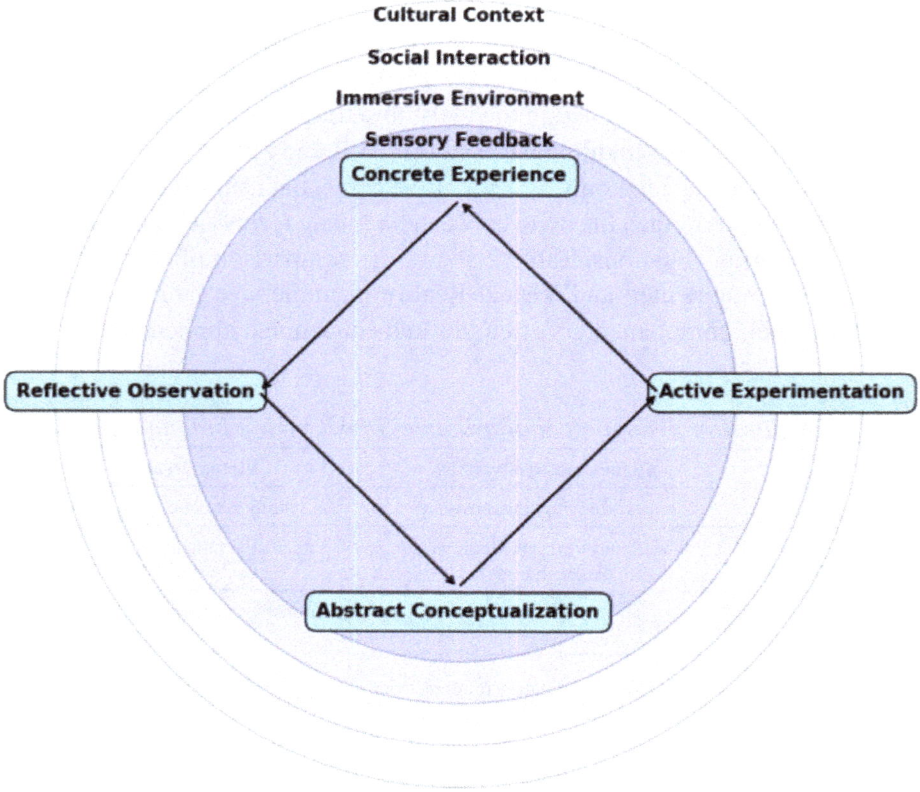

IMMERSIVE SCENARIOS IN LANGUAGE LEARNING

The most revolutionary impact of AR and VR on language learning is how the technology can bring forth a highly immersive and contextually sensitive experience, which simulates or even reinvents the real-world setting of communication. Not only do such scenarios give learners more repeated practice in using language, but they also cultivate their understanding of the culture and suit them to a higher level of pragmatic fluency. AR and VR applications create conditions in which language is an innate part of an environment, and with this shift, it changes the learning process to an active, task-oriented process. The fact that these scenarios are experienced promotes learning of vocabulary, syntax, and pronunciation, as well as exposing the learners to the socio-cultural laws that apply to authentic language use (Hayes et al., 2021; Lou, 2025; Shukla et al., 2023; Weng et al., 2024).

A popular example of this immersive possibility is apparent in the virtual marketplaces where learners get to practice the transactional language in a simulation that resembles real-life shopping or bargaining situations. Within a VR environment, a learner could explore a virtual Moroccan street market or a Japanese convenience store, deal with the seller, compare the goods, request them, and bargain in a second language. These activities introduce the learner to a practical lexicon that includes numbers, currency, adjectives describing size and color, and polite request forms in an emotionally understandable situation. Virtual transactions also differ from textbook drills in the sense that they allow learners to internalize language as a means of accomplishing communicative success and do not present vocabulary out of context, as do textbook drills. The increased time stress, environmental factors, and differentiation in the reactions of the interlocutors can mimic the unpredictability of real-life situations and assist in the development of fluency and adaptability. Figure 2 provides core immersive language learning scenarios enabled by AR and VR, demonstrating how virtual marketplaces, cultural overlays, and avatar-based dialogues support contextual, interactive, and culturally embedded language acquisition.

Figure 2. Types of immersive language learning scenarios in AR and VR

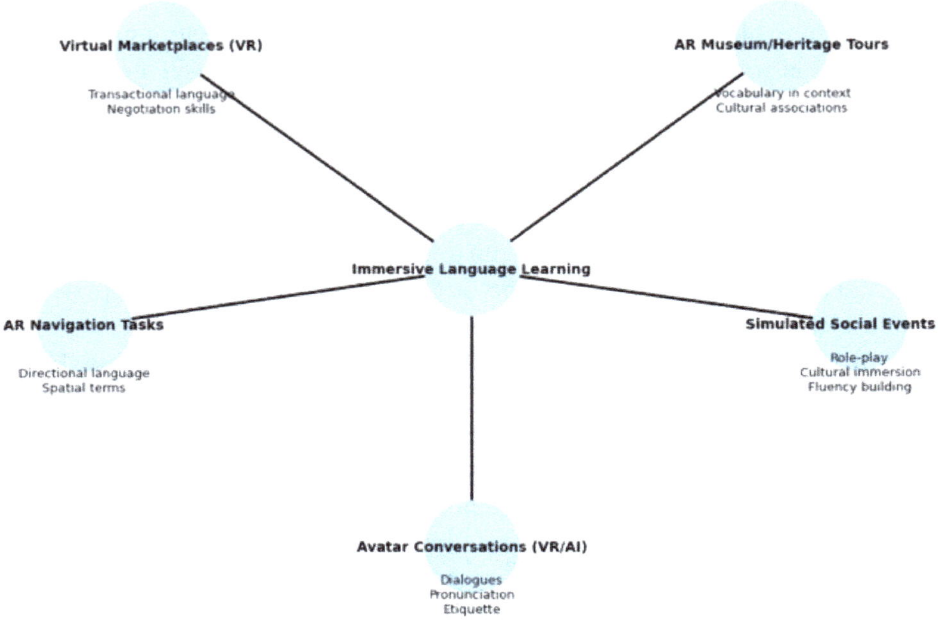

Augmented Reality can be an equally effective avenue to immersive education with its coating of language on top of culturally enriched settings. The overlays placed in museum or historical site environments can turn an ordinary visit into a multisensory linguistic experience. Shining a smartphone or AR headset at a historical artifact or work of art can enable the learner to read vocabulary labels, interpretive and cultural notes, all in the target language. The spatial, visual memory is used in such a type of learning, and words and other concepts are also encoded along with the physical places and sensory experiences. When one views a virtual gallery with AR that has a label of the French word le tableau, the word has become concrete through its visual and cultural context. A great deal of retention and understanding is enhanced by enhancing it with contextual encoding. What is more, AR can provide users with an interactive task that has to be done along the way, like solving language tasks at each of the stops; thus, vocabulary acquisition becomes purposeful and emotionally appealing.

The other major way of application is the simulation of conversations through avatars and cultural immersion activities that simulate social interactions within a range of environments. By means of VR or AI-based settings, students may interact with computerized characters who are programmed to mimic naturalistic speech, physiognomy, and cultural politeness. These simulations may occur in a park with a casual conversation, interviews, traveling, or even attending virtual cultural festivals. In another case, the learner may visit a virtual Spanish host family in one of the villages in the countryside, learn to cook alongside the family members, and discuss traditions and food habits. Those experiences have twofold use: they introduce learners to functional language use in various registers and make them better acquainted with the values, rituals, and social dynamics of the target culture. Moreover, they permit low-stakes repetition of social interaction, which eliminates much of the fear that speaking a foreign language in a real-world environment entail. Table 3 presents key affordances of AR and VR immersive scenarios in language learning, comparing how each technology contributes to vocabulary development, fluency, and cultural competence.

Table 3. Comparative affordances of AR and VR scenarios in language learning

Scenario Type	Augmented Reality (AR)	Virtual Reality (VR)
Transactional Practice	Moderate – via overlays	Highly simulated markets
Vocabulary Acquisition	High–object labeling	High–contextual exposure
Cultural Immersion	Moderate – museum guides	High festivals, customs
Conversational Fluency	Low – limited dialogues	High–avatar interactions
Real-time Feedback	High–AR–based cues	Moderate – AI-driven input
Mobility and Accessibility	High – via smartphones	Low – requires headset

In all these settings, in commercial, cultural, as well as conversational, the boat-rope of these is genuineness and relevance. Learners are now not so much practicing out-of-context phrases but are now engaged in an experience where language is required in order to play a part, to overcome problems, or to interact with others on a meaningful level. Such a strategy coincides with the principles of communicative language teaching and task-based language learning, where the real-world application, meaningful interaction, and purposeful use of language matter. By incorporating AR and VR into these kinds of immersive experiences, educators can help increase the vocabulary of their learners, their oral performance, as well as their cultural proficiency, all in a way that would be attention-catching and technologically rich.

LEARNING GAINS AND PEDAGOGICAL IMPACTS

The engagement of AR or VR in the setting of language learning is not only a tech upgrade, but it is a pedagogical shift. This means that they can show real improvements in language acquisition and user experience because of the surroundings these tools immerse the learner, which is always multi-sensory, active, and situated in a context of meaning. Increased interaction and motivation to augmented retention and personalized learning tracks, AR and VR are transforming not only the student experience of learning a foreign language, but also the design of teaching processes by educators. Both their pedagogical influence and their corresponding success in real-life case implementations are increasingly backed up by empirical evidence, which proves that immersive learning environments are often more effective than traditional learning methodologies in a number of ways (Cai et al., 2020; Nikimaleki & Rahimi, 2022).

The most frequent and observable effect of immersive technologies is that they provide a significant boost in learner engagement and motivation. Ideally, language learning involves persistent perseverance, repetition, and the ability to take communicative risks, and therefore may be difficult to perform in the traditional classroom. In VR and AR experiences, however, a feeling of presence, novelty, and emotionality is introduced to the learning process. The learning environment becomes task- and purpose-oriented when learners have to navigate a virtual restaurant, respond to an emergency call, and communicate with AI-powered characters in the form of avatars. Such scenarios provide intrinsic motivation because of their ability to give relevance and purpose to the language that is being communicated at the moment. The consistent findings are that students are more satisfied and more willing to engage in AR/VR-based language tasks than when using other materials like a textbook or worksheets. One of the comparative studies found an increase in attendance and participation rates of high school students learning French in the case of VR simulations of Parisian neighborhoods compared with those of their peers learning using non-immersive digital tools.

In addition to engagement, AR and VR can create much more retention and understanding by involving multi-sensory input. In teaching a language, what was traditionally done was to take the linguistic elements, like vocabulary lists or grammar lessons, and to take them out of the sensory and contextual information that supports real-life communication. Immersive technologies right this disconnection by excising, helping in stimulating several sensory pathways at once. Visual learners respond well to space and object indicators, auditory learners get to experience real-time discussions, and the kinesthetic learners are activated by gestures, movements, and manipulation of the virtual objects. These redundant inputs are not repetitive but synergistic; the brain can record the vocabulary more efficiently when a word is presented along with an image, a sound, an emotion, or an action. In case of a learner practicing the word knead in a VR bakery, he/ she has a much higher chance of remembering the word owing to the physical experience of the task. This can be attributed to dual-coding theory, whose foundation is the fact that memory formation is enhanced when verbal and non-verbal representations are combined.

This flexibility of the AR and the VR environments also enable diversified learning experiences to address individual preferences of cognition and other senses. In one VR language use experience, the visual learner can emphasize on interpretation of environmental signs, the auditory learner can focus on the dialogues and surrounding sound effects, whereas the kinesthetic learner can find it more accessible to perform actions as semantically as assembling objects or negotiating the physical obstacles. On top of that, there is also the possibility to personalize these platforms in real-time. For example, apps that use the AR language may increase or decrease the difficulty level depending on the learner's performance, or propose a review to vocabulary items that were mostly identified incorrectly by the learner. Inclusivity and diverse learning profiles reflected in the needs of neurodiverse students are encouraged by such responsiveness to unique individuals. Table 4 presents a comparison of key pedagogical benefits provided by immersive technologies in contrast to traditional language learning methods.

Table 4. Pedagogical impacts of AR/VR in language learning

Pedagogical Dimension	Traditional Methods	AR/VR-Enhanced Learning
Engagement & Motivation	Often passive, limited by repetition and abstraction	High interactivity, goal-oriented tasks, novelty
Retention & Recall	Text-intensive, low sensory encoding	Multi-sensory reinforcement via audio-visual-kinesthetic input
Adaptability to Learning Styles	One-size-fits-all; limited support for diverse learners	Real-time adjustments based on input and learner style
Cultural & Contextual Learning	Decontextualized exposure through isolated examples	Realistic cultural immersion in dynamic environments
Communicative Practice	Controlled, scripted drills or dialogues	Spontaneous, authentic conversation simulations

Such pedagogical merits are also justified by an increasing amount of empirical evidence. Şimşek, (2024) compared preschoolers' story retelling and comprehension using printed versus augmented reality (AR) storybooks. Post-test results showed that children using AR-enhanced books significantly outperformed those with printed books, suggesting AR's potential to improve narrative skills and understanding. Similarly, Geisen et al., (2023) explored the use of a VR-based rotation task in an extracurricular school dance class to enhance didactic, collaborative, and perceptual aspects of dance training. Findings from student feedback and observational analysis highlight the potential of VR as an innovative and integrative tool in sports education settings. These results are not only anecdotal but statistically relevant and thus provide a profound indication of the learning outcomes offered by immersive environments that can be expounded upon longer and stronger. Figure 3 presents convergence of visual, auditory, and kinesthetic modalities, each reinforcing different aspects of comprehension and memory in immersive environments.

Figure 3. Multi-sensory learning in AR/VR environments

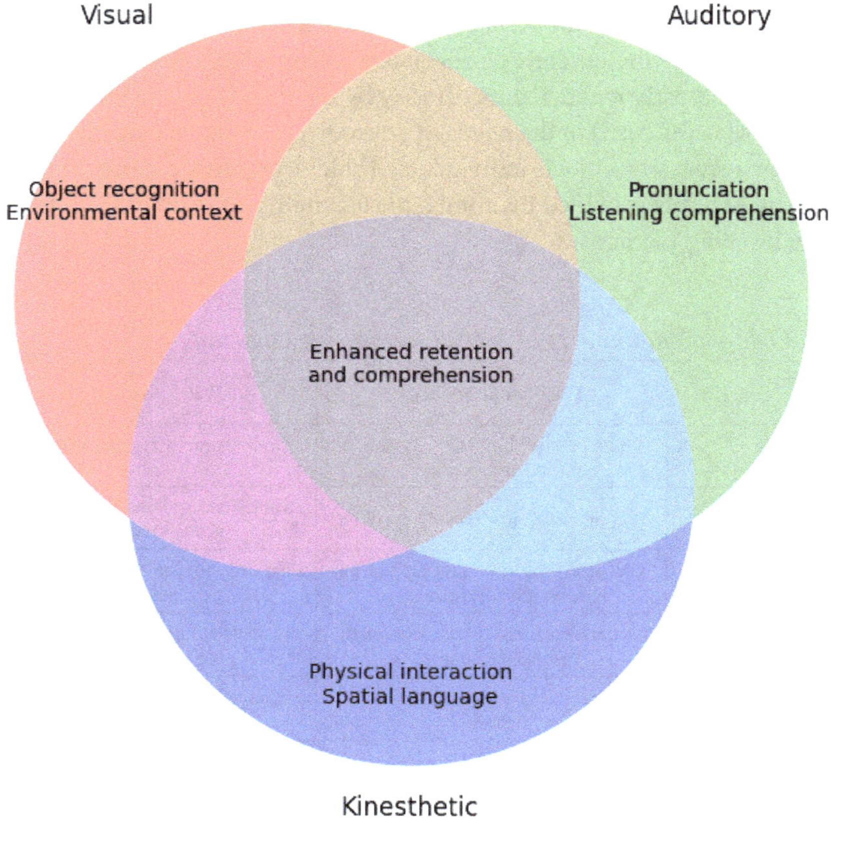

Summing it up, the pedagogical contribution of AR and VR to language teaching is both dimensional and full-fledged. These technologies foster a higher level of motivation by engaging participation, learning is more effective because of the integration of multiple senses, and instruction can be delivered at the learner level. Using more and more immersive technologies, educational institutions are not only modernizing their toolkits but moving toward a more responsive, inclusive, and efficient way of dealing with language learning altogether. Further study should persist in determining long-term educational achievement and principles of instructional design using AR/VR technologies, yet the current findings already indicate a paradigm shift that may have never been met by traditional methods as far as the rate of learning improvement.

CHALLENGES AND CONSIDERATIONS FOR IMPLEMENTATION

As the pedagogical value of AR and VR in language learning becomes increasingly evident, however, the road to full-scale implementation certainly has its twists and turns, especially in institutions that have limited funds to invest and knowledge to draw upon, or institutions with strict curricula to attend to. As is often the case in educational technology innovations, the step after demonstrating instructional utility is in ensuring that it can be practically and fairly, and in a sustainable way, implemented. Learning more about these impediments, as well as about the means of breaking them down, is a crucial step in working with immersive learning tools, as educators, administrators, and policymakers.

The main challenge is the budget associated with the implementation of AR/VR sceneries on a large level. Enterprise solutions like Meta Quest, HTC Vive, or Apple Vision Pro have high-ticket prices, and they require other hardware (PC, headsets, sensors, and accessory devices) to be compatible (Aros et al., 2025; Rocchi et al., 2025). Although AR experiences may work on smartphones or tablets, they still require an adequate internet connection, compatibility with the device, and regular software updates, which all may push the capabilities of schools in up-and-coming nations to the limit. Often, well-funded institutions also have logistical issues, either of equipping the classes with sufficient equipment or of dedicating space conducive to physical movements during immersion. Take the case of a university where there are 10 VR units to serve a class of 30 students as an example; this can limit the extent of access and also balkanize the course. In addition, technical problems, such as heated devices, program collapse, low Wi-Fi connection, and so on, may interrupt the learning process and cause disappointment to all involved in the process of learning process.

The other challenge that remains is the shortage of teacher training and the structural setup of the curriculum. The educators may not be familiar with AR or VR tools and feel uneasy or anxious about the technical nature of those tools. To the same extent that another technical staff exists, teachers might have issues mapping immersive content towards required learning outcomes or standardized testing styles. As another example, a VR tour of ancient Rome might be of interest to a teacher (to present the Latin language), but he/she might not be able to justify its use in a more traditional syllabus consisting of grammar drills and translation practice. Unless guided by specific training and presented with relevant materials, teachers are likely to use AR/VR learning only partially in their curricula or engage in shallow, one-time learning activities instead of embracing immersive learning and applying it to teach more comprehensively.

There is a significant challenge of accessibility and inclusion to the equitable implementation as well. There are chances that immersive technologies can end up excluding students with disabilities unless they are developed using universal access guidelines. Blind learners would not be able to learn anything useful because the world of VR is created using visual effects. Similarly, learners with ADHD or autism can also feel an overload in the over-stimulating premises unless the design contains a possibility of customized sensory input control. Devices and places to utilize AR/VR tools may not be readily available even when the tools are designed well. There are many communities where learners have to share mobile devices with other individuals in the family or do not have privacy at home, in which case, it is not possible to perform extended or asynchronous immersive practice. These disparities have the potential to jeopardize the inclusive nature of educational technology unless they are specifically overcome.

One aspect less written about but which needs to be mentioned is the issue of content relevance and localization. A large number of AR/VR applications are created in Western cultures and might not apply to the conditions of learners living in other places. An app to learn French developed in Paris can focus on city words and traditions that are foreign and not of interest to users in North Africa or in Southeast Asia. Unless the sociolinguistic background of immersive content adheres to their absorption by learners in their communication settings or routines, there is a hazard of disinterest or even bewilderment. This necessitates the desire to have culturally adaptive platforms that have customizable interactive material, which would allow local educators to redirect or edit immersive material to suit the environment of the learners.

Data privacy and ethical issues are becoming a very important consideration, especially when it comes to the integration of AI-based VR and cloud-based AR services. These systems can gather immense amounts of user information such as voice recordings, behaviour interactions, and biometric responses. In the absence of explicitly defined data governance policies, the privacy of students may be jeopardized. One such example would be an application used to record voice recordings to analyze the pronunciation that inadvertently captures the personal information that may be used to identify an individual. Developers and schools need to make sure that they meet ethical standards of data usage and provide them with opt-in/opt-out choices to learners and guardians.

The technical maintenance is a major challenge in the long-term sustainability of AR and VR technologies in learning institutions, besides being a costly endeavor and a pedagogical issue. Such a tool is not like conventional educational resources that would require regular updates, routine calibrations, and compatibility checks to run effectively. VR headsets can need the updating of firmware, adjusting of sensors, and battery management, whereas AR apps usually depend on the cell-specific features, including cameras, the processing power, and the versions of the operating system, which, in case they are outdated, may cause lags, app crashes, or inadequate performance. In addition, such technologies can be contingent on continuous access to the internet, cloud access, and access rights, and each institutional network may be configured differently. This can be very demanding, especially for schools and universities that do not have dedicated IT support teams. For example, a school that engages 20 VR headsets in language immersion applications can be exposed to downtimes in case of several uncharacterized and malfunctioning devices, which will amount to lost lesson plans and user frustrations. Educators can also become overwhelmed with them, having them fix them at the expense of classroom time when updates or installations break during live sessions. The effective running of immersive systems can be achieved only through investment in devices, technical staff, helpdesks, as well as simplified device maintenance processes, but these are the aspects that are frequently underinvested or at least not considered during the implementation process.

Despite such daunting obstacles, there are some encouraging trends in terms of technology and pedagogy that are setting the way for broader, more equitable, and more economical adoption of AR and VR in the language learning areas. All these new solutions help reduce the cost of entry and also solve a problem of scale, equity, and curriculum alignment, and therefore make use of immersive technologies an option even in more resource-limited educational contexts.

A remarkable trend is that of mobile-based VR and inexpensive AR applications, based on a consumer technology already in use, namely, the smartphone and tablet. Instead of using costly proprietary devices, now a great number of new tools can be used on Android and iOS-powered devices, which greatly increases their availability. They utilize device cameras and gyroscopic sensors to provide immersive experiences, which do not imply using specific VR headsets and spatial sensors. By way of example, user-interactive applications, such as Mondly VR, provide simulated conversations and travel scenarios using only a mobile phone and cheap cardboard viewer, whereas AR platform systems, such as Google Lens and Assemblr EDU, enable users to engage with vocabulary cards, cultural objects, and physical-location challenges using merely a phone camera. Such a movement to mobile accessibility implements an immersive language learning environment, a possibility no matter the socioeconomic background, which means it could be done in rural schools or other informal learning groups where advanced hardware may otherwise be unavailable.

At the same time, the VR web world is leveling the playing field by removing downloads and software requirements, even while specialized VR headsets remain popular and soaring in price. These platforms are web-based and can be hosted in the clouds so that learners can just plug into the virtual space using a URL and can do it from any web-connected device. As a language teacher, it creates a possibility to conduct real-time, scenario-based tasks, including virtual classroom discussions, culture tours, or even playing language games without significant investment in the new infrastructure. VR in the Web is also cross-platform friendly, so it can be utilized by hybrid classrooms or distant language exchange programs between learners in various countries.

On the other informative development is scalable modular AR content that can be readily integrated into customary learning environments. These applications are QR code-based, image recognition, or geolocation-driven to initiate bite-sized AR learning modules in textbooks, classroom materials, or items, or even field trips. As an example, a language learner who visits a historical site can scan a monument to activate an explanation of those vocabulary items, quizzes, or stories narrated in the target language, making passive observation active and contextual learning. Programs such as ZapWorks, BlippAR, and Metaverse Studio are being developed to enable teachers and even students to make their own AR projects with little or no programming knowledge (Belda-Medina, 2022). The localization, alignment, and promotion of student agency in creating materials by drawing on this democratization of content production means that curriculum integration is facilitated. Table 5 presents a comprehensive overview of the primary challenges to AR/VR adoption in language education, along with potential strategies to address them.

Table 5. Barriers to implementation and mitigation strategies

Challenge	Description	Potential Solution
Cost and Infrastructure	High hardware and setup costs; connectivity limitations	Mobile/web-based AR/VR; shared devices; cloud-based tools
Teacher Training	Lack of pedagogical training for immersive technologies	Targeted PD programs; integrated teacher toolkits
Curriculum Integration	Difficulty aligning immersive tasks with learning outcomes	Modular immersive content tied to curriculum standards
Accessibility and Inclusion	Barriers for students with disabilities or tech limitations	Multisensory design, custom controls, captions, and offline access
Content Localization	Culturally irrelevant or non-contextualized scenarios	Region-specific content creation; open authoring platforms
Ethical/Data Privacy Concerns	Collection of voice, biometric, and behavior data	Transparent data policies, opt-in consent, and GDPR compliance
Technical Maintenance	Need for updates, device management, and space adaptation	IT support, simple interface design, browser-based systems

On the teaching side, professional learning (PD) initiatives are also adapting to incorporate the use of immersive technologies with specific trainings designed to ensure that language teachers are prepared to use such technology in their very differentiated instruction. The initiatives go even further to provide more pedagogical support, focusing more on integration, content design, and assessment methods. As an example, such programs as EdTech Teacher Workshops or AR/VR in TESOL Training Modules provide scenario planning, analysis of case studies, and possibilities to allow teachers to draft and test AR/VR-infiltrated lessons. The training can become key not only to technical proficiency but also to an increase in teaching confidence to take on new roles as facilitators of interactive and immersive experiences focused on learners.

Notably, contemporary AR/VR platforms are increasingly designed with accessibility features in mind, which can ensure that such tools can be applied by all learners, even those with disabilities or neurodiverse conditions. Language learning tools are now more likely to feature real-time captions, text-to-speech translation, sign avatars, color-blind settings, and audio options to cater to a wider range of individuals with different wants and needs. The platforms have also provided some capabilities that enable the learners to have full control over themselves, depending on their sensory and cognitive load, where they can choose the speed of the avatar, the complexity of dialogues, and even the noise level in the surrounding environment. These affordances see the rise of a new understanding that inclusion has to be embedded at the ground level rather than an add-on.

All of these innovations are providing a more sustainable ecosystem of immersive language learning, one that is more accessible, adaptable, and scalable. Combining the low costs, scalability, modularity, teacher empowerment, and inclusive design, the practitioners are slowly transitioning the field beyond experimental pilots to realistic implementation. By working together, the education community can begin to integrate AR and VR in language teaching and learning, and make the application more than just a creative concept, but instead, a mainstream introduction that can enhance access to language and to cultures and broaden educational opportunities. Figure 4 shows a conceptual model showing the key components required for successful AR/VR integration in language education: aligning technology, pedagogy, accessibility, and institutional support.

Figure 4. AR/VR integration model in language education

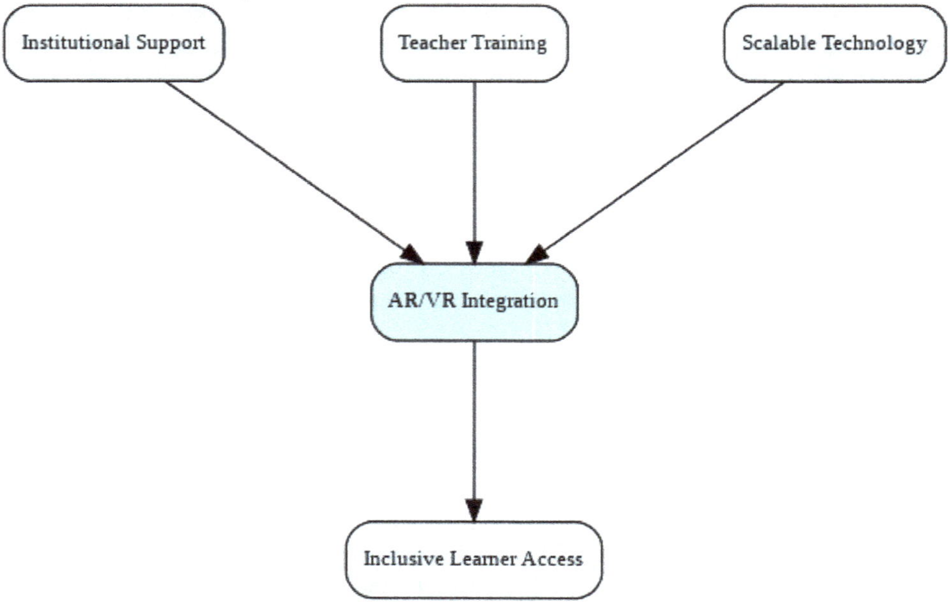

CONCLUSION

The chapter has addressed the language learning power and potential of AR and VR in immersive success, with a specific focus on the ability of the respective technologies to structure dynamic and environmentally rich conditions that reify a greater involvement in the process of language acquisition, better retention, and local practice of communication. As opposed to the teaching typically based on abstract rules and drills decontextualized far away in learning institutions, AR and VR provide experience-based ways of acquiring the language, i.e., the learners can study the language in how it is both lived, used, and comprehended by people in the actual social and cultural contexts. Virtual marketplaces, augmented reality-assisted museum visits, and avatar-enhanced simulations of dialogue are only some of the examples of immersive technologies that transform the way learners can interact with vocabulary, syntax, fluency, and pragmatics.

The learning assumptions of these technologies, which are predicated on experiential learning, situated cognition, and multimodal interaction, explain their capability to accommodate various personal learning requirements and audiences and improve motivation and memory. Nevertheless, the concept of implementing it is hampered by monetary, technical, curricular, and equity-oriented obstructions. Promisingly, evolutions like mobile-substrate VR, inherently browser-friendly languages, instructor-led training programmes, and accessible design practices are opening up access and viability faster and more than ever, and rendering immersive language learning more scalable and sustainable than ever before.

Moving into the future, future research and development should be on enhancing the intelligence and adaptability of such platforms. One potential future direction is the use of AI-powered virtual tutors that can provide real-time correction of speech, cultural feedback, and generation of personalized scenarios formed on the basis of the profile and state of the learner. Such smart agents might also enhance immersive experiences and hence be responsive and facilitating, particularly to autonomous or remote learners. Also, immersing multilingual learners into the same virtual world (in which individuals demonstrate language- and culturally-specific behavioral models) through immersive environments can also support the process of natural language acquisition and cross-cultural cooperation, as well as providing beneficial data to adaptive learning environments.

Still, the hybrid AR, VR, and AI in the end is about bringing a new era to language learning that is interactive, all-inclusive, and boundary-free. The increased availability and normalization of immersive technologies in educational systems across the globe are effectively poised to democratize language learning, enabling learners, no matter where they live or what their circumstances are, to be fully empowered members of the world in terms of making their communication meaningful and celebrated. With language being a competency as well as a bridge in the world, it is not only in the realms of immersive learning environments that the target would promise not just to ramp up linguistic competence but also promote empathy, cultural literacy, and global citizenship. Language learning does not only happen in a classroom or a book, but in the world of the virtual cafe, the virtual city street, and the interactive landscape of learning about another culture, where language is developed as an interactive tool rather than as an abstract concept.

REFERENCES

Al-Ansi, A. M., Jaboob, M., Garad, A., & Al-Ansi, A. (2023). Analyzing augmented reality (AR) and virtual reality (VR) recent development in education. In *Social Sciences and Humanities Open* (Vol. 8, Issue 1). DOI: 10.1016/j.ssaho.2023.100532

Anil, Ö., & Batdi, V. (2023). Use of augmented reality in science education: A mixed-methods research with the multi-complementary approach. *Education and Information Technologies*, *28*(5), 5147–5185. DOI: 10.1007/s10639-022-11398-6

Arena, F., Collotta, M., Pau, G., & Termine, F. (2022). An Overview of Augmented Reality. In *Computers* (Vol. 11, Issue 2). DOI: 10.3390/computers11020028

Aros, M., Arnold, H., & Chaparro, B. (2025). Exploring Device Usability: A Heuristic Evaluation of the Apple Vision Pro and Meta Quest 3. *Communications in Computer and Information Science, 2522 CCIS*, 3–12. DOI: 10.1007/978-3-031-94150-4_1

Belda-Medina, J. (2022). USING AUGMENTED REALITY (AR) AS AN AUTHORING TOOL IN EFL THROUGH MOBILE COMPUTER-SUPPORTED COLLABORATIVE LEARNING. *Teaching English with Technology*, *22*(2), 115–135.

Buchner, J., & Kerres, M. (2023). Media comparison studies dominate comparative research on augmented reality in education. *Computers & Education*, *195*, 104711. Advance online publication. DOI: 10.1016/j.compedu.2022.104711

Cai, S., Liu, E., Shen, Y., Liu, C., Li, S., & Shen, Y. (2020). Probability learning in mathematics using augmented reality: Impact on student's learning gains and attitudes. *Interactive Learning Environments*, *28*(5), 560–573. Advance online publication. DOI: 10.1080/10494820.2019.1696839

Cakmakci, G., Aydeniz, M., Brown, A., & Makokha, J. M. (2025). *Situated Cognition and Cognitive Apprenticeship Learning*. 293–311. DOI: 10.1007/978-3-031-81351-1_17

Chang, H. Y., Binali, T., Liang, J. C., Chiou, G. L., Cheng, K. H., Lee, S. W. Y., & Tsai, C. C. (2022). Ten years of augmented reality in education: A meta-analysis of (quasi-) experimental studies to investigate the impact. *Computers & Education*, *191*, 104641. Advance online publication. DOI: 10.1016/j.compedu.2022.104641

Chen, C., Wang, X. Y., Chen, J., Liu, Q., & Peng, S. L. (2022). An Active Security System Based on AR Smart Classes and Face Recognition Technology. *Wangji Wanglu Jishu Xuekan*, *23*(2), 245–253. Advance online publication. DOI: 10.53106/160792642022032302005

Garzón, J. (2021). An overview of twenty-five years of augmented reality in education. In *Multimodal Technologies and Interaction* (Vol. 5, Issue 7). DOI: 10.3390/mti5070037

Geisen, M., Fox, A., & Klatt, S. (2023). VR as an Innovative Learning Tool in Sports Education. *Applied Sciences 2023, Vol. 13, Page 2239, 13*(4), 2239. DOI: 10.3390/app13042239

Ghosheh Wahbeh, D., Najjar, E. A., Sartawi, A. F., Abuzant, M., & Daher, W. (2021). The role of project-based language learning in developing students' life skills. *Sustainability (Basel), 13*(12), 6518. Advance online publication. DOI: 10.3390/su13126518

Hamad, A., & Jia, B. (2022). How Virtual Reality Technology Has Changed Our Lives: An Overview of the Current and Potential Applications and Limitations. In *International Journal of Environmental Research and Public Health* (Vol. 19, Issue 18). DOI: 10.3390/ijerph191811278

Haoming, L., & Wei, W. (2024). A systematic review on vocabulary learning in AR and VR gamification context. *Computers & Education: X Reality, 4*, 100057. DOI: 10.1016/j.cexr.2024.100057

Hayes, A. T., Dhimolea, T. K., Meng, N., & Tesh, G. (2021). Levels of Immersion for Language Learning from 2D to Highly Immersive Interactive VR. In *Chinese Language Learning Sciences*. DOI: 10.1007/978-981-16-3416-1_4

Ikeda, S., Villagran, C. T., Fukuda, T., Okada, Y., Arai, F., Negoro, M., Hayakawa, M., & Takahashi, I. (2008). Patient-Specific IVR Endovascular Simulator with Augmented Reality for Medical Training and Robot Evaluation. *Journal of Robotics and Mechatronics, 20*(3), 441–448. Advance online publication. DOI: 10.20965/jrm.2008.p0441

Illés, É., & Akcan, S. (2017). Bringing real-life language use into EFL classrooms. *ELT Journal, 71*(1), 3–12. Advance online publication. DOI: 10.1093/elt/ccw049

Kamińska, D., Zwoliński, G., Laska-Leśniewicz, A., Raposo, R., Vairinhos, M., Pereira, E., Urem, F., Ljubić Hinić, M., Haamer, R. E., & Anbarjafari, G. (2023). Augmented Reality: Current and New Trends in Education. In *Electronics (Switzerland)* (Vol. 12, Issue 16). DOI: 10.3390/electronics12163531

Kovalchuk, O. I., Bondarenko, M. P., Okhrey, A. G., Prybytko, I. Y., & Reshetnyk, E. M. (2020). Features of using immersive technologies (virtual and augumented reality) in medical education and practice. *Morfolohiia, 14*(3), 158–164. Advance online publication. DOI: 10.26641/1997-9665.2020.3.158-164

Lebert, A., & Vilarroya, Ó. (2024). The links between experiential learning and 4E cognition. *Annals of the New York Academy of Sciences*, *1541*(1), 37–52. DOI: 10.1111/nyas.15238 PMID: 39383206

Lou, Y. (2025). The impact of virtual reality environments on English language acquisition: Innovative immersive learning technologies for communication skills development. *Journal of Computational Methods in Sciences and Engineering*. DOI: 10.1177/14727978251337950

Lui, T. W., & Goel, L. (2022). Learning effectiveness of 3D virtual reality in hospitality training: A situated cognitive perspective. *Journal of Hospitality and Tourism Technology*, *13*(3), 441–460. DOI: 10.1108/JHTT-03-2021-0091

Marougkas, A., Troussas, C., Krouska, A., & Sgouropoulou, C. (2023). Virtual Reality in Education: A Review of Learning Theories, Approaches and Methodologies for the Last Decade. In *Electronics (Switzerland)* (Vol. 12, Issue 13). DOI: 10.3390/electronics12132832

Martin, D., Malpica, S., Gutierrez, D., Masia, B., & Serrano, A. (2022). Multimodality in VR: A Survey. In *ACM Computing Surveys* (Vol. 54, Issue 10 s). DOI: 10.1145/3508361

Moesl, B., Schaffernak, H., Vorraber, W., Braunstingl, R., & Koglbauer, I. V. (2023). Multimodal Augmented Reality Applications for Training of Traffic Procedures in Aviation. *Multimodal Technologies and Interaction*, *7*(1), 3. Advance online publication. DOI: 10.3390/mti7010003

Nikimaleki, M., & Rahimi, M. (2022). Effects of a collaborative AR-enhanced learning environment on learning gains and technology implementation beliefs: Evidence from a graduate teacher training course. *Journal of Computer Assisted Learning*, *38*(3), 758–769. Advance online publication. DOI: 10.1111/jcal.12646

Pirker, J., & Dengel, A. (2021). The Potential of 360° Virtual Reality Videos and Real VR for Education - A Literature Review. In *IEEE Computer Graphics and Applications* (Vol. 41, Issue 4). DOI: 10.1109/MCG.2021.3067999

Rauschnabel, P. A., Babin, B. J., tom Dieck, M. C., Krey, N., & Jung, T. (2022). What is augmented reality marketing? Its definition, complexity, and future. In *Journal of Business Research* (Vol. 142). DOI: 10.1016/j.jbusres.2021.12.084

Rocchi, E., Ferrarotti, A., & Carli, M. (2025). A Comparison of the Meta Quest Pro and HTC Vive Focus 3 Eye-Tracking Systems: Analysis of Data Accuracy and Spatial Precision. *IEEE Access : Practical Innovations, Open Solutions*, *13*, 71995–72010. DOI: 10.1109/ACCESS.2025.3562672

Rojas-Sánchez, M. A., Palos-Sánchez, P. R., & Folgado-Fernández, J. A. (2023). Systematic literature review and bibliometric analysis on virtual reality and education. *Education and Information Technologies*, *28*(1), 155–192. Advance online publication. DOI: 10.1007/s10639-022-11167-5 PMID: 35789766

Rosendahl, P., & Wagner, I. (2024). 360° videos in education – A systematic literature review on application areas and future potentials. *Education and Information Technologies*, *29*(2), 1319–1355. Advance online publication. DOI: 10.1007/s10639-022-11549-9

Shadiev, R., Hwang, W. Y., Huang, Y. M., & Liu, T. Y. (2018). Facilitating application of language skills in authentic environments with a mobile learning system. *Journal of Computer Assisted Learning*, *34*(1), 42–52. Advance online publication. DOI: 10.1111/jcal.12212

Shukla, T., Duthade, A., & Deshbhratar, N. (2023). Improving Kids English Vocabulary Through Immersive Mobile Games. *INTERANTIONAL JOURNAL OF SCIENTIFIC RESEARCH IN ENGINEERING AND MANAGEMENT*, *07*(10), 1–11. Advance online publication. DOI: 10.55041/IJSREM26473

Şimşek, E. E. (2024). The effect of augmented reality storybooks on the story comprehension and retelling of preschool children. *Frontiers in Psychology*, *15*, 1459264. DOI: 10.3389/fpsyg.2024.1459264 PMID: 39439758

Tan, Y., Xu, W., Li, S., & Chen, K. (2022). Augmented and Virtual Reality (AR/VR) for Education and Training in the AEC Industry: A Systematic Review of Research and Applications. In *Buildings* (Vol. 12, Issue 10). DOI: 10.3390/buildings12101529

Vasarainen, M., Paavola, S., & Vetoshkina, L. (2021). A Systematic Literature Review on Extended Reality: Virtual, Augmented and Mixed Reality in Working Life. *Internationa Journal of Virtual Reality*, *21*(2), 1–28. DOI: 10.20870/IJVR.2021.21.2.4620

Weng, Y., Schmidt, M., Huang, W., & Hao, Y. (2024). The effectiveness of immersive learning technologies in K-12 English as second language learning: A systematic review. *ReCALL*, *36*(2), 210–229. Advance online publication. DOI: 10.1017/S0958344024000041

Wijnen-Meijer, M., Brandhuber, T., Schneider, A., & Berberat, P. O. (2022). Implementing Kolb's Experiential Learning Cycle by Linking Real Experience, Case-Based Discussion and Simulation. *Journal of Medical Education and Curricular Development*, *9*, 23821205221091511. Advance online publication. DOI: 10.1177/23821205221091511 PMID: 35592131

Wu, D., Yang, Z., Zhang, P., Wang, R., Yang, B., & Ma, X. (2023). Virtual-Reality Interpromotion Technology for Metaverse: A Survey. *IEEE Internet of Things Journal*, *10*(18), 15788–15809. DOI: 10.1109/JIOT.2023.3265848

Xu, P., Zhu, X., & Clifton, D. A. (2023). Multimodal Learning With Transformers: A Survey. *IEEE Transactions on Pattern Analysis and Machine Intelligence*, *45*(10), 12113–12132. DOI: 10.1109/TPAMI.2023.3275156 PMID: 37167049

Yuan, Y., Li, Z., & Zhao, B. (2025). A Survey of Multimodal Learning: Methods, Applications, and Future. *ACM Computing Surveys*, *57*(7), 1–34. Advance online publication. DOI: 10.1145/3713070

Zhao, X., Ren, Y., & Cheah, K. S. L. (2023). Leading Virtual Reality (VR) and Augmented Reality (AR) in Education: Bibliometric and Content Analysis From the Web of Science (2018–2022). *SAGE Open*, *13*(3), 21582440231190821. Advance online publication. DOI: 10.1177/21582440231190821

Chapter 9

The Role of VR and AR in Immersive Language Education

Ferdi Çelik
https://orcid.org/0000-0001-8272-4720
Ondokuz Mayıs University, Turkey

Volkan Duran
https://orcid.org/0000-0003-0692-0265
Iğdır University, Turkey

ABSTRACT

This chapter explores how virtual reality (VR) and augmented reality (AR) technologies transform language education through immersive, learner-centered experiences. Using a mixed-methods approach, bibliometric analyses of 634 publications (385 on VR, 249 on AR) and grounded theory, it finds that these technologies boost motivation, engagement, and contextual learning, though most studies prioritize technical development over pedagogical depth. VR simulates real-world and cultural scenarios, improving oral proficiency, vocabulary retention, and reducing anxiety. AR overlays digital content onto physical settings, enhancing interactivity and personalized feedback. However, challenges like hardware costs, teacher training, and scarce longitudinal data hinder adoption. Smaller research clusters (e.g., VR in healthcare language training or AR for special needs) highlight adaptability across contexts. Key next steps include integrating language acquisition theories, enhancing educator support, and pursuing large-scale comparative studies.

DOI: 10.4018/979-8-3373-3366-3.ch009

1. INTRODUCTION

In recent years, VR opened up new ways for learners to practice their language skills. VR can offer learners an immersive and interactive learning experience in which they can enhance their language skills through meaningful activities. Unlike traditional learning environments, VR allows learners to interact with digital objects and other people through the use of avatars (Creely, 2025). In VR environments, students can talk with speakers of English from various countries around the world in simulated environments. VR has been found to improve students' pronunciation, listening comprehension, and speaking skills (Sinthiya, 2023). In VR environments, the setting is usually appealing to the learners which can make learning more interesting and contribute to their retention and engagement levels (Wang, 2024).

Teachers can customize the VR so that it they can provide more tailored activities that may meet the diverse needs of learners (Lee et al., 2019). For instance, VR photo tours can function as an effective instrument for language educators to enhance the learning motivation of remote students (Figueroa, & Jung, 2025). In VR, learners can learn English without fear as it alleviates learners' anxiety (Meng et al., 2020) and sparks the learners' motivation (Palamar et al., 2023). These realistic digital setting when combined with meaningful scenarios (Chang et al., 2012), can improve vocabulary learning (Çelik & Baturay, 2024) and lift up reading skills (Syafri & Ramli, 2025). There are also VR platforms that are powered up by generative AI features (Uppoor et al., 2022) and it has been found that skill building in such platforms can support teamwork on global issues like climate change (Carmona-Galindo et al., 2025). Yet, it is not without flaw; costs can be steep (Flores et al., 2022), and teachers need training to use it in effective way (Algerafi et al., 2023; Familoni & Onyebuchi, 2024; Jang et al., 2024), and some feel queasy using it (Kaplan-Rakowski & Wojdynski, 2018). Design quirks, like avatars, can also trip up learning in English as a Foreign Language (EFL) settings (Muthmainnah et al., 2025). Still, VR can be regarded as good in breaking down language barriers that students face while learning the target languageAR overlays digital elements onto real-world settings, making abstract concepts tangible and enhancing comprehension (Jantanukul, 2024), while VR, especially in the metaverse, offers realistic scenarios for practicing communicative and cultural immersion (Sinthiya, 2023) and provides real-time feedback on pronunciation and grammar (Huang et al., 2021). Therefore, this study aimed at highlighting how AR and VR can transform language learning and informing educators with practical insights to bridge technology and pedagogy so that they can better prepare learners for a digitalized world. To this end, the research questions of the study are as follows:

1. How do VR and AR enhance language learning in terms of motivation, engagement, and contextual learning?
2. What are the dominant research trends and gaps in VR/AR-based language learning, as revealed by bibliometric and grounded theory analyses?
3. To what extent does current research prioritize technological development over pedagogical depth in VR/AR-assisted language education?
4. What are the key challenges that hinder the widespread adoption of VR and AR in language learning?
5. How can established language acquisition theories be integrated into VR/AR-based language education to enhance its pedagogical effectiveness?

2. METHODOLOGY

This study investigates the intersection of immersive technologies, specifically virtual reality (VR) and augmented reality (AR), with language learning, employing a mixed-methods approach that integrates bibliometric analysis and grounded theory. The methodology unfolds in two primary phases. The first phase involves a bibliometric analysis to systematically map the research landscape, identifying trends, themes, and disciplinary connections. The second phase applies grounded theory to interpret these quantitative findings, synthesizing recurring patterns into a cohesive theoretical framework that elucidates the role of VR and AR in language education.

2.1 Research Design

The study adopts an exploratory design, combining quantitative bibliometric methods with qualitative grounded theory techniques to provide a comprehensive understanding of the field. In the bibliometric phase, quantitative metrics such as document counts, citation distributions, and keyword frequencies are analyzed to uncover key trends, thematic clusters, and interdisciplinary intersections. Subsequently, the grounded theory phase interprets these quantitative insights through systematic coding, transforming raw data into a theoretical model that captures the essence of VR and AR applications in language learning.

2.1.1 Data Collection

2.1.1.1 Bibliometric Search Strategy

Data were sourced exclusively from the Web of Science Core Collection, a robust database for scholarly publications. Two distinct search queries were formulated to capture relevant literature. For VR, the search combined keywords “virtual reality” and "language learning," constrained to publications from 1995 to 2023 within the categories of Education, Educational Research, or Education Scientific Disciplines, yielding 385 documents. A parallel search for AR used the keywords "augmented reality" and "language learning," covering 1995 to 2017 within the same categories, retrieving 249 documents. Extracted data included meso- and micro-level citation topic distributions, author keywords, and prevalent terms from abstracts, providing a foundation for quantitative analysis.

2.1.1.2 Data Management

Retrieved records were systematically organized into datasets for analysis. Frequency counts of keywords and citation topics were computed to highlight prominent research areas, such as Human-Computer Interaction (HCI), Education & Educational Research, and Language & Linguistics. To facilitate interpretation, the data were visualized through tables and figures, including citation topic distribution tables and keyword frequency charts, which collectively illustrated the structure and scope of the research landscape.

2.1.2 Data Analysis

2.1.2.1 Bibliometric Analysis

The bibliometric analysis commenced with descriptive statistics, detailing document counts, category percentages, and keyword distributions across the corpus. Thematic mapping explored citation topic distributions at meso and micro levels, identifying dominant disciplinary orientations, such as the prevalence of HCI, alongside emerging foci like language policy and self-regulated learning. This mapping revealed major research clusters and nascent trends. Additionally, keyword frequency analysis extracted and quantified author and abstract keywords, such as "virtual," "reality," "learning," "language," and "augmented," providing a quantitative lens on core concepts and research priorities in VR/AR-based language learning.

2.1.2.2 Grounded Theory Analysis

To deepen the interpretation of bibliometric findings, grounded theory methods were applied through a structured coding process. Open coding involved reviewing bibliometric summaries, keyword frequencies, and topic distributions to identify initial concepts, such as technological foundations (e.g., "VR," "AR," "immersive"), language pedagogy (e.g., "language learning," "teaching"), learner-centered outcomes (e.g., "engagement," "motivation"), and interdisciplinary applications. Axial coding grouped these concepts into broader categories, such as Technological and Pedagogical Integration for technology-related codes and Learner Experience and

Outcomes for learner-focused codes, while exploring interrelationships among them. Selective coding synthesized these categories into a core theoretical narrative termed "Immersive Technological Transformation of Language Education."

2.1.3 Validity and Reliability

The study ensured robustness through triangulation, leveraging the complementary strengths of bibliometric and grounded theory analyses to validate findings. Data saturation was achieved through the consistent recurrence of codes and themes across the large corpus, bolstering the credibility of the emergent framework. To minimize bias, coding decisions and thematic interpretations underwent peer review by multiple researchers, enhancing inter-coder reliability.

2.1.4 Ethical Considerations

As the study utilized publicly available bibliometric data from the Web of Science Core Collection and involved no human subjects, it adhered to ethical standards for secondary data analysis, ensuring responsible data handling throughout the process.

2.1.5 Limitations

While the methodology offers a robust framework for mapping the VR/AR language learning landscape, limitations include potential bias from relying solely on the Web of Science, which may exclude relevant studies from other databases. Additionally, bibliometric data, while rich in quantitative insights, lack the contextual depth of classroom-based qualitative studies. Future research could address these gaps by incorporating additional databases, such as Scopus, and complementary qualitative investigations to provide a more holistic understanding of the field.

3. FINDINGS

3.1. Bibliometric Research on the Topic "Virtual Reality" and "Language Learning"

A preliminary bibliometric search using the Web of Science Core Collection on the topic "virtual reality" and "language learning" (refined to the publication years 1995–2023 and the categories "Education Educational Research" or "Education Scientific Disciplines") retrieved 385 documents (Figure 1). This result indicates a growing body of literature that explores the intersection of immersive technologies and language pedagogy over nearly three decades. The consistent appearance of publications across this timespan suggests that the application of VR in language learning has gained sustained scholarly attention, potentially driven by advancements in VR hardware and software as well as evolving pedagogical frameworks.

Figure 1. A preliminary bibliometric search using the Web of Science core collection on the topic "virtual reality" and "language learning"

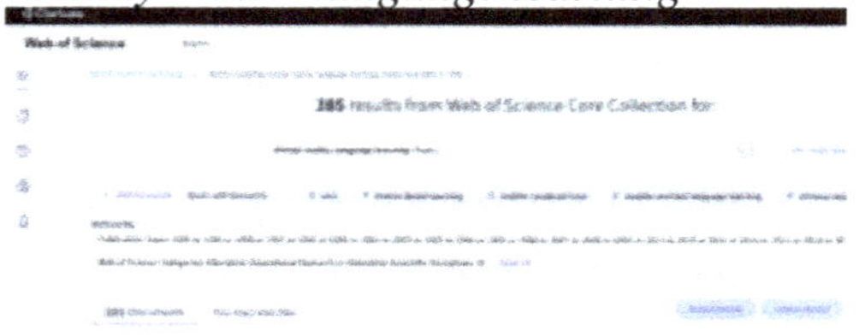

The citation topic distribution reveals that nearly half (49.61%) of the 385 documents fall under "Human Computer Interaction," indicating a strong focus on technological and user-interface dimensions of VR in language learning. Beyond this predominant area, a notable share is classified under "Education & Educational Research" (12.73%) and "Language & Linguistics" (11.69%), underscoring the pedagogical and linguistic foundations central to VR-based language instruction. Meanwhile, the presence of smaller clusters such as "Management," "Robotics," "Nursing," and "Neuroscanning" suggests that VR's applications in language learning extend into interdisciplinary domains, ranging from organizational contexts and automation to healthcare and cognitive research. Although these areas each comprise a relatively small percentage of the overall corpus, their inclusion highlights the breadth of interest in VR-enhanced language learning across diverse fields. Taken together, these data point to a research landscape dominated by technology-oriented and educational frameworks, with growing intersections in specialized and cross-disciplinary sectors. This distribution signals that while the foundational conversation revolves around the design, usability, and instructional effectiveness of VR for language acquisition, scholars are also increasingly exploring its implications in various professional and scientific contexts.

Table 1. The citation topic distribution

Citation Topics Meso	Record Count	% of 385
4.284 Human Computer Interaction	191	49.610
6.11 Education & Educational Research	49	12.727
6.69 Language & Linguistics	45	11.688
6.3 Management	9	2.338
4.116 Robotics	8	2.078
1.14 Nursing	7	1.818
1.7 Neuroscanning	7	1.818
4.322 Remote Research & Education	6	1.558
1.136 Autism & Development Disorders	5	1.299
4.48 Knowledge Engineering & Representation	3	0.779
6.317 Risk Assessment	3	0.779
6.73 Social Psychology	3	0.779
10.126 Philosophy	2	0.519
10.240 Music	2	0.519
6.185 Communication	2	0.519
6.86 Human Geography	2	0.519
1.172 Sports Science	1	0.260
1.44 Nutrition & Dietetics	1	0.260
1.6 Immunology	1	0.260
1.66 Hiv	1	0.260
1.72 Obstetrics & Gynecology	1	0.260
10.290 Art	1	0.260
10.325 Translational Studies	1	0.260
3.35 Zoology & Animal Ecology	1	0.260
4.237 Safety & Maintenance	1	0.260

The micro-level citation topic distribution underscores the breadth and interdisciplinary nature of research on "virtual reality" and "language learning." Nearly half of the documents (49.61%) are associated with "Augmented Reality," suggesting that many studies conflate or integrate VR and AR in educational contexts. A sizable portion also pertains to "Language Policy" (9.35%) and "Self-regulated Learning" (4.42%), reflecting a pedagogical and linguistic focus.

Beyond these main clusters, there is notable diversity: from "Science Education" and "Technology Acceptance Model" to specialized domains like "Medical Education," "Computational Thinking," "Human-Robot Interaction," and "Phonological Awareness." Smaller but still significant categories, such as "Music Psychology,"

"Semantic Web," "Sustainable Development," and "Social Marketing," highlight that VR/AR in language learning has become relevant in various professional, disciplinary, and societal contexts.

Overall, this distribution suggests a research landscape where immersive technologies intersect with language education, cognitive sciences, policy studies, and numerous applied fields. While AR/VR-based language instruction remains the primary focus, the wide range of smaller topic clusters indicates a growing recognition of how these technologies can be adapted and studied in diverse educational, professional, and socio-cultural settings.

Table 2. The micro-level citation topic distribution

Citation Topics Micro	Record Count	% of 385
4.284.1027 Augmented Reality	191	49.610
6.69.342 Language Policy	36	9.351
6.11.31 Self-regulated Learning	17	4.416
6.11.295 Science Education	8	2.078
6.3.368 Technology Acceptance Model	7	1.818
4.322.1428 Remote Laboratory	6	1.558
6.11.1094 Medical Education	6	1.558
6.11.1526 Computational Thinking	6	1.558
1.14.849 Surgical Education	5	1.299
4.116.1415 Human-robot Interaction	5	1.299
1.136.283 Autism	4	1.039
1.7.637 Action Observation	3	0.779
4.116.1097 Biped Robot	3	0.779
6.11.190 Teacher Education	3	0.779
6.69.218 Phonological Awareness	3	0.779
10.126.1995 Peirce	2	0.519
10.240.657 Music Psychology	2	0.519
4.48.322 Semantic Web	2	0.519
6.11.2332 Higher Education	2	0.519
6.11.2525 Open Education Resources	2	0.519
6.11.333 Digital Learning	2	0.519
6.185.1004 Internet Addiction	2	0.519
6.317.2518 Sustainable Development	2	0.519
6.69.1116 Academic Writing	2	0.519
6.73.2468 Social Marketing	2	0.519

The most frequent words in the corpus of author keywords can be given as follows: *virtual* (59); *reality* (58); *learning* (55); *language* (44); *education* (10); *augmented* (8); *VR* (6); *foreign* (6); *assisted* (6); *technology* (5); *computer* (5); *based* (5); *teaching* (4); *immersive* (4); *immersion* (4); *game* (4); *worlds* (3); *mixed* (3); *languages* (3); *gamification* (3); *environment* (3); *English* (3); *distance* (3); *communicative* (3); *Chinese* (3); *artificial* (3); *student* (2); *strategies* (2); *sign* (2); *second* (2); *review* (2); *programming* (2); *pedagogical* (2); *oral* (2); *multimodal* (2); *model* (2); *mobile* (2); *media* (2); *intelligence* (2); *environments* (2); *engagement* (2); *embodiment* (2); *embodied* (2); *educational* (2); *digital* (2); *communication* (2); *analysis* (2); *acquisition* (2); *writing* (1); *world* (1); *VRALL* (1); *vocabulary* (1); *videos* (1); *video* (1); *VECAR* (1); *VEC3D* (1); *user* (1); *understanding* (1); *ubiquitous* (1).

The list of most frequent words in the corpus of author keywords provides valuable insights into the research trends, technologies, and pedagogical focuses within the field of VR and language learning.

1. Core Research Focus: Technology and Language Learning

The most frequent keywords "virtual" (59), "reality" (58), "learning" (55), and "language" (44) form the foundation of this research domain. These terms indicate that studies are centered on leveraging VR as a technological tool to enhance language learning. Their high frequencies suggest they are the defining concepts, appearing consistently across publications.

Technological Emphasis

Keywords like "VR" (6), "technology" (5), "computer" (5), "augmented" (8), "immersive" (4), and "immersion" (4) highlight a strong focus on immersive technologies. The presence of "augmented" points to the significant role of AR often explored alongside or in comparison to VR, reflecting a broader interest in XR solutions.

Learning Context

"Learning" (55) and "education" (10) emphasize the educational applications of these technologies, positioning VR as a tool to support academic and instructional goals.

2. Pedagogical Applications and Approaches

Several keywords reveal how VR is being applied in educational settings, particularly for language learning:

Teaching and Pedagogy

"Teaching" (4), "pedagogical" (2), and "educational" (2) suggest that researchers are investigating how VR can be integrated into teaching practices and instructional design.

Assisted Learning

"Assisted" (6) and "based" (5) likely relate to phrases such as "technology-assisted language learning" or "VR-based education," indicating the supportive role of technology in pedagogy.

Student-Centered Focus

"Student" (2) and "strategies" (2) point to studies examining learner experiences and learning strategies within VR environments.

3. Linguistic and Communicative Goals

The keywords also reflect a strong linguistic orientation, with a focus on specific language skills and contexts:

Language Acquisition

"Language" (44), "languages" (3), "foreign" (6), "English" (3), "Chinese" (3), and "acquisition" (2) indicate that VR is widely used for foreign language learning, with particular attention to globally prominent languages like English and Chinese.

Communicative Skills

"Communicative" (3), "oral" (2), and "communication" (2) suggest an emphasis on developing interactive and spoken language abilities, aligning with communicative language teaching approaches.

Specialized Skills

Less frequent terms like "vocabulary" (1) and "writing" (1) hint at targeted studies on lexical acquisition and written language skills, though these appear to be less dominant areas.

4. Immersion and Engagement

A key strength of VR lies in its immersive capabilities, which are well-represented in the keyword list:

Immersive Environments

"Immersive" (4), "immersion" (4), "worlds" (3), and "environments" (2) underscore the importance of creating virtual spaces where learners feel fully engaged. These terms suggest research on how simulated environments enhance the learning experience.

Gamification and Motivation

"Game" (4), "gamification" (3), and "engagement" (2) indicate that gamified elements are being explored to boost learner motivation and participation, a trend that leverages VR's interactive potential.

Mixed Reality

"Mixed" (3) likely refers to mixed reality (MR), blending VR and AR, pointing to innovative approaches that combine multiple immersive technologies.

5. Emerging Trends and Interdisciplinary Connections

The keyword list also reveals emerging areas and interdisciplinary influences:

Artificial Intelligence

"Artificial" (3) and "intelligence" (2) suggest that AI is being integrated with VR, possibly for adaptive learning systems or natural language processing to enhance language instruction.

Embodied Learning

"Embodiment" (2) and "embodied" (2) reflect interest in how physical presence in VR impacts cognitive processes, potentially drawing on embodied cognition theories.

Mobile and Ubiquitous Access

"Mobile" (2) and "ubiquitous" (1) indicate efforts to make VR language learning accessible via portable devices, enabling flexible, anytime-anywhere education.

Multimodal Approaches

"Multimodal" (2) and "media" (2) point to the use of multiple communication modes (e.g., audio, visual, haptic) in VR, enriching the learning experience.

6. Niche and Specialized Areas

Keywords with lower frequencies (e.g., 1 or 2 occurrences) highlight niche or emerging topics:

Specific Technologies

Terms like "videos" (1), "video" (1), "vecar" (1), and "vec3d" (1) may relate to specific VR tools, such as 360-degree videos or 3D modeling, used in language learning.

Sign Language

"Sign" (2) could indicate research on VR applications for sign language education, a specialized subfield.

Programming and Analysis

"Programming" (2) and "analysis" (2) suggest technical studies on VR system development or data-driven evaluations of learning outcomes.

User Experience

"User" (1) and "understanding" (1) might reflect usability studies or research on comprehension in VR contexts.

The most frequent author keywords in this corpus paint a comprehensive picture of research on VR and language learning. The field is heavily driven by technological innovation, with VR and AR serving as core tools to create immersive, engaging learning environments. Educationally, the focus is on enhancing language acquisition particularly foreign languages through communicative and gamified approaches. Emerging trends, such as AI integration, embodied learning, and mobile accessibility, signal a broadening scope that intersects with other disciplines. Meanwhile, niche areas like sign language and specific skill development (e.g., vocabulary, writing) suggest potential directions for future exploration. Together, these keywords highlight a dynamic research landscape that blends technology, pedagogy, and linguistics to advance language education.

Most Frequent Words in the Corpus for the Keywords

Most frequent words in the corpus for the keywords can be given as:

- reality (15); virtual (8); education (8); environments (6); augmented (6); language (5); immersion (5); technology (4); life (4); 2nd (4); students (3); environment (3); english (3); communication (3); worlds (2); vocabulary (2); technologies (2); speaking (2); motivation (2); game (2); future (2); design (2); challenges (2); world (1); web (1); use (1); tool (1); thinking (1); theory (1); telecollaboration (1); styles (1); sense (1); self (1); science (1); program (1); plus (1); play (1); platform (1); performance (1); participation (1); multiliteracies (1); motor (1); modality (1); learning (1); learners (1); keywords (1); instruction (1); insights (1); input (1); humans (1); gestures (1); games (1); framework (1); experience (1); determination (1); computational (1); classroom (1); chinese (1); autonomy (1)

The most frequent keywords in the corpus on virtual reality (VR) and language learning provide insight into the primary research trends and focuses in this field.

1. Immersive Technologies as the Core

The high frequency of technology-related keywords such as reality (15), virtual (8), augmented (6), environments (6), immersion (5), and technology (4) underscores the central role of immersive technologies in this field. These terms suggest that research is heavily focused on leveraging VR and augmented reality (AR) to create engaging, interactive environments for language learning. The emphasis on environments and immersion indicates a priority on simulating realistic settings where learners can practice language skills, reflecting the technological foundation of this research.

2. Education and Language Learning Context

Keywords like education (8), language (5), students (3), and learning (1) highlight the educational framework of this research. The prominence of these terms shows that VR is being explored as a tool to enhance language education, with a specific focus on how it benefits students and learners. This suggests a practical orientation toward improving teaching and learning outcomes in classroom or autonomous settings, as further hinted by terms like classroom (1) and autonomy (1).

3. Specific Language Skills and Applications

The corpus includes keywords tied to particular language skills and languages, such as vocabulary (2), speaking (2), English (3), and Chinese (1), alongside communication (3). These terms indicate that researchers are investigating how VR can target measurable linguistic outcomes, like vocabulary acquisition and oral proficiency. The mention of widely taught languages like English and Chinese suggests a focus on globally relevant educational applications, aiming to address practical needs in language instruction.

4. Gamification and Engagement

Terms like game (2), motivation (2), and play (1) point to the integration of gamified elements in VR language learning. This trend reflects an interest in making learning more engaging and enjoyable, likely to boost learner motivation. The presence of design (2) further suggests that researchers are exploring how to thoughtfully craft these experiences to maximize educational impact, blending fun with effective pedagogy.

5. Future-Oriented Perspective

The keywords future (2) and challenges (2) reveal a forward-looking approach. Researchers appear to be not only studying current applications of VR in language learning but also anticipating its evolution and the obstacles to its widespread adoption. This indicates a field that is proactive about addressing implementation issues and envisioning long-term potential.

6. Emerging and Niche Areas

Less frequent but notable keywords like gestures (1), motor (1), computational (1), and telecollaboration (1) hint at emerging interdisciplinary directions. For instance, gestures and motor may relate to embodied learning, where physical actions enhance language acquisition, while computational could point to the role of AI or adaptive systems. These niche terms suggest that the field is beginning to explore innovative, specialized applications of VR.

Most Frequent Words in the Corpus of Abstracts

Most frequent words in the corpus of abstracts can be given as:

- learning (243); language (193); virtual (144); reality (118); vr (94); students (60); technology (50); foreign (40); study (37); based (36); research (34); results (32); using (31); learners (31); environment (30); education (27); use (26); immersive (26); chinese (24); skills (23); technologies (21); used (19); teaching (19); paper (19); design (19); vocabulary (18); english (18); development (18); tools (17); studies (17); real (17); languages (17); applications (17); experience (15); educational (15); cultural (15); communication (15); second (14); new (14); environments (14); digital (14); enhanced (13); current (13); ar (13); world (12); immersion (12); game (12); engagement (12); data (12); augmented (12); application (12); years (11); potential (11); pedagogical (11); motivation (11); learn (11); interactive (11); interaction (11); analysis (11)

Key Research Trends

Research in this field centers on language learning, with learning (243) and language (193) as dominant themes. VR emerges as the pivotal technology, virtual (144), reality (118), and VR (94) collectively underscore its transformative role in education. The focus is practical and learner-centered, targeting students (60) and learners (31) in educational (15) settings, where VR enhances teaching and learning processes.

VR's Strengths

The technology's immersive (26) and interactive (11) qualities stand out, fostering engagement (12) and motivation (11) through rich, virtual environments (30). These features make learning experiential, contrasting with real (17) settings.

Language Learning Focus

Studies often explore foreign (40) language acquisition, spotlighting Chinese (24) and English (18), and honing skills (23) like vocabulary (18) and communication (15). This aligns with second (14) language learning priorities.

Research Approach

Empirical evidence drives the field, with study (37), research (34), and results (32) reflecting a rigorous, data-backed approach. Design (19) hints at both VR environments and study structures.

Innovation and Trends

VR is seen as a cutting-edge tool, tied to new (14) developments and potential (11). Emerging trends like AR (13) and game (12)-based strategies, alongside cultural (15) dimensions, suggest a forward-looking scope. Broader implications point to immersive digital (14) spaces shaping contemporary education.

This snapshot reveals a dynamic field where VR empowers language learning, blending innovation with practical, learner-focused applications.

The most frequent words in this corpus paint a picture of a dynamic research field centered on using virtual reality to enhance language learning, particularly for foreign languages like Chinese and English. The focus is on students and learners, with VR leveraged as an immersive, interactive, and motivating tool to develop skills such as vocabulary and communication. Studies are often empirical, exploring how VR environments, sometimes gamified or culturally enriched, improve educational outcomes. The presence of terms like AR and cultural aspects suggests a broadening scope, while words like potential and development indicate optimism about VR's future in language education. In essence, this corpus reflects a blend of cutting-edge technology and pedagogical innovation aimed at transforming how languages are taught and learned.

3.2. Understanding Bibliometric Research Results on the Topic "Virtual Reality" and "Language Learning" Based on Grounded Theory

Open Coding (Identifying Initial Concepts)

In grounded theory, open coding entails reading through the data (in this case, the bibliometric summaries, keyword frequencies, and topic distributions) and labeling recurring concepts. Several initial concepts emerged from the data:

1. Technological Foundations

Codes: "VR," "AR," "immersive," "technology," "virtual environments," "human-computer interaction" Description: References to hardware, software, user interfaces, and immersive digital platforms.

2. Language Pedagogy and Educational Frameworks

Codes: "language learning," "foreign language," "communicative," "teaching," "education," "pedagogical approaches" Description: Studies focusing on how VR/AR aligns with language teaching methodologies (e.g., communicative language teaching, self-regulated learning).

3. Learner-Centered Focus

Codes: "student," "learner," "engagement," "motivation," "experience," "interaction," "communication" Description: Emphasis on how learners respond to VR-based interventions (e.g., affective outcomes, cognitive engagement, skill development).

4. Interdisciplinary Expansion

Codes: "management," "nursing," "robotics," "neuroscanning," "music psychology," "semantic web," "human-robot interaction" Description: VR language applications extending into specialized fields (healthcare, organizational contexts, computational domains).

5. Skills and Competencies

Codes: "vocabulary," "oral skills," "communication," "phonological awareness," "writing," "reading," "cultural competence" Description: Specific linguistic outcomes and the role of VR in enhancing them.

6. Research Orientation

Codes: "study," "results," "data," "analysis," "design," "empirical," "quantitative," "qualitative," "mixed methods". Description: Focus on how research is conducted, reported, and validated (e.g., measuring effectiveness, user attitudes, learning outcomes).

7. Engagement and Gamification

Codes: "gamification," "game," "motivation," "immersive worlds," "interactive," "embodiment" Description: Game-based or gamified VR environments to increase learner motivation, sense of presence, and autonomy.

8. Emerging Trends and Future Directions

Codes: "artificial intelligence," "mobile," "ubiquitous," "sustainable development," "risk assessment," "digital learning".Description: New technologies and frameworks that intersect with VR-based language learning, extending potential future research paths.

These open codes represent the diverse aspects of VR-based language learning literature.

Axial Coding (Grouping Concepts into Categories and Subcategories)

Next, axial coding involves clustering the open codes into overarching categories and exploring relationships among them. From the above open codes, we can form the following major categories:

1. Technological and Pedagogical Integration

Core Subcategories: Technological Foundations (VR, AR, Human-Computer Interaction), Instructional Design (Teaching strategies, pedagogical frameworks), Learning Environment Characteristics (Immersion, gamification, engagement). Studies show how VR platforms are designed and implemented to meet pedagogical needs, merging technology with language teaching theories.

2. Learner Experience and Outcomes

Core Subcategories: Learner-Centered Focus (motivation, engagement, interaction), Skill Development (vocabulary, oral communication, phonological awareness), Assessment and Evaluation (results, data, analysis, performance measurements). Key Relationship: VR-based instruction influences learners' motivation, interaction quality, and language proficiency gains.

3. Interdisciplinary Applications and Contexts

Core Subcategories: Sector-Specific Uses (healthcare, robotics, management), Emerging Cross-Disciplinary Topics (music psychology, autism, neurosurgery, semantic web). Key Relationship: VR language learning extends beyond traditional classroom contexts, suggesting collaborative or domain-specific applications.

4. Future Directions and Innovations

Core Subcategories: AI Integration and Adaptive Systems, Ubiquitous and Mobile Learning, Sustainability and Global Education. Key Relationship: Researchers are looking beyond current VR setups to explore next-generation technologies (AI, mixed reality, mobile VR) and broader societal impacts (sustainability, policy, global reach).

Within axial coding, these categories interrelate around the central phenomenon of "Using VR to enhance language learning through immersive and technologically advanced methods." Each category addresses a different slice of the phenomenon, pedagogical design, learner experience, interdisciplinary expansion, and future prospects.

Selective Coding (Identifying the Core Phenomenon and Integrating the Theory)

Selective coding refines the axial categories into a more cohesive theoretical narrative. The overarching theme that ties these categories together could be framed as:

"Immersive Technological Transformation of Language Education"

1. Core Phenomenon

VR-based language instruction is increasingly recognized as a transformative educational strategy. It leverages immersive environments, gamification, and user-centered interfaces to foster deeper engagement and communicative competence.

2. Context/Conditions

Advances in VR hardware/software and pedagogical research create fertile ground for experimentation. Educational institutions seeking innovative methods, along with global demand for foreign language proficiency, fuel interest in VR solutions.

3. Actions/Strategies

Researchers develop VR language curricula emphasizing immersion, collaboration, and authentic communicative tasks. They incorporate gamified elements and adaptive AI features to personalize learning experiences. Studies measure effectiveness via learning outcomes, engagement metrics, and user feedback.

4. Intervening Variables

Cost, accessibility, and technological infrastructure can either facilitate or hinder VR adoption. Institutional support and educator training also shape the success of implementation. Interdisciplinary collaboration opens new avenues but requires alignment of different expertise and goals.

5. Consequences/Outcomes

When effectively implemented, VR-based language programs show promising gains in oral proficiency, vocabulary, cultural awareness, and learner motivation. Interdisciplinary investigations link VR-supported language learning to cognitive development, social interaction, and professional skill-building.

6. Future Directions

Emerging technologies (mobile VR, AR, AI-driven analytics, embodied cognition) are likely to deepen or diversify applications. Studies increasingly explore VR's broader societal impact, addressing sustainability, policy, and inclusive education.

Thus, the core grounded theory proposition is that VR's immersive, interactive qualities, combined with evolving pedagogical strategies, drive a transformation in language education, with outcomes contingent on technological, institutional, and learner variables.

5. Emergent Theoretical Framework

Bringing these insights together, the grounded theory analysis suggests a theoretical model where:

Central Phenomenon: *"Immersive Technological Transformation of Language Education."*

Driving Forces (Conditions): Technological advancements in VR/AR, global language education demands, pedagogical interest in experiential and communicative methods.

Mediators (Intervening Conditions): Resource availability, educator training, institutional buy-in, interdisciplinary collaboration.

Strategies/Actions: Curriculum design integrating VR, interactive tasks, gamification, AI-driven personalization, cross-domain partnerships.

Outcomes/Consequences: Enhanced language proficiency, motivation, cultural competence, emerging fields of research and practice, and broader acceptance of immersive approaches in formal and informal learning contexts.

Through this lens, VR-based language learning is not just a new technological tool but a pedagogical shift that is shaped by, and in turn shapes how educators and learners conceptualize language instruction. While challenges remain (costs, access, robust empirical evidence across contexts), the ongoing research trajectory is one of growing adaptation and interdisciplinary innovation.

3.3. Bibliometric Research on the Topic "Augmented Reality" and "Language Learning"

In the context of bibliometric research, this screenshot from the Web of Science Core Collection represents a systematic query focused on the intersection of augmented reality (AR) and language learning as a research topic. The retrieved results (249 documents) indicate the academic interest in this field, as filtered by specific publication years (ranging from 1995 to 2017) and categorized under education, educational research, and education scientific disciplines.

Figure 2. A preliminary bibliometric search using the Web of Science Core Collection on the topic "augmented reality" and "language learning"

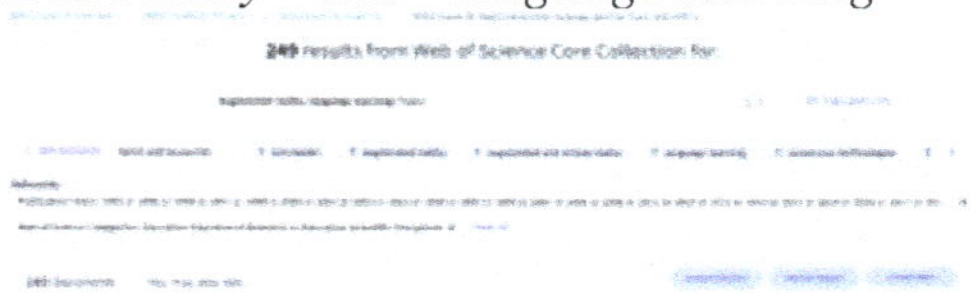

The bibliometric analysis of augmented reality (AR) and language learning research in the Web of Science Core Collection (Figure 2) reveals a strong dominance of human-computer interaction (HCI) (61.04%), highlighting a technology-driven focus. Language & Linguistics (12.85%) and Education & Educational Research (10.04%) are significantly lower, suggesting a gap in pedagogical and linguistic studies. Interdisciplinary contributions from management (4.42%), robotics (2.41%), and telecommunications (0.40%) indicate emerging applications in EdTech, AI tutors, and communication technologies. Additionally, niche areas like neuroscanning (0.80%), autism & developmental disorders (1.61%), and hearing loss (0.40%) show interest in AR's cognitive and accessibility benefits. However, social psychology (0.40%) and educational theories are underrepresented, pointing to a need for deeper exploration of learning engagement, cognitive impact, and instructional frameworks in AR-based language learning.

Table 3. Citation topics meso

Citation Topics Meso	Record Count	% of 249
4.284 Human Computer Interaction	152	61.044
6.69 Language & Linguistics	32	12.851
6.11 Education & Educational Research	25	10.040
6.3 Management	11	4.418
4.116 Robotics	6	2.410
1.136 Autism & Development Disorders	4	1.606
1.7 Neuroscanning	2	0.803
1.150 Hearing Loss	1	0.402
1.172 Sports Science	1	0.402
10.290 Art	1	0.402
4.13 Telecommunications	1	0.402
4.17 Computer Vision & Graphics	1	0.402
6.223 Hospitality, Leisure, Sport & Tourism	1	0.402
6.73 Social Psychology	1	0.402

The micro-level citation analysis of augmented reality (AR) and language learning research shows that AR (61.04%) is the most prominent topic, emphasizing its technological focus. Language policy (9.24%) and self-regulated learning (4.82%) indicate interest in educational frameworks and learner autonomy. The Technology Acceptance Model (4.42%) suggests research on user adoption of AR tools. Niche topics like phonological awareness (1.61%), conversation analysis (1.21%), and computational thinking (1.61%) point to specialized language learning applications. Autism (1.61%), medical education (1.61%), and deaf studies (0.40%) highlight accessibility and cognitive aspects. Additionally, fields like biped robots (1.61%), human-robot interaction (0.40%), and reinforcement learning (0.40%) reflect emerging AI and robotics applications in AR-based language learning. However, creativity, metaphor use, and higher education (each 0.40%) remain underexplored, suggesting potential research gaps in cognitive and pedagogical applications of AR.

Table 4. Citation topics micro

Citation Topics Micro	Record Count	% of 249
4.284.1027 Augmented Reality	152	61.044
6.69.342 Language Policy	23	9.237
6.11.31 Self-regulated Learning	12	4.819
6.3.368 Technology Acceptance Model	11	4.418
1.136.283 Autism	4	1.606
4.116.1097 Biped Robot	4	1.606
6.11.1094 Medical Education	4	1.606
6.11.1526 Computational Thinking	4	1.606
6.69.218 Phonological Awareness	4	1.606
6.69.610 Conversation Analysis	3	1.205
6.11.295 Science Education	2	0.803
1.150.1148 Deaf	1	0.402
1.172.1331 Sport Psychology	1	0.402
1.7.1043 Multisensory Integration	1	0.402
1.7.637 Action Observation	1	0.402
10.290.2219 Art Education	1	0.402
4.116.1415 Human-robot Interaction	1	0.402
4.116.862 Reinforcement Learning	1	0.402
4.13.696 Indoor Localization	1	0.402
4.17.1432 Digital Topology	1	0.402

continued on following page

Table 4. Continued

Citation Topics Micro	Record Count	% of 249
6.11.1248 Creativity	1	0.402
6.11.1859 Assurance Of Learning	1	0.402
6.11.2332 Higher Education	1	0.402
6.223.247 Tourism	1	0.402
6.69.1855 Metaphor	1	0.402

Most frequent words in the corpus on the topic in author keyfords for "augmented reality" and "language learning":

reality (57); *learning* (55); *language* (49); *augmented* (48); *english* (11); *technology* (10); *mobile* (10); *virtual* (8); *teaching* (8); *second* (7); *foreign* (7); *education* (6); *educational* (5); *based* (5); *technologies* (4); *design* (4); *communication* (4); *assisted* (4); *application* (4); *systematic* (3); *skills* (3); *review* (3); *pedagogical* (3); *motivation* (3); *mixed* (3); *learner* (3); *information* (3); *games* (3); *computer* (3); *chinese* (3); *ar* (3); *vocabulary* (2); *ubiquitous* (2); *strategies* (2); *school* (2); *research* (2); *recognition* (2); *programming* (2); *phonics* (2); *metaverse* (2); *learners* (2); *gamification* (2); *game* (2); *environment* (2); *early* (2); *children* (2); *character* (2); *aurasma* (2); *artificial* (2); *acquisition* (2); *years* (1); *visuospatial* (1); *vecar* (1); *usability* (1); *unit* (1); *training* (1); *things* (1); *thematic* (1); *telugu* (1)

The list provided represents the most frequent words in a corpus (a collection of texts) related to the topics "augmented reality" and "language learning." These words give insight into the key themes, concepts, and areas of focus within this domain.

1. Core Themes

Augmented Reality (AR) and Related Concepts.

Words like *"reality"* (57), *"augmented"* (48), *"virtual"* (8), *"mixed"* (3), *"metaverse"* (2), and *"artificial"* (2) highlight the prominence of AR and related immersive technologies. This suggests that the corpus frequently discusses how AR is integrated into language learning environments. Terms like *"technology"* (10), *"technologies"* (4), and *"mobile"* (10) indicate the importance of technological tools and platforms in implementing AR-based solutions.

Language Learning.

Words such as *"learning"* (55), *"language"* (49), *"English"* (11), *"vocabulary"* (2), *"phonics"* (2), and *"acquisition"* (2) emphasize the focus on teaching and acquiring languages, particularly English but also other languages like *"Chinese"* (3) and *"Telugu"* (1).

The inclusion of *"second"* (7) and *"foreign"* (7) suggests discussions around second language acquisition and foreign language education.

2. Pedagogical and Educational Contexts

Teaching and Education.

Words like "teaching" (8), "education" (6), "educational" (5), "pedagogical" (3), and "school" (2) reflect the role of AR in formal educational settings. The corpus likely explores how AR enhances traditional teaching methods or supports pedagogical innovation. Terms such as "learner" (3), "learners" (2), and "skills" (3) indicate a focus on the needs and development of students using AR tools.

Motivation and Engagement.

The presence of "motivation" (3), "gamification" (2), and "games" (3) suggests that AR is being used to increase learner engagement and motivation through interactive and game-like experiences.

3.Applications and ToolsApplications of AR

Words like "application" (4), "design" (4), "based" (5), and "assisted" (4) point to the practical implementation of AR in language learning. For example, AR might be used in apps or systems designed to assist learners. Specific tools or platforms like "Aurasma" (2) are mentioned, indicating case studies or examples of AR software being used.

Technological Features.: Terms such as "recognition" (2), "usability" (1), and "visuospatial" (1) suggest discussions about the technical aspects of AR, including its ability to recognize objects or gestures and its overall user experience.

4. Broader Contexts

Interdisciplinary Connections: Words like communication (4), information (3), computer (3), and programming (2) show that AR in language learning intersects with fields like computer science, information technology, and communication studies. The mention of early (2) and children (2) indicates interest in using AR for early childhood education or foundational language learning.

Research and Reviews: Terms such as review (3), research (2), and systematic (3) suggest that the corpus includes academic papers or systematic reviews analyzing the effectiveness of AR in language learning

5. Emerging Trends

Gamification and Immersive Environments: Words like game (2), environment (2), and gamification (2) indicate a growing trend of using AR to create immersive, game-based learning environments.

Future Directions: The inclusion of metaverse (2) and ubiquitous (2) hints at futuristic applications of AR, where language learning might occur in fully immersive virtual worlds or seamlessly integrated into everyday life.

The corpus reflects a multidisciplinary exploration of how augmented reality is transforming language learning. Key themes include: The integration of AR into educational settings to enhance language teaching and learning. The use of gamification and immersive technologies to engage learners. A focus on specific languages (e.g., English, Chinese) and broader pedagogical strategies. Discussions about the technical design, usability, and effectiveness of AR tools.

This analysis highlights the potential of AR to revolutionize language education by making it more interactive, engaging, and accessible. The corpus underscores the intersection of technology, education, and linguistics, with AR serving as a bridge to innovative and effective language learning methodologies.

Most frequent words in the corpus on the topic in keywords for "augmented reality" and "language learning":

education (12); english (10); vocabulary (8); technology (7); students (6); game (6); technologies (5); reality (5); design (4); challenges (4); augmented (4); motivation (3); achievement (3); virtual (2); pedagogy (2); mobile (2); load (2); instruction (2); environment (2); context (2); cognitive (2); children (2); beliefs (2); worlds (1); university (1); trends (1); text (1); systems (1); support (1); styles (1); strategies (1); software (1); simulation (1); science (1); research (1); reading (1); plus (1); phones (1); performance (1); pedagogical (1); multiliteracies (1); moocs (1); model (1); media (1); mathematics (1); loci (1); learning (1); language (1); keywords (1); insights (1); improvement (1); immersion (1); gamification (1); games (1); future (1); framework (1); experiences (1); experience (1); environments (1)

The list of most frequent words in the corpus related to the topics "augmented reality" and "language learning" provides valuable insights into the key themes, concepts, and areas of focus within this interdisciplinary field.

AR is reshaping language education, particularly for English, by enhancing vocabulary acquisition and fostering student engagement. It is being integrated into pedagogical frameworks, though challenges such as technical barriers and financial constraints remain. AR employs gamification and simulations to craft immersive, contextually rich learning environments. Its influence extends beyond language instruction, intersecting with fields like STEM and online education. Scholars are examining AR's cognitive benefits and its potential to elevate student performance. This analysis highlights AR's transformative role, merging technology, pedagogy, and learner engagement to drive innovative and effective educational experiences.

The corpus emphasizes the role of augmented reality in transforming language learning by enhancing vocabulary acquisition, motivation, and engagement through gamified and immersive experiences. It addresses challenges, pedagogical strategies, and the integration of AR into diverse educational contexts, reflecting its potential to revolutionize language education

Most frequent words in the corpus on the topic in abstracts for "augmented reality" and "language learning":

learning (229); language (168); ar (147); reality (104); augmented (86); students (68); study (51); teaching (49); based (43); english (40); research (38); technology (37); mobile (37); application (37); use (34); learners (34); foreign (33); virtual (31); used (27); education (27); using (25); skills (25); group (25); activities (25); technologies (23); results (22); environment (22); applications (22); teachers (20); new (19); findings (19); potential (17); design (17); vr (16); second (16); motivation (16); future (16); experimental (16); educational (16); vocabulary (15); studies (15); real (15); paper (15); languages (15); game (15); strategies (14); games (14); enhanced (14); data (14); pedagogical (13); experience (13); early (13); different (13); development (13); world (12); tools (12); support (12); provide (12); framework (12)

The list of most frequent words in the corpus related to "augmented reality" (AR) and "language learning" provides a comprehensive view of the key themes, concepts, and trends within this interdisciplinary field.

Core Themes

Learning and Language:

Words like learning (229), language (168), english (40), foreign (33), and second (16) indicate that the corpus focuses heavily on how AR is being used to enhance language acquisition and proficiency, particularly for English as a second or foreign language. The prominence of vocabulary (15) suggests that AR is frequently applied to teach and reinforce word knowledge, a foundational aspect of language learning.

Augmented Reality and Technology:

Terms such as ar (147), reality (104), augmented (86), technology (37), and technologies (23) highlight the central role of AR and related technological advancements in transforming traditional educational practices. Words like mobile (37), application (37), and applications (22) suggest that AR tools are often delivered through mobile devices or software platforms, making them accessible and scalable.

Pedagogical and Educational Contexts

Teaching and Learning Strategies:

Words like teaching (49), based (43), pedagogical (13), and strategies (14) reflect discussions about how AR tools are integrated into instructional frameworks to support effective teaching and learning. The mention of teachers (20) and learners (34) highlights the dual focus on both educators and students, emphasizing the need for tools that cater to diverse needs.

Student-Centered Focus:

The prominence of students (68), group (25), and activities (25) indicates that AR is often designed with collaborative and interactive learning experiences in mind. Terms like motivation (16) and enhanced (14) suggest that AR is seen as a tool to boost learner engagement and improve learning outcomes.

Applications and Tools

Games and Gamification:

The frequent use of game (15) and games (14), along with gamification (implied), indicates that AR is often leveraged to create game-like experiences that make language learning fun and engaging. Words like virtual (31), vr (16), and world (12) further emphasize the immersive nature of AR, which can simulate real-world language use or create virtual environments for practice.

Learning Environments:

Words such as environment (22), real (15), and world (12) suggest that AR is used to create interactive and context-rich learning spaces, both physical and virtual, where students can practice language skills.

Research and Academic Contexts

Studies and Findings:

The presence of study (51), research (38), findings (19), and results (22) indicates that the corpus includes a significant amount of empirical research, likely evaluating the effectiveness of AR in language learning. Terms like experimental (16) and data (14) suggest that many studies employ experimental designs or quantitative methods to assess AR's impact.

Future Directions:

Words like future (16), potential (17), and new (19) reflect discussions about the untapped possibilities of AR in language education, including emerging trends and innovations.

Emerging Insights

Cognitive and Contextual Factors:

Words like skills (25), development (13), and early (13) point to an interest in understanding how AR impacts cognitive processes and supports early language development. The term framework (12) suggests that researchers are developing theoretical models or frameworks to guide the design and implementation of AR tools.

Support and Tools:

The inclusion of support (12), tools (12), and provide (12) indicates that AR is seen as a supportive mechanism, offering resources and scaffolding for learners and educators alike. The corpus reflects a comprehensive exploration of how augmented reality is reshaping language learning, particularly in educational contexts. Key themes include:The integration of AR into language teaching to enhance vocabulary acquisition, motivation, and student engagement. The use of gamification and immersive environments to create interactive and enjoyable learning experiences. A focus on overcoming challenges and designing effective AR tools for diverse learners, including children and university students. Discussions about the cognitive, contextual, and technological aspects of AR, as well as its broader implications for education and research.

The corpus underscores the transformative potential of augmented reality in language education. It highlights the intersection of technology, pedagogy, and learner engagement, with AR serving as a powerful tool to create innovative, immersive, and effective learning experiences. The findings suggest that AR is not only enhancing traditional language learning methods but also paving the way for future advancements in educational technology. The corpus emphasizes the role of augmented reality in transforming language learning by enhancing vocabulary acquisition, motivation, and engagement through gamified and immersive experiences. It addresses challenges, pedagogical strategies, and the integration of AR into diverse educational contexts, reflecting its potential to revolutionize language education.

3.4. Understanding Bibliometric Research Results on the Topic "Augmended Reality" and "Language Learning" Based on Grounded Theory

Open Coding (Identifying Initial Concepts)

In open coding, we label recurring ideas and themes from the bibliometric results and keyword analyses. The codes that emerged:

1. Technology-Centric Foundations

Codes: "augmented reality," "human-computer interaction," "mobile," "virtual," "metaverse," "reinforcement learning," "robotics" *Description:* The strong prevalence of HCI and AR tools points to a research landscape driven by technological innovation and user-interface design.

2. Language Acquisition & Instruction

Codes: "language policy," "foreign language," "second language," "phonological awareness," "conversation analysis," "vocabulary". *Description:* Studies focus on specific linguistic aspects (e.g., vocabulary building, phonology) and broader policy or curricular frameworks for AR-based language teaching.

3. Educational Context & Pedagogy

Codes: "education," "teaching," "pedagogical," "self-regulated learning," "motivation," "gamification," "cognitive impact" *Description:* Many publications discuss how AR might fit into formal education, particularly its motivational and self-directed learning benefits, yet the volume is relatively smaller compared to the technical HCI angle.

4. Learner Experience & Engagement

Codes: "students," "learner autonomy," "accessibility," "autism & developmental disorders," "hearing loss," "early childhood," "engagement," "motivation" *Description:* Research often highlights AR's potential to engage learners interactively and support diverse needs, from early childhood settings to specialized interventions.

5. Interdisciplinary Extensions

Codes: "management," "telecommunications," "human-robot interaction," "medical education," "computational thinking" *Description:* The presence of these codes underscores AR's expanding role beyond traditional language classrooms, reaching into corporate training, robotics, and healthcare contexts.

6. Adoption & Acceptance

Codes: "technology acceptance model," "user adoption," "implementation challenges," "feasibility," "instructor training". *Description:* References to user acceptance frameworks emphasize the need to understand how teachers and learners perceive and integrate AR tools.

7. Future Directions & Gaps

Codes: "creativity," "social psychology," "metaphor," "higher education," "art education".*Description:* The relatively low frequency of these codes indicates uncharted territory. While AR shows promise, deeper exploration of cognitive-linguistic theories, creativity, socio-emotional learning, and higher education frameworks is needed.

Axial Coding (Grouping Concepts into Categories and Subcategories)

In axial coding, we group the open codes into larger conceptual categories, examining how they relate to one another:

1. Technological Integration in Language Pedagogy

Subcategories: AR Tools & HCI Focus (dominant technology orientation). *Pedagogical Adaptations* (designing AR-based tasks, gamification, scaffolding). *Relationships:* The bulk of research is driven by technology design and usage, often adapting or layering AR onto existing pedagogical approaches.

2. Learner Engagement & Cognitive Outcomes

Subcategories: Motivation & Immersion (gamification, self-regulated learning), *Accessibility & Specialized Needs* (autism, hearing loss, early childhood). *Relationships:* AR's potential is consistently linked to higher engagement, interactive feedback, and diverse learner support, though in-depth long-term cognitive studies remain less common.

3. Interdisciplinary & Contextual Applications

Subcategories: Sector-Specific Uses (management training, robotics, healthcare), *Methodological Extensions* (computational thinking, reinforcement learning, conversation analysis). *Relationships:* AR-based language research intersects with multiple fields, suggesting collaborative innovation but also hinting at fragmented or niche studies.

4. Challenges & Adoption Factors

Subcategories: Technology Acceptance Model & Adoption (stakeholder perceptions, training). *Gaps in Pedagogical Frameworks & Theories* (social psychology, creativity, advanced language pedagogy). *Relationships:* While AR's technical feasibility garners interest, many educators and institutions remain uncertain about practical implementation, especially regarding cost, efficacy, and teacher preparedness.

From these axial groupings, a central phenomenon emerges: *"Using AR to enhance language learning, but with a disproportionately technical emphasis compared to pedagogical and linguistic underpinnings."*

Selective Coding (Identifying the Core Phenomenon and Integrating the Theory)

Selective coding refines the categories into a unified explanation or "storyline." Based on the axial coding above, we might define the overarching phenomenon as:

"Augmented reality's role in transforming language learning through immersive, technology-driven environments, while facing underexplored pedagogical, linguistic, and socio-emotional dimensions."

1. Core Phenomenon

AR is presented as a transformative tool in language learning, leveraging HCI breakthroughs to create immersive, interactive experiences.

2. Context/Conditions

Rapid technological development, growing EdTech markets, and global demand for effective language learning solutions set the stage for AR's emergence. Lack of robust pedagogical frameworks to guide AR integration in everyday instruction remains a critical context.

3. Strategies/Actions

Researchers and educators experiment with gamification, mobile AR apps, interactive tasks, and multimodal content to boost language skills. Specialized interventions (e.g., for autism, hearing impairment) leverage AR's adaptability for accessibility.

4. Intervening Variables

Technology acceptance, *institutional support*, *cost*, *infrastructure*, and *instructor readiness* can either propel or hinder successful AR adoption. The presence or absence of theoretical grounding in *learning sciences*, *linguistics*, and *socio-cultural frameworks* significantly shapes outcomes.

5. Consequences/Outcomes

Preliminary studies often report increased motivation, engagement, and vocabulary retention. Yet, the field sees a gap in large-scale, longitudinal evidence and deeper theoretical alignment, especially regarding advanced language competencies (e.g., pragmatics, discourse) and social-emotional learning.

6. Future Directions

Opportunities for AI-driven personalization, extended reality (combining VR, AR, MR), and integrated pedagogical models. Research bridging AR with more robust frameworks, *social psychology*, *adult learning theory*, *creativity*, and *critical pedagogy*, could fill existing theoretical voids.

Hence, the grounded theory suggests that while AR's technological promise has captured the research spotlight, its full pedagogical and linguistic impact requires more cohesive, theory-driven approaches. Addressing these gaps will likely accelerate AR's effectiveness and legitimacy in mainstream language education.

5. Emergent Theoretical Framework

We can consolidate these findings into a theoretical model:

- Central Phenomenon: *"Technologically Immersive Augmented Reality for Language Learning."*
- Driving Forces (Conditions): Technological innovations in AR, emphasis on interactive/immersive learning, global push for language proficiency.
- Mediators (Intervening Conditions): Institutional readiness, educator training, technology acceptance, availability of robust pedagogical frameworks.
- Strategies/Actions: Creation of AR-enhanced curricula (e.g., gamification, mobile AR apps, specialized interventions), focus on user engagement, exploratory integration of interdisciplinary methodologies.
- Consequences/Outcomes: Positive engagement and motivation in short-term studies, potential for enhanced language skills and accessibility; however, theoretical underdevelopment may limit long-term, scalable adoption.

Applying a grounded theory lens to the bibliometric data on "augmented reality" and "language learning" reveals a field strongly influenced by technology-centric innovation, where AR's immersive capabilities are seen as a key to engaging learners. Nonetheless, a pedagogical and linguistic gap emerges: while many researchers champion AR's potential, fewer delve deeply into theoretical instructional design, language acquisition theory, or longitudinal pedagogical frameworks. Hence, the theoretical proposition arising from these data is:

AR offers promise as a catalyst for more immersive, engaging language learning experiences, but sustained impact requires stronger pedagogical foundations, interdisciplinary synergies, and systematic implementation approaches. In other words, *AR has the capacity to revolutionize language education*, but to fully realize its potential, it must be integrated with robust instructional models, well-trained educators, and evidence-based language teaching principles, an area ripe for future inquiry.

4. DISCUSSION

The bibliometric analyses presented in Sections 3.1 and 3.2 shed light on how VR and AR are increasingly explored within the domain of language learning. Both searches, one focusing on VR (385 documents) and the other on AR (249 documents), underscore a rapidly growing interdisciplinary research landscape driven by advances in educational technology and global demands for effective language instruction.

A notable finding is the strong dominance of HCI across both VR and AR studies. This signals a marked emphasis on the technological underpinnings, hardware, software, interface design, potentially at the expense of more in-depth pedagogical, linguistic, and socio-emotional considerations. Although sizeable clusters in Education & Educational Research and Language & Linguistics exist, they are still overshadowed by technology-driven orientations. This suggests that, while VR/AR tools are gaining traction in language learning contexts, their integration is often guided by emerging technologies rather than robust pedagogical frameworks.

Learner-centered approaches do figure prominently, with a significant focus on engagement, motivation, and immersive experiences. Game-based and gamified applications are commonly cited, reflecting the belief that immersive and playful environments can support deeper language acquisition. However, despite the enthusiasm for interactive affordances, ranging from real-time simulations to collaborative virtual worlds, many studies stop short of investigating longitudinal impacts on linguistic development. Consequently, questions remain about how VR/AR might facilitate complex skills (e.g., advanced grammar, discourse pragmatics) or address varied learner profiles (e.g., young children, adult learners, special needs populations).

Interdisciplinary expansions into fields like robotics, management, healthcare, and cognitive research indicate the adaptability of VR/AR-based language applications to specialized contexts. These smaller clusters highlight promising but fragmented directions, such as AR for autism interventions or VR for nursing communication training, yet also underline a need for broader frameworks. If researchers and practitioners aim to scale these targeted interventions, systematic studies of pedagogical effectiveness and contextual adaptions are essential.

Both sets of bibliometric results consistently point to emerging trends: mobile and ubiquitous learning, integration of AI-driven personalization, and mixed or extended reality solutions. Yet, the emphasis on user acceptance models and feasibility studies highlights the challenges of implementation, cost, access, educator training, and institutional support. These intervening variables can significantly shape whether VR/AR remains a niche novelty or becomes a mainstream tool in language education.

Overall, the grounded theory analyses for both VR and AR underscore similar core themes:

1. Technological and Pedagogical Integration – How immersive tools intersect with instructional design and learning theories.

2. Learner Experience and Outcomes – Motivation, engagement, and the measurable linguistic benefits of immersive environments.
3. Interdisciplinary Applications – The expansion of VR/AR language solutions to niche and cross-domain contexts.
4. Challenges and Future Directions – Addressing gaps in theoretical grounding, cost-effectiveness, and multi-stakeholder adoption.

In short, the discussion points to a field where the *promise* of VR/AR is consistently acknowledged, but sustained impact and theoretical rigor still warrant deeper exploration. In the light of these key points. Virtual Reality (VR) has seen a steady increase in research interest in language learning, as evidenced by bibliometric analysis from 1995 to 2023. The recent studies align with historical trends, reflecting both technological advancements and educational adaptations. The bibliometric search revealed that "Human-Computer Interaction" dominates VR-related language learning research at 49.61%, followed by "Education & Educational Research" (12.73%) and "Language & Linguistics" (11.69%). This suggests a primarily technology-driven focus with pedagogical considerations being secondary. Stark (2025) complements these findings by examining VR's role in speech and language rehabilitation, demonstrating its potential in cognitive-linguistic interventions. However, the study emphasizes accessibility concerns, aligning with the challenges noted in previous VR adoption trends, such as cost and scalability.

At a micro-level, augmented reality (AR) was identified as the leading citation topic within VR and language learning at 49.61%, reinforcing the notion that AR and VR are frequently researched in tandem. This is supported by Wu et al. (2025), who explored the role of AR/VR in digital game-based language learning, highlighting its effectiveness in engagement and skill development. Bibliometric data shows "Language Policy" (9.35%) and "Self-Regulated Learning" (4.42%) as key VR topics, suggesting VR supports both immersion and independent learning. Lu (2025) combined VR with deep learning for an adaptive, gamified setup that improved retention and engagement. VR is also growing in fields like nursing, neuroscience, and autism education, especially post-COVID-19, per Lapka and Shcherbak (2025). Niche areas like "Neuroscanning" (1.82%) and "Medical Education" (1.56%) highlight VR's expanding role beyond traditional classrooms.

Keyword analysis emphasizes immersion and interaction, with "virtual" (59 uses), "learning" (243 uses), and "immersion" (4 uses) standing out. Lutfi et al. (2025) found VR videos outperformed traditional methods in language learning, while Wu et al. (2025) showed game-based VR boosts engagement. Overall, VR has evolved into a versatile educational tool, enhancing engagement and accessibility, though challenges like cost and unproven long-term effects persist. For AR, analysis of 249 documents (1995–2017) shows growing interest, especially in Human-Computer

Interaction (61.04%, 152 papers). Language studies (12.85%, 32 papers) and education (1.04%, 25 papers) are rising, pointing to AR's benefits for vocabulary and motivation (Fan et al., 2020). Smaller clusters like autism education (1.61%) suggest broader potential (Bryant et al., 2024). AR excels in self-regulated learning (4.82%) and user adoption (4.42%), but more classroom-focused research is needed.

Author keyword and abstract analysis indicate a strong emphasis on educational technology and learner outcomes, with high frequencies for terms such as 'learning' (229 mentions), 'language' (168 mentions), and 'students' (68 mentions). This reflects the prevailing research interest in AR's pedagogical applications and aligns with meta-analytic findings showing AR's significant impact on linguistic and affective learning gains (Cai et al., 2022). However, the relatively low frequency of terms related to complex language skills, such as reading comprehension or syntax acquisition, suggests a need for expanded research on AR's impact beyond vocabulary learning.

Overall, this bibliometric analysis complements existing empirical studies by mapping the research landscape of AR in language learning. While AR has demonstrated substantial promise in vocabulary acquisition, motivation enhancement, and engagement, further research is needed to explore its efficacy in higher-order language skills, interdisciplinary applications, and long-term learning outcomes. Future studies should also aim to bridge the gap between technological development and pedagogical best practices to optimize AR's potential as an effective language learning tool.

5. CONCLUSION

The findings demonstrate that research on VR and AR in language learning has evolved over nearly three decades, transitioning from initial experiments to increasingly diverse implementations. While early studies may have focused on the novelty of immersive experiences, the growing number of publications in education, linguistics, and human-computer interaction indicates that these technologies are edging closer to mainstream acceptance.

At a macro level, VR- and AR-based studies share a technology-centric approach. Most investigations are grounded in the design and evaluation of specific tools, interfaces, or prototypes. Even so, the collected data consistently reflect an interest in how VR/AR can facilitate communicative, collaborative, and context-rich language practice. This alignment with constructivist, learner-centered paradigms suggests that, when thoughtfully integrated, immersive technologies could address key pedagogical goals, especially motivation, authenticity, and active engagement.

However, the bibliometric analyses also highlight underexplored facets, for instance, rigorous longitudinal assessments, robust theoretical models of language acquisition, and systematic frameworks for teacher training. Smaller but notable clusters in areas like healthcare, robotics, and special education underscore the versatility of immersive learning, yet they also underscore the fragmentation in the field. Put simply, *VR/AR in language learning is no longer an emerging trend but remains a discipline needing more cohesive theoretical grounding and empirical depth.*

6. RECOMMENDATIONS

Future research should move beyond proof-of-concept studies and incorporate established language acquisition theories (e.g., socio-cultural theory, communicative language teaching, task-based learning) to more systematically evaluate VR/AR's impact on diverse linguistic skills (e.g., speaking fluency, discourse pragmatics, literacy skills). To establish efficacy and guide best practices, larger-scale, longitudinal research is needed. Comparative studies (VR/AR interventions vs. traditional or blended methods) can offer insights into whether immersive approaches significantly outperform existing pedagogical models over time. Integrating VR/AR effectively requires teacher preparation and institutional buy-in. Programs that provide practical training in VR/AR pedagogy, covering design, classroom management, and technology troubleshooting, will be critical for scaling innovations. Researchers and practitioners should explore low-cost or mobile-based VR/AR solutions to broaden accessibility, especially in under-resourced contexts. Partnerships with technology providers and open-source communities may help reduce financial barriers and expand equitable access. The smaller clusters (e.g., robotics, nursing, autism interventions) reveal a potential for cross-domain synergy. Collaborations between language educators, HCI experts, cognitive scientists, and industry partners can catalyze more robust solutions and share lessons learned across fields. Future work should explore the intersection of AI-driven personalization with VR/AR to support real-time feedback, learner analytics, and adaptive language tasks. This fusion can enhance motivation and allow tailored learning paths for individual students. Beyond language proficiency, VR/AR can simulate cultural contexts and foster social-emotional learning, areas ripe for investigation. Understanding how immersive environments influence cultural competence, empathy, and collaboration can deepen the value proposition of VR/AR-based language learning.

REFERENCES

Algerafi, M., Zhou, Y., Oubibi, M., & Wijaya, T. (2023). Unlocking the potential: A comprehensive evaluation of augmented reality and virtual reality in education. *Electronics (Basel)*, *12*(18), 3953. DOI: 10.3390/electronics12183953

Bryant, L., Bailey, B., & Hemsley, B. (2024). Developing and evaluating an immersive augmented reality application for children with developmental communication disability: InterPlay for language learning. *International Journal of Speech-Language Pathology*. Advance online publication. DOI: 10.1080/17549507.2024.2361734 PMID: 39135408

Bryant, L., Bailey, B., & Hemsley, B. (2024). Developing and evaluating an immersive augmented reality application for children with developmental communication disability: InterPlay for language learning. *International Journal of Speech-Language Pathology*. PMID: 39135408

Cai, Y., Pan, Z., & Liu, M. (2022). Augmented reality technology in language learning: A meta-analysis. *Journal of Computer Assisted Learning*, *38*(4), 929–945. DOI: 10.1111/jcal.12661

Carmona-Galindo, V. D., Velado-Cano, M. A., & Groat-Carmona, A. M. (2025). The ecology of climate change: Using virtual reality to share, experience, and cultivate local and global perspectives. *Education Sciences*, *15*(3), 290. DOI: 10.3390/educsci15030290

Çelik, F., & Baturay, M. H. (2024). The effect of metaverse on L2 vocabulary learning, retention, student engagement, presence, and community feeling. *BMC Psychology*, *12*(1), 58. DOI: 10.1186/s40359-024-01549-4 PMID: 38308350

Chang, B., Sheldon, L., Si, M., & Hand, A. (2012). Foreign language learning in immersive virtual environments. In *Proceedings of SPIE 8289, The Engineering Reality of Virtual Reality 2012* (p. 82890E). SPIE. https://doi.org/DOI: 10.1117/12.909835

Creely, E. (2025). Innovating language learning in higher education: Creativity, creative pedagogies and technologies in the 21st century. In *2024 International Graduate Research Symposium.*

Familoni, B., & Onyebuchi, N.Babajide Tolulope FamiloniNneamaka Chisom Onyebuchi. (2024). Augmented and virtual reality in U.S. education: A review: Analyzing the impact, effectiveness, and future prospects of AR/VR tools in enhancing learning experiences. *International Journal of Applied Research in Social Sciences*, *6*(4), 1043. Advance online publication. DOI: 10.51594/ijarss.v6i4.1043

Fan, M., Antle, A., & Warren, J. L. (2020). Augmented reality for early language learning: A systematic review of augmented reality application design, instructional strategies, and evaluation outcomes. *Journal of Educational Computing Research*, *58*(6), 1059–1100. DOI: 10.1177/0735633120927489

Figueroa, R. B.Jr, & Jung, I. (2025). Integrating VR tours in online language learning: A design-based research study. *Australasian Journal of Educational Technology*, *41*(1), 70. DOI: 10.14742/ajet.9498

Flores, J., Huffman, L., Lozada, V., & Dávila, E. (2022). Adopting VR in the classroom: Perceptions of bilingual and ESL education pre-service teachers. *EDMETIC*, *11*(1), 13610. DOI: 10.21071/edmetic.v11i1.13610

Huang, X., Zou, D., Cheng, G., & Xie, H. (2021). A systematic review of AR and VR enhanced language learning. *Sustainability (Basel)*, *13*(9), 4639. DOI: 10.3390/su13094639

Jang, B., Yunus, M., & Hashim, H. (2024). Virtual reality (VR) and its acceptance among English language teachers. *International Journal of Academic Research in Progressive Education and Development*, *13*(3), 21546. DOI: 10.6007/IJARPED/v13-i3/21546

Jantankul, W. (2024). Immersive reality in education: Transforming teaching and learning through AR, VR, and mixed reality technologies. *Journal of Education and Learning Reviews* . https://doi.org/DOI: 10.60027/jelr.2024.750

Kaplan-Rakowski, R., & Wojdyński, T. (2018). Students' attitudes toward high-immersion virtual reality assisted language learning. In *Proceedings of the 26th International Conference on Computers in Education* (pp. 824–829). Asia-Pacific Society for Computers in Education. https://doi.org/DOI: 10.14705/rpnet.2018.26.824

Lapka, O., & Shcherbak, A. F. (2025). *Innovative approaches to learning foreign languages: The impact of virtual reality on overcoming the language barrier*. Gamification and Augmented Reality., DOI: 10.56294/gr202590

Lee, M. J., Spryszynski, A., & Nersesian, E. (2019). Personalizing VR educational tools for English language learners. *CEUR Workshop Proceedings*, •••, 2327.

Lu, J. (2025). Research on interactive teaching of college English based on deep learning and virtual reality. *International Journal of High Speed Electronics and Systems*, 2540240. Advance online publication. DOI: 10.1142/S0129156425402402

Lutfi, N., Sunardi, A. I., Utami, A. N. E. W., & Nafiah, R. (2025). Integrating virtual reality into English language instruction: Assessing efficacy and student motivation. *Journal of Education*, *7*(2), 10975–10989. Advance online publication. DOI: 10.31004/joe.v7i2.8154

Meng, N., Li, Q., Knight, L., & Hayes, A. (2020). Panel - Improve foreign language speaking proficiency in immersive environment. In *Proceedings of the 2020 IEEE International Conference on e-Learning, e-Management, and e-Services* (pp. 411–412). IEEE. https://doi.org/DOI: 10.23919/iLRN47897.2020.9155192

Muthmainnah, M., Cardoso, L., Marzuki, A. G., & Al Yakin, A. (2025). A new innovative metaverse ecosystem: VR-based human interaction enhances EFL learners' transferable skills. *Discover Sustainability*, *6*(1), 156. Advance online publication. DOI: 10.1007/s43621-025-00913-7

Palamar, S., Brovko, K., & Semerikov, S. (2023). Enhancing foreign language learning in Ukraine: Immersive technologies as catalysts for cognitive interest and achievement. *CEUR Workshop Proceedings*, *3624*, 69–81.

Sinthiya, B. (2023). English language learning in the metaverse: Exploring the potential of AR and VR. *Shanlax International Journal of English*, *12*(S1-Dec, S1), 52. DOI: 10.34293/rtdh.v12iS1-Dec.52

Stark, B. C. (2025). Speech and language rehabilitation: The potential role of immersive VR interventions. *[Preprint]*. https://doi.org/DOI: 10.31219/osf.io/rhz3c

Syafri, N. A., & Ramli, S. (2025). Design development of Arabic-Rumi transliteration module with virtual reality concept to improve primary school students' Arabic reading skills: Design and development. *Multidisciplinary Applied Research and Innovation, 6* (2), 53–61. Retrieved from https://publisher.uthm.edu.my/periodicals/index.php/mari

Uppoor, N., Banerjee, D., Shah, D., Mishra, P., & Saha, I. (2022). Interactive language learning with VR and NLP assistance. In *2022 IEEE 7th International Conference for Convergence in Technology (I2CT)* (pp. 1–6). IEEE. https://doi.org/DOI: 10.1109/I2CT54291.2022.9824754

Wang, L. (2024). A study on the application of virtual reality technology in English language education relying on immersive learning environments and interactive experience design. *Journal of Electrical Systems*. https://doi.org/DOI: 10.52783/jes.2626

Wu, J. G., Zhang, D., Lee, S. M., & Xian, J. (2025). Game on! Digital gaming and augmented reality/virtual reality in language learning. *International Journal of Computer-Assisted Language Learning and Teaching*, *15*(1), 1–15. Advance online publication. DOI: 10.4018/IJCALLT.367601

Yang, S., & Mei, B. (2018). Understanding learners' use of augmented reality in language learning: Insights from a case study. *Journal of Education for Teaching*, *4*(4), 511–513. DOI: 10.1080/02607476.2018.1450937

Chapter 10
The Future of Language Learning: A Blend of Reality and Virtuality

Ali Rahmanipur
https://orcid.org/0009-0000-2417-2833
Bu-Ali Sina University, Iran

Moein Shokri
https://orcid.org/0009-0001-5762-2367
University of Salento, Italy

ABSTRACT

This chapter explores the transformative potential of augmented reality (AR) and virtual reality (VR) in mobile language education, arguing that these technologies address critical gaps in situational learning and cognitive load management. Drawing on theories of multimedia cognition and situated learning, it examines how AR/VR environments replicate authentic language contexts while optimizing mental engagement. The chapter analyzes current applications across formal education (e.g., classroom integration) and self-directed informal learning (e.g., gamified apps), highlighting their distinct advantages. Findings reveal that AR/VR not only enhances vocabulary retention and conversational fluency but also reduces learner anxiety through immersive, low-stakes practice. However, challenges like accessibility and teacher training persist. By synthesizing empirical evidence and practical implementations, this chapter underscores AR/VR's role in democratizing language acquisition and proposes frameworks for scalable adoption.

DOI: 10.4018/979-8-3373-3366-3.ch010

INTRODUCTION

Imagining the future of language learning while considering recent technological advancements such as Artificial intelligence, Digital Education, and Personalized Learning is extremely promising. These recent advancements have fundamentally changed the landscape of language learning by integrating cultural elements and practical situations into the educational experience. Moreover, by doing so, students and teachers are going beyond the borders of classroom language learning.

One of the most significant flaws of learning a new language in a classroom environment for years was the lack of real-world experience. When confronted with actual scenarios, students struggled to utilize the knowledge they had learned in their language courses. This issue was one of the most critical problems with traditional language teaching methodology. No matter how hard the student was studying and practicing, and no matter how engaging and active the teacher tried to be, there was always something missing in the process. Conventional methods fail because they detach words from the worlds they inhabit.

However, what significant change will revolutionize the future of language teaching and learning? How will this fill the huge gap starting at the classroom and books all the way to the situation and context?

Augmented Reality (AR) and Virtual Reality (VR) are cutting-edge technologies that have been broadly implemented in the educational landscape of various subjects (Huang et al., 2021). Augmented reality (AR) technology first appeared in the 1990s, when it was predominantly applied for training in aircraft engineering and surgical training purposes (Caudell & Mizell, 1992; Bajura et al., 1992). However, this technology did not gain the desired attention due to its arduous application. The evolution of technology has resulted in the creation of gadgets and devices that are more practical and easier to operate, thereby enhancing the user-friendliness of AR and VR technologies. Today, everyone can use Extended Reality. A range of industries and organizations are incorporating them into their training programs for various subjects and skills, and it was not long before language-teaching classrooms joined this movement.

Augmented Reality (AR) encompasses a three-dimensional technology that displays digital content within the context of the real world (Abad-Segura et al., 2020). Employing graphic computing and object recognition technologies could facilitate a richer user visualization experience, blending real-world elements with virtual objects (Redondo et al., 2020). This approach possesses a distinctive capability to establish engaging hybrid learning settings that integrate digital and physical elements, promoting the enhancement of processing skills, including critical thinking, problem-solving, and communication through collaborative tasks that rely on interdependence.

In virtual reality (VR), users are presented with a three-dimensional simulated world that provides visual experiences, creating an immersive sensation free from the boundaries of time and space (Chang et al., 2020). Virtual reality in language education offers various benefits, including visual assistance, heightened interest in learning, and chances for genuine learning experiences (Shadiev & Yang, 2020). Students will be far more motivated to learn the new language by being free to experience and navigate real-life situations, and teachers will no longer be bound to repetitive methods and contents since they can transport students to virtual foreign cities or simulated social interactions.

The application of AR and VR not only facilitates the process of language teaching and learning but also helps second-language students understand the cultural context. Language is a multifaceted phenomenon that goes beyond the basic elements of grammar, vocabulary, sentence structure, and reading and listening practices. Unlike established language education techniques, modern approaches leverage innovative technologies like augmented and virtual reality to support understanding of cultural components and offer real-life experiences. The new wave of language learners often finds traditional teaching strategies unengaging, describing them as repetitive and uninspiring. Their engagement with new technologies is notably positive, as their daily lifestyles integrate a variety of technological uses, particularly in the realm of gamified experiences. We believe teachers, policymakers, and instructional designers should embrace the latest technological advancements, particularly AR and VR, in their language teaching curricula to ensure that the new generation of students remains actively involved in their learning.

This chapter delves into the pivotal role of Augmented Reality (AR) and Virtual Reality (VR) as essential tools for the future of language learning. It will discuss how these technologies allow learners to engage in realistic conversations within simulated environments, enhancing retention, engagement, and cultural comprehension. The focus will be on the theoretical underpinnings of AR and VR, highlighting their ability to create immersive, interactive, and engaging experiences while critically analyzing the application of extended reality (particularly AR and VR) in educational settings. Given the challenges of utilizing these technologies, a comprehensive review of prior research will be presented to facilitate a problem-solving discussion. Ultimately, we will formulate a cohesive and effective conclusion regarding incorporating these technologies into language teaching practices by drawing on two theoretical frameworks—situated learning and constructivism—and empirical evidence from recent studies.

BACKGROUND

Real-World Learning

Academic subjects from schools to universities and beyond are often visually taught to learners so that the material may be from different curricula, Mathematics (Bertrand et al., 2024), Science (Tsichouridis et al., 2020), History (Remolar et al., 2021), Geography (Bos et al., 2022), and specifically English language teaching (Huang et al., 2021; Rudnik, 2023). They try to visualize the material in learners' minds to make the content more understandable and coherent. Until now, and before the emergence of various educational technologies, this was done in a conventional way and with the help of teaching methods such as lectures, creative demonstrations, videos, or websites developed by teachers to create the necessary motivation and engagement in students to learn that material. Also, accessing some educational bases to mobile museum complexes or centers equipped with various learning systems made it easier for teachers to create a learning situation similar to the real world for learners.

Given the technological developments and the emergence of various educational technologies, are conventional teaching methods and teacher creativity sufficient to implement situational learning or an education mixed with engagement and motivation?

Naturally, no, because newly born technologies have brought a supernatural feature that creates double attraction and engagement for learners. Today, with the emergence of newly born technologies such as VR and AR, most schools are equipped with situational learning and simulation of learning with the real world to fill this gap using this technology. However, this issue also faces shortcomings, which we will discuss in the rest of the chapter. In light of what has been said so far in this chapter, we will look at the newly born technologies of VR and AR in language teaching, specifically English language teaching. Additionally, we will discuss the theories and learning methods in this field and their obstacles. Although introducing these tools to language teaching and helping facilitate learning is a time-consuming task, the chapter's authors tried to include the necessary material in a comprehensive and obstacle-free manner.

Augmented reality and virtual reality technologies offer promising possibilities in language teaching. Research shows that these technologies can create deep and engaging educational experiences that increase learner motivation and improve the quality of learning (Huang et al., 2021). The unique feature of these technologies is their ability to create a link between the real environment and the virtual space, which provides dynamic and interactive conditions for language learners, a possibility that does not exist in conventional educational methods (Alizadeh, 2019; Dobrova et al., 2018).

Studies indicate that the use of these technologies can positively impact various aspects of language learning, including vocabulary learning, improving pronunciation, and strengthening communication skills (Rudnik, 2023). Although both technologies have valuable capabilities, augmented reality is currently considered a more suitable option for training teachers and language classes due to its ease of access and implementation. In contrast, the technical limitations of virtual reality have challenged its widespread use (Rudnik, 2023).

To effectively utilize these technologies, teachers must receive the necessary training and consider the individual characteristics of learners (Huang et al., 2021). Augmented and virtual reality are effective tools for creating attractive and effective educational environments in the field of language learning.

Situational Learning

To properly understand knowledge, teaching it in real spaces and situations related to the subject is necessary. Social interactions and teamwork are key elements of this learning method. For effective learning, the learning environment should be designed and aligned with the learners' practical tasks and prepare them for success in real-life situations (Mei & Sheng, 2011). Such an environment can include interactive experiences with the social, psychological, and physical environment in which the individual is located. Also, educational content linked to the learner's daily life provides a platform for developing deep and reflective thinking. It can be said that situational learning is one of the notable educational approaches in the discussion of language teaching with the help of augmented reality and virtual reality because the necessary infrastructure in this approach is similar to the practical operation of these two technologies. Therefore, teachers and instructional designers should prepare environments similar to situational learning environments to use these technologies to provide the necessary measures for more effective and motivated learning.

Key Principles of Real-Life Learning Environments

Effective learning environments should:

Ensure Authenticity:

Incorporate authentic contexts reflecting real-world knowledge application.
Design authentic activities mirroring practical use cases.

Facilitate Expert Engagement:

Provide access to expert practices and process modeling.
Offer mentoring and scaffolding during critical learning phases.

Encourage Diverse Perspectives & Collaboration:

Support multiple roles and perspectives to enrich understanding.
Foster collaborative knowledge construction (aligned with constructivist principles).

Promote Reflection & Articulation:

Enable reflection to aid abstract conceptualization.
Encourage expression to make tacit knowledge explicit.

Integrated Assessment:

Implement embedded assessment to evaluate learning authentically.

Consequently, instructional designers, educators, and educational technologists are anticipated to create conditions that enable learners to engage with augmented and virtual realities within the language learning context. This engagement should allow them to experience a simulated environment that parallels the real world while also empowering them to become active creators of their knowledge through the interplay between these two environments (Amineh & Asl, 2015).

Cognitive Load in Multimedia Learning

Technologies such as augmented reality and virtual reality are now being used to change traditional education's paradigm and its conventional and mundane methods (Johnson et al., 2016). This use of multimedia technologies has led to the emergence of an important field of study (Çeken & Taşkın, 2025). Multimedia, for example, such as AR and VR, combine visual and verbal elements to present information and enhance the learning process through educational messages that combine visual and verbal information (Richter et al., 2016). Meyer (1997) first proposed the idea of multimedia learning as a process of acquiring knowledge through dual processing channels. With the increasing use of multimedia technologies in education, extensive empirical studies have evaluated the effectiveness of this approach. This systematic research ultimately led to Meyer's formulation of the Cognitive Theory of Multimedia Learning (CTML) (2009). This theoretical framework explains the cognitive mechanisms underlying effective learning through multiple media. With the rapid development of multimedia technologies, the scope of application of multimedia learning principles has expanded beyond traditional frameworks. It has expanded to computer-based learning systems, online learning environments, especially virtual learning spaces (Klingenberg et al., 2023), and augmented reality learning environments (Lai et al., 2019). These developments illustrate the increasing dynamism of educational platforms and the need to rethink learning strategies to accommodate these emerging technologies. However, despite the significant expansion of research in multimedia learning, significant research challenges, especially in augmented reality (AR) and virtual reality (VR) implementation, still require deeper investigation.

In this section of the present chapter, we aim to draw the attention of three key groups of instructional designers, teachers, and curriculum planners to the critical issue of cognitive load management, especially when utilizing language teaching technologies that have two essential characteristics:

- Integration with multimedia elements
- Simultaneous engagement of multiple sensory channels

Cognitive Load Theory, as an educational framework, is rooted in a fundamental understanding of human cognitive mechanisms (Sweller et al., 2011). By focusing on the concept of cognitive load, this theory provides valuable theoretical and practical insights to improve the quality of learning experiences. In modern educational environments based on augmented reality (AR) and virtual reality (VR), paying attention to the additional cognitive load resulting from the simultaneous processing of visual and auditory stimuli is particularly important. This phenomenon is directly related to the methods of presenting educational content and educational design patterns. It reveals the need to optimize educational materials in order to prevent inefficient designs and, as a result, achieve more effective learning (Kalyuga, 2009). Accordingly, this theory suggests several practical strategies to reduce additional cognitive load and improve the efficiency of the learning process. Redesigning the physical learning environment with the aim of reducing distracting environmental stimuli can significantly reduce external cognitive load. This, through the optimal allocation of cognitive resources to the learning process, leads to improved learning quality at two levels: deeper conceptual understanding and long-term content retention. The design and organization of the learning environment based on these technologies play a significant role in influencing cognitive load and, consequently, learning outcomes, which is more important than merely using these technologies.

CURRENT APPLICATIONS

After examining the theoretical foundations and context surrounding the use of augmented reality and virtual reality in educational settings, particularly in language acquisition, it is pertinent to explore the practical applications of this technology among modern language learners. This section analyzes how language institutes, educational establishments, universities, and individual language learners leverage AR and VR. Undoubtedly, it is anticipated that each of the discussed environments and applications has its limitations and problems, in addition to their respective advantages and benefits. These elements will be thoroughly examined and discussed in detail.

The global outbreak of the COVID-19 pandemic has led to extensive institutional and behavioral disruptions in various domains, particularly in education. The effects on students are extraordinary: On April 9, 2020, over 1.5 billion individuals from primary to higher education could not attend school (UNESCO, 2020) (Teräs et al.., 2020). According to Williamson et al. (2020), During the COVID-19 pandemic, media and technology have the potential and responsibility to enhance knowledge and practice significantly. The shift to online and digital education formats, alongside the growth of remote teaching and learning, has emerged due to the extensive closure of schools, colleges, and universities. In this politically charged environment, where various entities are engaged in debates over crisis management, distance education has become a pressing issue for political figures, educational organizations, charitable groups, teachers, parents, and students (Williamson et al., 2020). The circumstances prompted governments worldwide to invest in digitalizing educational systems, encompassing language education as well. In the final analysis, the trajectory of technological progress has brought about the current use of AR and VR, which embody many of the key features that are in demand within the realm of digital and modern education, including personalized language learning, student engagement, accessibility, and the enhancement of motivation and interaction. These technologies are still evolving and have much to accomplish, yet they have already traveled a considerable distance.

Formal Education

Augmented Reality: Liu, T. Y. published a report in 2009 regarding a study that focused on creating a sensor and a handheld AR-integrated ubiquitous learning (u-learning) environment named the Handheld English Language Learning Organization (HELLO). In this study, students practiced English language skills across the campus utilizing a handheld ARLE (Augmented Reality Learning Experience) prototype. The lessons featured vocabulary and sentence patterns relevant to different locations, such as the classroom, gallery, library, gym, laboratory, cafeteria, and health clinic. In this scenario, students observe and learn while attempting to generate language within familiar settings and environments. This familiarity facilitates their ability to connect with the context and situations where the language is applied (Santos et al., 2013). In this context, the described augmented reality technology alters the classroom setting to simulate real-life environments, enabling students to practice language in situations where it is applicable. This technology has demonstrated its capacity to transcend the limitations of real-world experiences, enabling the creation of scenarios that are less probable in everyday life. A research study titled "Study of Learning System Based on Augmented Reality and Mobile Learning" by Tarng and Ou (2012) presents a virtual ecological environment for butterflies, allowing students to engage with butterfly ecology. Through smartphones, students can nurture virtual caterpillars on host plants and observe their development, thereby gaining insights into the butterfly life cycle. The findings of this study indicate that when specific subjects are challenging to experience in a classroom setting due to accessibility or financial constraints, AR technology can be employed to simulate these experiences for learners. While Tarng and Ou's research addresses explicitly the teaching of ecology in elementary and junior high schools, the application of this technology is not confined to that discipline, as it can generate a vast array of virtual objects. In language instruction, educators often focus on real-life scenarios that students are likely to encounter in the future; however, incorporating imaginative and fictional contexts can enhance student engagement and foster creativity.

While AR offers numerous advantages and significant contributions, it also presents particular challenges. A prevalent concern among students is the " Usability " issue (M. Akçayır & G. Akçayır, 2017). Research conducted globally indicates that many students struggle to engage with AR technology effectively. This challenge is particularly critical when considering the age group of students learning a new language, as it significantly narrows the potential user base by excluding younger learners. Kaufmann and Dünser (2007) found that although users expressed satisfaction with the controllability and learnability of the technology, they were frustrated by various technical issues related to its application.

Additionally, Kerawalla et al. (2006) reported that teachers shared similar concerns, particularly regarding the rigidity of the content. Given that AR content is often limited, and many educators lack the training to create and adapt new materials suitable for their teaching methods, this has led to a reluctance among some teachers to utilize AR. Nevertheless, despite these challenges, the future of AR applications remains promising, as the identified issues are addressable and are anticipated to be resolved for the next generation of language educators and learners.

Virtual Reality: Ma and Zheng (2011) categorize virtual reality systems into three distinct types: immersive, semi-immersive, and non-immersive. A non-immersive

VR system typically consists of a desktop computer equipped with a 3D graphical interface, which users interact with via a keyboard, mouse, and a small monitor. In contrast, a semi-immersive system features a larger display that enhances the graphical experience and may incorporate gesture recognition technology to facilitate more natural user interactions. This type of system offers a broader field of view, resulting in a greater sense of immersion compared to non-immersive systems. The final category, fully immersive VR, utilizes head-mounted displays that surround entirely the user's field of vision, thereby providing an unparalleled sense of presence. Immersive VR systems are arguably the most recognized among the various types of VR technologies.

Table 1. Tools of each VR category

Immersive	Head-Mounted Display (HMD), HeadCoupled Display (HCD), Visual display units and speakers within a HMD unit, Data gloves, Body suits, Joystick, Space ball, 3D mouse, Hand-held sensors, Cameras
Semi-Immersive	Either a large screen monitor, or a large screen projector, or multiple television projection systems,
Non-Immersive	A monitor, A keyboard and A mouse

The application of various categories of augmented reality (AR) systems is influenced by multiple factors that educational institutions or systems consider. The primary objective of employing virtual reality (VR) is to enhance language acquisition by immersing students in relevant scenarios, enabling them to respond appropriately based on their learning. This approach necessitates depicting real-world and, in some instances, imaginative events and situations. Implementation can occur at different levels, utilizing diverse VR technologies. For instance, immersive systems require a sophisticated array of tools that may not be readily available to all, and many of these tools can be prohibitively expensive.

Additionally, the technical knowledge required to operate these technologies necessitates trained educators, presenting both financial and time-related challenges. However, these systems offer intricate and advanced functionalities that significantly enhance the learning experience. Conversely, semi-immersive technologies are more affordable and user-friendly, yet they still require specialized environments for installation, making them less suitable for standard classrooms. Lastly, non-immersive VR systems are both accessible and cost-effective, but they lack the engagement and sophistication found in the other two categories.

In summary, following the pandemic, educational systems globally began allocating funds and resources towards modern and digital teaching methodologies, including language acquisition. This surge of innovation has also extended to integrating AR and VR. While these emerging technologies are still nascent, their development progresses swiftly and holds great promise. Current research indicates that the use of AR and VR is primarily confined to experimental exercises and has not yet become a widespread phenomenon. The implementation and practices associated with these technologies differ according to the financial and educational strategies of various schools, universities, and institutions. In the realm of language education, there are challenges related to usability and content tailored for advanced learning technologies; however, there are also significant benefits. The integration of AR and VR, particularly in language teaching and learning, has made substantial strides and is rapidly evolving to shape the future of education.

SELF-DIRECTED AND INFORMAL LEARNING

The integration of AR and VR in education is not solely emphasized by formal institutions; rather, the true potential of these technologies may lie beyond traditional educational frameworks. Self-directed learning (SDL) and informal language acquisition often occur by incorporating language practice into daily leisure activities, gaming, and social interactions. The widespread popularity of Pokémon GO during the summer of 2016 drew significant public interest in AR. As is typical with emerging technological trends, the educational community quickly began to investigate how this game could be utilized in teaching and learning, particularly in the context of second language acquisition (Godwin-Jones, 2016). Furthermore, a study by Symonenko et al. (2020) examined various popular VR applications for foreign language learning, including Mondly, VRSpeech, VR Learn English, Gold Lotus, AltSpaceVR, and VirtualSpeech. They concluded that VR applications present substantial opportunities for engaging students in the foreign language learning process while successfully achieving three primary objectives: enhancing language acquisition, preparing undergraduates for real-world and professional scenarios outside their native language, and improving communication skills. Consequently, as users of these technologies, language learners interact with new vocabulary, contextual language practice, grammar exercises, and other activities that typically require significant effort from teachers and students but are now rendered enjoyable, making learning a new language more engaging.

Self-directed learning offers modern language learners a flexible schedule and tailored content, particularly for those who struggle with conventional teaching methods. However, it presents challenges related to motivation, consistency in self-paced learning, and a lack of structured feedback, which are crucial in language education. Private tutors can play a vital role in addressing these challenges by monitoring and assessing the language acquisition processes that integrate AR and

VR. The rise of remote and online meetings has made finding and hiring such tutors easier. Nonetheless, a significant concern is that these educators must be proficient in the aforementioned technologies. Unfortunately, locating teachers trained in extended reality teaching methodologies can be tricky. Nevertheless, given the rapid advancement of these technologies and their increasing application in language education globally, this issue will likely resolve itself soon.

The current utilization of augmented reality (AR) and virtual reality (VR) in language learning settings, whether formal or informal, is not yet widespread but is experiencing rapid growth. As the demand for extended reality technologies increases, the industry is expanding, increasing the availability of related tools, equipment, usability options, and training programs. These developments not only enhance accessibility to these technologies but also promote their application, particularly in language education. In recent years, it has become evident that language learners are eager to adopt innovations that facilitate a more engaging, enjoyable, and personalized learning experience. The recent widespread integration of artificial intelligence further supports this observation. It is not an unfounded assertion that augmented and virtual reality represents the future of language learning.

CASE STUDIES

Recent research shows that virtual reality (VR) and augmented reality (AR) technologies provide better results than traditional language teaching methods, especially in strengthening listening and reading skills for learners of different levels (Pataquiva & Klimova, 2022). However, researchers believe that in order to utilize these technologies optimally in the language learning process, it is essential that teachers undergo appropriate training courses and that the factors affecting learners' learning are further investigated (Huang et al., 2021).

In this section of the chapter, we will analyze several case studies to examine the successes, challenges, and innovations related to the use of augmented reality and virtual reality technologies in language teaching. We will also compare the practical applications of these technologies in language learning to find out how they can help improve the language learning process for learners.

Case Study No. 1

With the development of new educational technologies, Augmented Reality (AR) learning has emerged as an emerging approach in second language teaching. The present study examines the practical experience of augmented reality in improving English as a foreign language writing skills within the framework of mobile-based learning. This study is based on a research paper titled Using Augmented Reality-based Mobile Learning Material in EFL English Composition: An Exploratory Case Study, (Liu & Tsai, 2013).

Research Context

In 2011, a study was conducted on five 20-year-old English language students at Kainan University in Taiwan. The students had no prior experience using mobile-based learning. The main objective was to investigate the effect of using augmented reality content on the process of writing a descriptive essay in English.

Instruments and Execution Process

The educational content was delivered through an application designed for smart mobile devices. The application used GPS to identify the students' location and then used the mobile phone camera to combine images of the real environment with digital information. The augmented information included the buildings' names, uses, history, and visual descriptions of prominent university landmarks.

Students moved around the campus environment, receiving textual and visual information related to different locations. In the end, they were required to write a descriptive essay about one of the locations or routes they observed. Qualitative analysis of the content written by the students showed that they effectively used the vocabulary and linguistic structures presented in the augmented reality content.

Participant Feedback

Feedback collected through a questionnaire showed that most students reported the experience of using augmented reality as engaging, practical, and effective in language learning. They emphasized that this tool increased their active participation, cognitive engagement, and motivation in writing.

Overview

The present study shows that augmented reality, especially in mobile learning environments, can be used as an effective tool to enhance writing skills in a second language. By providing real-world learning experiences and situational learning, this technology transforms writing from an abstract activity into a meaningful, tangible, and relevant experience in everyday life. Also, integrating visual, spatial, and textual elements in the learning content provides a good opportunity to strengthen semantic memory, functional vocabulary, and writing structures. This study can be a valuable reference for administrators and teachers who want to design mobile-based English curricula or courses that encourage students to learn the language in different environments.

Case Study No. 2

In the era of new educational technologies, Augmented Reality (AR) has been considered a promising tool in second language teaching, especially for younger age groups. This study provides a practical example of the successful use of augmented reality in teaching the English alphabet to kindergarten children. It is based on a study titled "The Effectiveness of Using Augmented Reality Apps in Teaching English Alphabet to Kindergarten Students" (Safar et al., 2016).

Research Context

This study was conducted in Kuwait in 2016, with the participation of 42 kindergarten children. Participants were selected from two kindergartens and randomly divided into two control and experimental groups. This quasi-experimental study compared the performance of the two groups in learning the English alphabet.

Instruments and Execution Process

In the experimental group, learning was carried out using the applications "AR Flashcards Animals-Alphabet" and "AR Flashcards Alphabet." These applications displayed English letters in 3D, colorful, and interactive forms to engage children's visual and auditory memory simultaneously. In contrast, the control group experienced education in the traditional way using flashcards and a classroom board. The duration of the education for both groups was 7 weeks.

Participant Feedback

Children in the experimental group showed high interaction and excitement towards using the augmented reality application. They participated in the sessions with great motivation and interest and were active in learning by spontaneously repeating letters and words along with 3D objects. Educators also reported that the concentration and interactive participation level during the education had increased significantly.

Overview

The present study shows that using augmented reality applications in language teaching to young children can not only improve their cognitive performance in learning the alphabet but also increase motivation, enjoyment of learning, and active involvement in the education process. Augmented reality transforms the learning space from an abstraction to a sensory and engaging experience by linking language content to visual and tangible objects, making it a powerful tool for language teaching in preschool.

Challenges and Considerations

Despite the positive results, researchers noted that the success of using AR technology at this age requires the provision of hardware infrastructure (tablet or smartphone) and technical support. Also, teachers and trainers must receive the necessary training to use this technology correctly so that they can use it purposefully in the teaching-learning process (Safar et al., 2016).

Case Study No. 3

With the advancement of interactive technologies, foreign language learning has moved away from traditional methods and towards simulated and immersive environments. In this study, advanced Chinese language students at a US public university gained a unique experience of learning the target language and culture by using interactive virtual reality tools such as Google Cardboard and Google Expeditions. The study, "Using Interactive Virtual Reality Tools in an Advanced Chinese Language Class: A Case Study," investigated students' perceived benefits and challenges of using VR tools to learn the Chinese language and culture (Xie et al., 2019).

Research Context

Twelve Chinese language students (four females and eight males) from various academic disciplines in the United States participated in this study. Ten were native English speakers, and two were Chinese language learners. Most of the students had previously taken several Chinese language courses, and eight had a history of study trips to China.

Instruments and Execution Process

Over a semester (15 weeks), students were paired up and given six "virtual tour guides" to give oral presentations about cultural and historical sites in China. Four of these tours were designed using the Google Expeditions app. Each group chose a real-world location, such as the Great Wall of China, Beijing, or Guilin, and used the app's 360-degree video data and text descriptions to create a complete educational tour in Chinese.

Students extracted the necessary content from the app and then wrote the presentation. On the presentation day, one group member acted as a guide using an iPad while the other students virtually walked around the presented location using Google Cardboard.

Participant Feedback

Most participants described the experience of using VR tools as "exciting," "useful," and "motivating." They stated that seeing real places helped them better learn vocabulary and language structures and made them more interested in Chinese history and culture. One student stated:

"I felt like I was in the place I was talking about; this immersive experience deepened my connection with the Chinese language."

Students also noted that using VR reduced the cognitive load associated with preparing the presentation, allowing them to focus their energy on more accurate language production and pronunciation. In addition, many reported increased confidence in presenting in Chinese compared to the beginning of the semester.

Overview

The study showed that combining virtual reality with second language teaching can transform learning from a theoretical and classroom-based mode to a real-world cultural experience. Combining realistic images, historical data, and an interactive environment helped students learn more advanced vocabulary, feel less anxious about speaking, and establish a deeper connection with the target culture. Of course, the success of this method depends on careful preparation, technical support, and innovative design of learning activities.

Challenges and Considerations

Despite the positive aspects, some students also faced challenges. These challenges include dizziness caused by using headsets, technical problems such as internet connection, and the inability to follow virtual tours due to incorrect pronunciation by some students. Also, some students felt anxious when faced with new technologies or were confused in coordinating with applications.

Suggestions include allocating more time for initial technical training, holding individual practice sessions with the teacher, and providing a vocabulary list with pinyin pronunciation to make the learning experience more consistent and efficient for all students (Xie et al., 2019).

Case Study No. 4

One of the main challenges of second language learning in non-English speaking countries is the lack of real-world environments for language practice. The present study investigates the empirical impact of augmented reality (AR) technology in creating a simulated classroom for English language teaching in schools. The research focuses on the impact of AR-based instruction on increasing students' motivation, attention, self-confidence, satisfaction, and overall learning performance and is analyzed using the ARCS motivational model. This study is based on "Enhancing English-Learning Performance through a Simulation Classroom for EFL Students Using Augmented Reality—A Junior High School Case Study" (Chang et al., 2020).

Research Context

In this study, 40 junior high school students from Da-li District, Taichung City, Taiwan, participated. The students were divided into experimental (n=21) and control (n=19) groups. Both groups were taught by the same teacher over 18 weeks. However, they experienced different teaching methods: the experimental group used the HP Reveal Aurasma Studio augmented reality app, while the control group was taught in a traditional PowerPoint-based format.

Instruments and Execution Process

To create realistic learning situations, the teacher designed eight AR instructional videos using the Aurasma Studio platform about practical airport conversations (such as getting a ticket, boarding a plane, and talking to airport agents). Students in the experimental group used their smartphones to scan specific images in the textbook and play the instructional videos on their devices.

Each video included dialogues accompanied by 3D animations and audio content that helped students interact in a simulated English environment. Meanwhile, the control group received the same material through traditional methods without

AR interaction. This new approach to language learning can significantly impact students' learning experience and increase their motivation and engagement.

Participant Feedback

Statistical analysis showed that the experimental group's scores on the post-test were significantly higher than the control group. This difference was examined in the four skills of listening, speaking, reading, and writing, and it was found that the use of augmented reality positively improved learning performance. In addition, the questionnaire responses of the experimental group students showed that they:

were more focused while learning.
enjoyed the interactive content.
gained more confidence in speaking English.
were satisfied with their learning experience.

Based on the ARCS model analysis, the variables of interest, focus, confidence, and learning satisfaction were all positively and significantly correlated with learning performance.

Overview

This case study demonstrates a successful example of using AR in English language teaching in middle schools. Augmented reality technology created an engaging and interactive environment and moved language teaching away from the theoretical level and into a practical and meaningful experience. Designing simulated classes with multimedia content can increase students' intrinsic motivation and smooth the path to learning a second language.

Challenges and Considerations

Although the results were positive, some limitations were also observed; including the need to equip schools with AR technologies, prepare educational content by teachers, and the need for students to be initially familiar with how to use the application for instructional design. Also, some students initially experienced anxiety in using technology, which decreased over time (Chang et al., 2020).

Case Study No. 5

Given the increasing role of technology in second language teaching, this study examines the impact of integrating Virtual Reality (VR) technology on developing English language speaking skills among ESL learners in an independent Chinese high school in Malaysia. The tool used was the ImmerseMe platform, an application that provides simulations of real-life situations such as going to a coffee shop, booking a hotel, or reporting a theft with spoken interactions in 3D space. This study was conducted within the context of a research project titled Exploring the Impact of VR Integration on ESL Learners' English Verbal Communication Skills: A Case Study in a Malaysian High School (Raman et al., 2024).

Research Context

Fifteen students whose first language was Chinese were purposively selected. According to the official SPM test, all of them had an intermediate level of speaking skills and had initial experience using VR headsets. The training was conducted over 8 weeks in the 2022/2023 academic year.

Instruments and Execution Process

The ImmerseMe app was used alongside smartphones and VR goggles to implement the intervention. The teacher selected topics relevant to the curriculum and set them up in the platform environment. The topics, in order of the 8-week training course, included:

1. Hotel Reservations 2. Ordering Coffee 3. Reporting a Theft 4. Movie Recommendations 5. Making Vacation Decisions 6. Describing Traditional Ceremonies 7. Encouraging Environmental Actions 8. Conclusions in Conversations

Participant Feedback

Each student spoke to avatars in a simulated environment that acted out different scenarios. The app automatically analyzed and scored the students' articulation, pronunciation, and fluency. The teacher could also track each student's progress through a dashboard. This study's findings showed that using virtual reality in conversational environments can significantly improve learners' fluency, accuracy, confidence, and motivation.

Overview

The present study shows that virtual reality can play a key role in developing second language speaking skills, especially when combined with interactive activities, realistic scenario design, and automated feedback. Platforms such as ImmerseMe have not only created realistic simulated environments but also provided the opportunity to learn in the context of practical language situations. This educational method can be used as a new model for language teaching in non-English-speaking schools.

Challenges and Considerations

- Although the overall experience was successful, some challenges were reported:
- Technical issues in some sessions (resolved using ICT lab computers)
- Test anxiety in one student, which caused a lack of progress.
- Difficulty in some subjects for students with limited vocabulary (Raman et al., 2024).

FUTURE DIRECTIONS

This section aims to explore future trajectories for AR/VR in relation to Technological Innovations, Pedagogical Developments, Research Priorities, and Accessibility or Implementation. We anticipate that this discussion will serve as a valuable resource for language educators, researchers, policymakers, and instructional designers, assisting them in advancing the integration of these rapidly evolving technologies. The recommendations presented here are derived from a range of scholarly articles. We contend that insights gained from practical applications of these technologies may offer greater advantages than our initial expectations. Nonetheless, we have also outlined some anticipated challenges that are prevalent in contemporary language education.

Educators play a crucial role as both facilitators and gatekeepers in deciding what content is introduced into the classroom. Consequently, the preparedness of teachers is the primary prerequisite for ensuring sustainable education, which in turn offers equitable opportunities for students to engage in ongoing learning in the face of rapid technological advancements (Lee & Hwang, 2022). At present, the availability of trained educators equipped to meet students' needs through AR/VR tools is quite restricted. This limitation is somewhat justifiable, as the use of these technologies is also constrained. Consequently, educational institutions and organizations often deem it financially impractical to invest in teacher training initiatives. Given that the development of AR teaching materials necessitates significant time for curriculum design and system creation (Chang et al., 2020), it is essential to have dedicated teachers for classes utilizing AR/VR. However, very few educational systems are capable of offering classes specifically focused on AR/VR.

A significant challenge in both research and practical applications is that Augmented and Virtual Reality technologies have only recently been integrated into language education methodologies. Consequently, researchers must engage in long-term studies. The success of language acquisition hinges on sustained practice to enhance communication skills (Chang et al., 2020). Therefore, further investigation into the long-term effects of these technologies in language learning is essential.

In their 2013 study, Liu and Tsai indicated that while the majority of participants acknowledged certain benefits of the learning materials—such as user-friendliness, engaging interactions, and helpful support—they expressed a desire for more comprehensive information about the scenic locations. One participant noted that they did not utilize the mobile device during the activity due to a lack of familiarity with information technology. This instance underscored the importance of having adequate familiarity with AR-based mobile learning resources. Future research could specifically identify these issues to inform subsequent developments.

CONCLUSION

The transformative potential of augmented reality (AR) and virtual reality (VR) in mobile language learning is not merely a technological evolution but a pedagogical revolution. These technologies effectively connect theoretical knowledge with practical experience, addressing two essential components of successful education: situated learning and cognitive load management. As illustrated in this chapter, integrating these concepts with immersive technologies is transforming the methods of language instruction, facilitating learning in vibrant, contextually relevant settings that reflect real-life interactions rather than being confined to conventional classroom environments.

Central to the process of language acquisition is the necessity for contextual and socially integrated practice. Conventional approaches—such as textbooks, scripted conversations, or video lessons—frequently do not capture the spontaneity and cultural subtleties inherent in authentic communication. AR and VR address this gap by creating immersive environments where learners can engage in meaningful interactions, whether negotiating in virtual cafés in Paris, maneuvering through crowded train stations in Tokyo, or working collaboratively with peers in global teamwork exercises.

Significantly, these technologies engage with the social and psychological facets of language learning. For instance, virtual reality's feature of user anonymity diminishes the fear of being judged, which empowers anxious learners to practice their speaking skills without the burden of social repercussions. This functionality illustrates that augmented and virtual reality do more than merely simulate environments; they also cultivate emotionally safe spaces that encourage exploration, mistakes, and personal growth.

Incorporating AR and VR into language education is consistent with Mayer's (2009) Cognitive Theory of Multimedia Learning (CTML), which suggests that processing information through visual and verbal channels improves memory retention. Conventional multimedia resources, such as videos and slides, frequently inundate learners with fragmented information. In contrast, AR and VR effectively organize information spatially, integrating vocabulary hints into interactive three-dimensional objects (for instance, labels that appear above items in a virtual marketplace) or situating grammar rules within the context of dialogues. This approach aligns with the findings of Lai et al. (2019), which indicate that AR settings diminish unnecessary cognitive load by delivering information in naturally segmented and relevant task-oriented segments.

Nevertheless, obstacles remain. Klingenberg et al. (2023) emphasize that inadequately designed virtual reality experiences can increase cognitive load due to overwhelming stimuli or confusing interfaces. A case study presented in this chapter illustrated how users of an initial augmented reality language application faced difficulties with overly complex visual overlays, resulting in frustration. These challenges underscore the importance of a learner-centered design approach, which focuses on simplicity, flexibility, and adherence to the evidence-based principles of the Cognitive Theory of Multimedia Learning (CTML).

The implementation of augmented and virtual reality in language education is expanding, yet it is not uniform across the board. Traditional educational institutions and universities that have set up VR language laboratories are leading the way in language education. They are typically at the forefront of this innovation. However, financial constraints and inadequate infrastructure prevent many from participating. On the other hand, independent learners are finding greater access to cost-effective AR applications like Mondly and Immerse, which enhance vocabulary learning through gamification via smartphone cameras. This contrast highlights a significant issue: Although the potential of AR/VR is widespread, its accessibility varies greatly.

The five case studies mentioned in the chapter collectively demonstrate that AR and VR enhance language learning by providing immersive, context-rich experiences that improve engagement, retention, and confidence. From Taiwanese university students using AR for writing practice to Malaysian ESL learners developing fluency through VR conversations, these technologies bridge the gap between classroom instruction and real-world application. Key successes include increased motivation, reduced anxiety, and improved performance across skills like vocabulary acquisition and speaking fluency. Challenges such as technical barriers, teacher training needs, and equitable access emphasize that effective implementation necessitates pedagogical alignment, user-friendly design, and institutional support. These cases underscore AR/VR's potential to transform language education—not by replacing traditional methods but by creating dynamic, interactive learning environments that prepare learners for honest communication.

The future of mobile language learning is poised to integrate the physical and digital realms harmoniously. As AR glasses become lighter and VR platforms evolve to be more engaging, the distinction between "learning" a language and "living" it will diminish. Policymakers and developers must focus on equity during this transition, ensuring that these technologies do not deepen existing inequalities but act as tools for equal opportunity. Ultimately, AR and VR represent innovations and the cornerstone of a new framework in which language education transcends traditional classroom settings and becomes an integral part of daily life.

REFERENCES

Abad-Segura, E., González-Zamar, M. D., Rosa, A. L. D. L., & Cevallos, M. B. M. (2020). Sustainability of educational technologies: An approach to augmented reality research. *Sustainability (Basel)*, *12*(10), 4091. DOI: 10.3390/su12104091

Akçayır, M., & Akçayır, G. (2017). Advantages and challenges associated with augmented reality for education: A systematic review of the literature. *Educational Research Review*, *20*, 1–11. DOI: 10.1016/j.edurev.2016.11.002

Alizadeh, M. (2019). Augmented/virtual reality promises for ELT practitioners. *Diversity and Inclusion*, 360–368.

Amineh, R. J., & Asl, H. D. (2015). Review of constructivism and social constructivism. *Journal of Social Sciences. Literature and Languages*, *1*(1), 9–16.

Bajura, M., Fuchs, H., & Ohbuchi, R. (1992). Merging virtual objects with the real world: Seeing ultrasound imagery within the patient. *Computer Graphics*, *26*(2), 203–210. DOI: 10.1145/142920.134061

Bertrand, M. G., Sezer, H. B., & Namukasa, I. K. (2024). Exploring AR and VR tools in mathematics education through culturally responsive pedagogies. *Digital Experiences in Mathematics Education*, *10*(3), 462–486. DOI: 10.1007/s40751-024-00152-x PMID: 39497730

Bos, D., Miller, S., & Bull, E. (2022). Using virtual reality (VR) for teaching and learning in geography: Fieldwork, analytical skills, and employability. *Journal of Geography in Higher Education*, *46*(3), 479–488. DOI: 10.1080/03098265.2021.1901867

Caudell, T. P., & Mizell, D. W. (1992). Augmented reality: An application of heads-up display technology to manual manufacturing processes. In Nunamaker, J. F., & Sprague, R. H. (Eds.), *Proceedings of the Twenty-Fifth Hawaii International Conference on System Sciences* (Vol. 2, pp. 659–669). IEEE Computer Society Press. DOI: 10.1109/HICSS.1992.183317

Çeken, B., & Taşkın, N. (2025). Examination of multimedia learning principles in augmented reality and virtual reality learning environments. *Journal of Computer Assisted Learning*, *41*(1), e13097. DOI: 10.1111/jcal.13097

Chang, S. C., Hsu, T. C., Chen, Y. N., & Jong, M. S. Y. (2020). The effects of spherical video-based virtual reality implementation on students' natural science learning effectiveness. *Interactive Learning Environments*, *28*(7), 915–929. DOI: 10.1080/10494820.2018.1548490

Chang, Y.-S., Chen, C.-N., & Liao, C.-L. (2020). Enhancing English-Learning Performance through a Simulation Classroom for EFL Students Using Augmented Reality—A Junior High School Case Study. *Applied Sciences (Basel, Switzerland), 10*(21), 7854. DOI: 10.3390/app10217854

Godwin-Jones, R. (2016). Augmented reality and language learning: From annotated vocabulary to place-based mobile games. *Language Learning & Technology, 20*(3), 9–19. DOI: 10.64152/10125/44475

Huang, X., Zou, D., Cheng, G., & Xie, H. (2021). A systematic review of AR and VR enhanced language learning. *Sustainability (Basel), 13*(9), 4639. DOI: 10.3390/su13094639

Johnson, L., Becker, S. A., Cummins, M., Estrada, V., Freeman, A., & Hall, C. (2016). *NMC horizon report: 2016 higher education edition.* The New Media Consortium.

Kalyuga, S. (2009). *Cognitive load factors in instructional design for advanced learners.* Nova Science Publishers.

Kaufmann, H., & Dünser, A. (2007). Summary of usability evaluations of an educational augmented reality application. In *International Conference on Virtual Reality* (pp. 660–669). Springer. DOI: 10.1007/978-3-540-73335-5_71

Kerawalla, L., Luckin, R., Seljeflot, S., & Woolard, A. (2006). "Making it real": Exploring the potential of augmented reality for teaching primary school science. *Virtual Reality (Waltham Cross), 10*(3-4), 163–174. DOI: 10.1007/s10055-006-0036-4

Klingenberg, S., Fischer, R., Zettler, I., & Makransky, G. (2023). Facilitating learning in immersive virtual reality: Segmentation, summarizing, both or none? *Journal of Computer Assisted Learning, 39*(1), 218–230. DOI: 10.1111/jcal.12741

Lai, A. F., Chen, C. H., & Lee, G. Y. (2019). An augmented reality-based learning approach to enhancing students' science reading performances from the perspective of the cognitive load theory. *British Journal of Educational Technology, 50*(1), 232–247. DOI: 10.1111/bjet.12716

Lave, J., & Wenger, E. (1991). *Situated learning: Legitimate peripheral participation.* Cambridge university press. DOI: 10.1017/CBO9780511815355

Lee, H., & Hwang, Y. (2022). Technology-Enhanced Education through VR-Making and Metaverse-Linking to Foster Teacher Readiness and Sustainable Learning. *Sustainability (Basel), 14*(8), 4786. DOI: 10.3390/su14084786

Liu, P. H. E., & Tsai, M. K. (2013). Using augmented-reality-based mobile learning material in EFL English composition: An exploratory case study. *British Journal of Educational Technology*, *44*(1), E1–E4. DOI: 10.1111/j.1467-8535.2012.01302.x

Liu, T. Y. (2009). A context-aware ubiquitous learning environment for language listening and speaking. *Journal of Computer Assisted Learning*, *25*(6), 515–527. DOI: 10.1111/j.1365-2729.2009.00329.x

Ma, M., & Zheng, H. (2011). Virtual reality and serious games in healthcare. In *Advanced Computational Intelligence Paradigms in Healthcare 6* (pp. 169–192). Springer. DOI: 10.1007/978-3-642-17824-5_9

Mayer, R. E. (2002). Multimedia learning. [). Academic Press.]. *Psychology of Learning and Motivation*, *41*, 85–139. DOI: 10.1016/S0079-7421(02)80005-6

Mayer, R. E. (2009). *Multimedia learning* (2nd ed.). Cambridge University Press. DOI: 10.1017/CBO9780511811678

Mei, H. H., & Sheng, L. S. (2011). Applying situated learning in a virtual reality system to enhance learning motivation. *International Journal of Information and Education Technology (IJIET)*, *1*(4), 298–302.

Meyer, R. H. (1997). Value-added indicators of school performance: A primer. *Economics of Education Review*, *16*(3), 283–301. DOI: 10.1016/S0272-7757(96)00081-7

Pataquiva, F. D. P. F., & Klimova, B. (2022). A systematic review of virtual reality in the acquisition of second language. [iJET]. *International Journal of Emerging Technologies in Learning*, *17*(15), 43–53. DOI: 10.3991/ijet.v17i15.31781

Raman, K., Hashim, H., & Ismail, H. H. (2024). *Exploring the impact of VR integration on ESL learners' English verbal communication skills: A case study in a Malaysian high school.* Arab World English Journal.

Redondo, B., Cózar-Gutiérrez, R., González-Calero, J. A., & Ruiz, R. S. (2020). Integration of augmented reality in the teaching of English as a foreign language in early childhood education. *Early Childhood Education Journal*, *48*(2), 147–155. DOI: 10.1007/s10643-019-00999-5

Remolar, I., Rebollo, C., & Fernández-Moyano, J. A. (2021). Learning history using virtual and augmented reality. *Computers*, *10*(11), 146. DOI: 10.3390/computers10110146

Richter, J., Scheiter, K., & Eitel, A. (2016). Signaling text-picture relations in multimedia learning: A comprehensive meta-analysis. *Educational Research Review*, *17*, 19–36. DOI: 10.1016/j.edurev.2015.12.003

Rudnik, Y. (2023). The use of augmented reality and virtual reality technologies in teaching foreign languages. *Educological Discourse, 1*(40), 165–183. DOI: 10.28925/2312-5829.2023.110

Safar, A. H., Al-Jafar, A. A., & Al-Yousefi, Z. H. (2016). The effectiveness of using augmented reality apps in teaching the English alphabet to kindergarten children: A case study in the State of Kuwait. *Eurasia Journal of Mathematics, Science and Technology Education, 13*(2), 417–440. DOI: 10.12973/eurasia.2017.00624a

Santos, M. E. C., Chen, A., Taketomi, T., Yamamoto, G., Miyazaki, J., & Kato, H. (2013). Augmented reality learning experiences: Survey of prototype design and evaluation. *IEEE Transactions on Learning Technologies, 7*(1), 38–56. DOI: 10.1109/TLT.2013.37

Shadiev, R., & Yang, M. (2020). Review of studies on technology-enhanced language learning and teaching. *Sustainability (Basel), 12*(2), 524. DOI: 10.3390/su12020524

Sweller, J. (1988). Cognitive load during problem solving: Effects on learning. *Cognitive Science, 12*(2), 257–285. DOI: 10.1207/s15516709cog1202_4

Sweller, J., Ayres, P., & Kalyuga, S. (2011). *Cognitive load theory.* Springer. DOI: 10.1007/978-1-4419-8126-4

Symonenko, S., Zaitseva, N., Osadchyi, V., Osadcha, K., & Shmeltser, E. (2020). Virtual reality in foreign language training at higher educational institutions.

Tarng, W., & Ou, K. L. (2012). A study of campus butterfly ecology learning system based on augmented reality and mobile learning. In *2012 IEEE Seventh International Conference on Wireless, Mobile and Ubiquitous Technology in Education* (pp. 62–66). IEEE. DOI: 10.1109/WMUTE.2012.17

Teräs, M., Suoranta, J., Teräs, H., & Curcher, M. (2020). Post-Covid-19 education and education technology 'solutionism': A seller's market. *Postdigital Science and Education, 2*(3), 863–878. DOI: 10.1007/s42438-020-00164-x PMID: 40477107

Tsichouridis, C., Batsila, M., Vavougios, D., & Ioannidis, G. (2020). Virtual and augmented reality in science teaching and learning. In *The Impact of the 4th Industrial Revolution on Engineering Education* (pp. 193–205). Springer. DOI: 10.1007/978-3-030-40274-7_20

Williamson, B., Eynon, R., & Potter, J. (2020). Pandemic politics, pedagogies and practices: Digital technologies and distance education during the coronavirus emergency. *Learning, Media and Technology, 45*(2), 107–114. DOI: 10.1080/17439884.2020.1761641

Xie, Y., Ryder, L., & Chen, Y. (2019). Using interactive virtual reality tools in an advanced Chinese language class: A case study. *TechTrends*, *63*(3), 251–259. DOI: 10.1007/s11528-019-00389-z

KEY TERMS AND DEFINITIONS

Augmented Reality (AR): integrates digital elements such as text, images, and 3D models into the physical environment, thereby enriching user engagement. In the context of language education, AR offers immersive and contextualized experiences, including interactive labels and virtual scenarios, which enhance learner engagement, improve vocabulary retention, and facilitate practical application in real-world settings.

Cognitive Load: pertains to the extent of mental resources utilized by working memory when engaging in cognitive activities. This notion was first proposed by John Sweller in the 1980s and laid the groundwork for Cognitive Load Theory (CLT). The theory posits that human working memory has a finite capacity, suggesting that educational programs should be structured to minimize cognitive overload and enhance the learning experience (Sweller, 1998).

Extended Reality (XR): serves as a comprehensive term that includes Augmented Reality (AR), Virtual Reality (VR), and Mixed Reality (MR), integrating physical and digital environments with different degrees of immersion. In the context of language education, XR provides versatile experiences, ranging from AR enhancements for vocabulary to complete VR immersion, facilitating adaptive, interactive, and contextually enriched language learning.

Multimedia Learning: an instructional approach that integrates various forms of media, including text, audio, images, animations, and videos, to communicate ideas effectively. The primary goal of this approach is to improve understanding, engage learners, and support the educational process. Mayer posits that individuals utilize two cognitive pathways—verbal and visual—to assimilate information. When these pathways are engaged concurrently, learning can be enhanced, as long as appropriate cognitive design principles are adhered to. Thus, the amalgamation of diverse media components can significantly enrich the educational experience and deepen comprehension of the subject matter (Mayer, 2002).

Self-Directed Learning (SDL): refers to a method in which individuals assume responsibility for their educational journey by establishing objectives, selecting resources, and overseeing their advancement. In the context of language acquisition and instruction, SDL enables learners to customize their study techniques—such as utilizing applications, engaging in media immersion, and practicing conversations—according to their personal requirements, thereby enhancing their independence and motivation.

Situated learning: a fundamental principle within social and context-oriented learning theories, positing that effective learning transpires when individuals engage actively in authentic or simulated scenarios relevant to specific situations. This concept was introduced by Lave and Wenger in 1991, highlighting that learning is not solely a cognitive process but also involves interactions with the social environment and practical activities. In essence, individuals learn more effectively when they are immersed in real-life contexts and can leverage their experiences from those situations (Lave & Wenger, 1991).

Virtual Reality (VR): envelops users in a completely digital realm, effectively isolating them from the physical environment. In the context of language education, VR facilitates the creation of simulated environments, such as virtual classrooms and international cities, which provide opportunities for immersive practice, cultural engagement, and immediate interaction in the target language.

Chapter 11
Emotional Dynamics of EFL Students in Eportfolio-Based Speaking Assessment:
An In-Depth Exploration of Positive and Negative Emotions

I. Putu Indra Kusuma
https://orcid.org/0000-0002-1574-6070
Universitas Pendidikan Ganesha, Indonesia

Luh Indrayani
https://orcid.org/0009-0006-5852-6162
Universitas Pendidikan Ganesha, Indonesia

Gres Desika Br Kaban
Universitas Pendidikan Ganesha, Indonesia

Belia Damayanti Br Sembiring
Universitas Pendidikan Ganesha, Indonesia

ABSTRACT

Previous studies of eportfolios in speaking courses suggest that this digital assessment technique significantly improves speaking skill improvement. However, what kind of positive and negative emotions experienced by the English as a foreign language (EFL) students remains unclear as previous studies did not profoundly discuss them. This study therefore purposed to explore the positive and negative

DOI: 10.4018/979-8-3373-3366-3.ch011

emotions experienced by EFL students during the implementation of eportfolios in speaking courses. This study employed a phenomenology approach and recruited 20 Indonesian EFL students. The data were collected through semi-structured interviews and analyzed using inductive thematic analysis. The findings revealed that while eportfolios supported emotional growth through enjoyment, pride, motivation, and relief, they also triggered various negative emotions, including anxiety, embarrassment, frustration, and exhaustion. Four implications were drawn to effectively implement eportfolios in speaking courses.

INTRODUCTION

The pioneered work of Huang and Hung (2010) with a video blog application to support the implementation of eportfolios in speaking courses has inspired later researchers to conduct similar research in different educational context and with varied digital tools (e.g. Cepik & Yastibas, 2013; Kusuma et al., 2021; Kusuma & Waluyo, 2023; Sun & Yang, 2015; Yastibas & Cepik, 2015; Zheng & Barrot, 2022). For example, Kusuma and Waluyo (2023) employed social media to support the implementation of eportfolios to Indonesian students where they instructed them to upload the speaking clips to YouTube, give feedback to their friends' videos, and did self-reflections after reading the comments. This process not only encouraged students' speaking practice, but also facilitated metacognitive awareness and learners autonomy. Thus, the advent of Web 2.0 technology has revolutionized the implementation of portfolio assessment in teaching speaking skills by making the speaking process becoming more interactive, accessible and collaborative.

The literature of eportfolios in speaking courses suggests that this digital assessment technique brings significant cognitive advantages for speaking skill improvement. Through the structured nature of eportfolio tasks, students are encouraged to engaged in regular speaking practice, monitor their own progress and make improvement based on feedback. Research indicates that eportfolio implementation in speaking courses could improve EFL students' vocabulary and sentences mastery (Hsu, 2016). Moreover, the opportunity to revise speaking tasks contributes to improved pronunciation (Cabrera-Solano, 2020; Cepik & Yastibas, 2013; Huang & Hung, 2010; Sun & Yang, 2015), greater fluency in speaking discourse (Cabrera-Solano, 2020; Rabbani Yekta & Kana'ni, 2020), content, and performance (Kusuma, Mahayanti, Adnyani, et al., 2021; Kusuma & Waluyo, 2023; Zheng & Barrot, 2022). Given the benefits it brings to language learning through regular practices, peer feedback, and self-reflection, eportfolios has now been implementing in many speaking courses around the globe.

In addition to cognitive development, numerous studies have also indicated that eportfolios brought positive contributions to the affective domain (Bensalem & Hamed Alenazi, 2023; Hung, 2011; Kusuma et al., 2022; Kusuma, Mahayanti, Gunawan, et al., 2021; Kusuma & Waluyo, 2023; Sun & Yang, 2015; Zheng & Barrot, 2022). For example, the process of recording and receiving feedback through eportfolios has been shown to contribute to EFL students' motivation to practice speaking skills (Cabrera-Solano, 2020) which later can lead to a notable increase in their self-confidence when using the target language (Cepik & Yastibas, 2013; Sun & Yang, 2015). This increased confidence can reduce speaking anxiety and foster more active classroom participation. In addition, Eportfolios also improve EFL students' self-efficacy (Kusuma & Waluyo, 2023).

However, several studies' minor findings revealed that the implementation of eportfolios in speaking courses could influence students' negative emotions. The minor findings include the stressful situation when creating the videos (Kusuma, Mahayanti, Gunawan, et al., 2021), social pressure, and the reluctance to upload speaking clips to social media (Zheng & Barrot, 2022). But these studies did not explore in much detail about the negative emotions experienced by the EFL students. Meanwhile, EFL teachers need more information about how eportfolios in speaking courses influence students' positive and negative emotions to provide better understanding about this assessment technique and the considerations about the do's and don't's when implementing it in speaking courses.

Looking the above gaps and to guide the investigation, this study posed the following research questions:

1. What positive emotions were experienced by the English as a foreign language students during the implementation of eportfolios in speaking courses?
2. What negative emotions were experienced by the English as a foreign language students during the implementation of eportfolios in speaking courses?

LITERATURE REVIEW

Emotional Well-Being

Emotion is often linked to human beings' development of thinking and behavior (Cong-LEM, 2022). Emotional well-being is commonly defined as an individual's perception of overall life satisfaction and emotional state, reflecting the equilibrium between positive and negative emotions. This emotional balance plays a vital role in shaping life satisfaction, resilience, and even longevity (Zhu & Fan, 2018). Although emotions are fundamental to language learning and teaching, they have historically received limited attention in applied linguistics research (Dewaele et al., 2019). Positive emotions extend beyond feelings of joy, interest, and love to include cognitive states and prosocial behaviors, such as expressing gratitude, all of which contribute to emotional well-being and overall health (Ching & Chan, 2020). In contrary, negative emotions including fear, sadness, and anger can disrupt a person's mental well-being and overall psychological functioning (Erbas et al., 2014).

In educational contexts, emotions are internal states that have significant effects on how well students do and how engaged they are at school (Pekrun, 2006). Positive emotions, such as enjoyment, pride, and hope can enhance students' motivation, give them a sense of success and support ongoing engagement in learning activities (Namaziandost et al., 2022; Richards, 2020). As a result, it enables students to stay involved in learning activities on a regular basis and in a meaningful way (Jaber Rafidi & Wagner, 2023; Mihai et al., 2022). On the other hand, negative emotional experiences, such as anxiety, shame, or frustration might make people disengage, avoid work and lose confidence in academic performance (Ningrum & Rokhuma, 2023). When students become emotionally disengaged, they may start to do poorly and lose faith in their talents, especially when they have to do things that are emotionally draining, like speaking (Richards, 2020). That's why it's crucial for educators to be sensitive and respond to how their pupils are feeling. This will help to create a supportive learning environment that encourages both involvement and academic success.

In conclusion, emotional well-being plays a foundational role in shaping students as the language learning outcomes. Positive emotions not only support motivation and performance, but it is also contribute to a learner's overall health and resilience. Negative emotions, if left unaddressed, can lead to avoidance behaviors and long-term disengagement from language use. For that reason, fostering emotional well-being in EFL contexts should not be seen as a secondary concern but as an essential component of effective pedagogy.

Speaking Performance

Speaking plays an important role in language learning, as it contributes significantly to learners' ability to communicate effectively (Harmer, 2007). Through speaking, students can convey their thoughts, feelings, and intentions in both every day and formal situations. Interestingly, as Brown (2004) points out, speaking is not merely about saying words, it also involves listening actively and making sense of what is heard. This indicates that good speaking skills rely on both language knowledge and the capacity to apply that knowledge in real-time exchanges.

There are several aspects that shape a learner's speaking performance. Ilinawati et al. (2021) identify five main components, namely: pronunciation, fluency, grammatical accuracy, vocabulary, and content. Pronunciation helps listeners understand what is being said, while fluency reflects how smoothly one can speak without unnecessary pauses. Grammatical accuracy contributes to clarity and coherence in sentence construction. Vocabulary allows for more precise expression. Finally, the quality of content is closely related to how ideas are structured and connected which also plays as a key role in making speech meaningful (Ilinawati et al., 2021).

In the context of Indonesian learners, where English is used as a foreign language, speaking often presents multiple challenges. These are not limited to linguistic issues alone. Emotional factors, such as anxiety and self-doubt, as well as classroom dynamics can interfere with learners' ability to speak confidently. For instance, Pratolo et al., (2019) observed that many students face ongoing struggles with pronunciation, fluency, grammar, and vocabulary. These challenges are often compounded by limited speaking practice and a lack of supportive learning environments. At the same time, they face anxiety, low self-confidence, and limited opportunities for practice. These difficulties reflect what Goh and Burns (2012) describe as the cognitive and emotional complexity of speaking. This situation is when learners must manage their thoughts, language skills, and social interactions all at once. For example, when speaking topics is lack of relevance or students have little peer interaction, then their ability to produce coherent and meaningful speech can also be affected. Therefore, speaking instruction must not only address the five technical aspects of speaking (Ilinawati et al., 2021) but also reduce emotional barriers and provide a supportive environment for speaking practice.

From a teaching perspective, speaking instruction should move beyond isolated drills or scripted tasks. Research emphasizes that real engagement, task-based learning and guided support are all important for helping learners improve their communicative competence (Richards, 2008). Learners need frequent and meaningful opportunities to speak, receive feedback, reflect on their progress to gradually improve their speaking performance. Without these supports, it becomes difficult to balance fluency with accuracy, especially in high-pressure or anxiety-inducing situations. As a result, careful planning needs to be considered both for linguistics and emotional aspect in order to teach speaking effectively.

Eportfolios in Speaking Courses

Originally, portfolios were made as a tool to collect and display students' classroom work for assessment purposes (Efe, 2016; Johnson & Johnson, 2002; O'Malley & Pierce, 1996). As the digital technology has grown, these old portfolios have turned into electronic version, known as eportfolios (Georgi & Crowe, 1998). Digital technologies are used in eportfolios to arrange many types of students' works, such as: audio recordings, presentations, and images. These benefits make eportfolios easier to be utilized in tracking learning progress over time (Cepik & Yastibas, 2013; Dougherty & Coelho, 2017; Namaziandost et al., 2020). Eportfolios highlight the value of self-reflection and peer feedback, like traditional portfolio does (Kwak & Yin, 2018). These help students evaluate their own learning and development (Cepik & Yastibas, 2013). Thus, eportfolios fit well with student-centered learning and active learning methods because it encourages students to think on what they have learned and allows feedback from peers and teachers (Yastibas & Yastibas, 2015).

When eportfolios used to assess speaking skills, it allows students to record and store different types of oral performances. For instance, students might upload audio responses to a personal blog (Huang & Hung, 2010; Hung & Huang, 2015) or post video presentations to platforms like QQ or YouTube (Zheng & Barrot, 2022; Cepik & Yastibas, 2013; Kusuma & Waluyo, 2023). These various platforms enable students to show their speaking abilities in a more dynamic and engaging ways. Research shows that students can reflect on and evaluate their performances outside classroom. These activities support independent learning and ongoing improvement (Cepik & Yastibas, 2013; Kusuma et al., 2021; Sun & Yang, 2015). Due to its flexibility, e-portfolios become a practical option for conducting speaking courses fully online. The cycle of eportfolios, namely: creating, reviewing, and improving help students take charge of their own learning. Students can look back on past recordings, increase their awareness of personal growth and help them focus on areas that need improvement (Sun & Yang, 2015; Kusuma et al., 2021). Moreover, studies suggest that using e-portfolios can boost students' motivation, confidence, and sense of independence (Bensalem & Hamed Alenazi, 2023; Kusuma & Waluyo, 2023).

Previous Studies

Numerous studies have indicated that eportfolios brought positive contributions to the affective domain. For example, Cabrera-Solano (2020) conducted an eportfolios study by recruiting 42 Ecuador EFL students and found that this assessment technique motivated students to practice speaking skills. The regular practices done by the students could help the EFL students improve their self-confidence as found by Cepik and Yastibas (2013) who conducted a similar study with 17 EFL students in Turkey as well as Sun and Yang (2015) who conducted the study with 14 EFL students in Taiwan. In addition, eportfolios, through regular practices, feedback from peers, and self-reflections become the sources of EFL students' self-efficacy as found by Kusuma and Waluyo (2023) who conducted a study with 55 Indonesian EFL Students. in another study with 60 Indonesian EFL students, Kusuma et al. (2022) found that eportfolios contributed to the reduced foreign language anxiety. Similarly, Bensalem and Hamed Alenazi (2023) found that e-portfolios significantly enhanced students' speaking fluency and confidence in a university EFL context. Learners appreciated the constructive feedback from both peers and instructors, which helped them track their progress and identify areas for improvement. Moreover, students reported that e-portfolio use reduced their shyness and increased their motivation to speak in English, especially after gaining more control over the speaking process through repeated video submissions. These findings suggest that the integration of eportfolios not only enhances language performance and decrease anxiety but also fosters positive emotional experiences, such as increased motivation, confidence, and self-efficacy, which are essential for sustained engagement in EFL learning.

However, several studies found that the implementation of eportfolios in speaking courses could influence students' negative emotions. Even though a few participants in Sun and Yang's (2015) study felt that they improved their speaking skills, however, in the beginning of the implementation of eportfolios, they stated that felt stresful with creating the videos and even wanted to drop the class. Quite similar findings were revealed by Kusuma, Mahayanti, Gunawan, et al. (2021) who did a study with ten Indonesian EFL students where a few participants stated that they could not remember all lines they had to say when recorded the speaking clips. Also, they spent much time to record, edit, and uploaded the videos which sometimes demotivated them to do the activities in eportfolios. In addition, Zheng and Barrot (2022), who conducted a study with 42 Chinese EFL students, found that a few participants felt social pressure because their videos would be watched by public as they uploaded the speaking clips on social media. Similarly, Bolliger and Shepherd (2010), in a study involving online graduate students, found that some learners viewed e-portfolios as an additional burden rather than a useful learning tool. One of the participants shared feeling frustrated while creating content, which in turn reduced their motivation and emotional engagement. Although previous studies have pointed out feelings of stress, demotivation, and social anxiety associated with using e-portfolios in speaking classes, they tend to touch on these emotional experiences only at the surface level. As a result, there remains a need for deeper exploration into the emotional challenges EFL students may face.

RESEARCH METHODS

Research Design

This study used a phenomenological research design to explore how senior high school students emotionally responded to the use of eportfolios. A phenomenological approach was chosen because it allows researchers to understand the essence and meaning behind people's lived experiences, especially focusing on their emotions and personal reflections as they interact with specific phenomena (Creswell & Poth,

2018). In this case, the phenomenon under investigation was the implementation of eportfolios in a speaking class. This design made it possible to capture both the positive and negative feelings students experienced as they engaged with eportfolios in a classroom context.

Participants and Context

This research was conducted in an Indonesian public senior high school where eportfolio had recently been implemented. The researchers applied for an IRB prior to conducting the study. The consent from the headmaster to approach the potential participants was also sought after the IRB approval. The researchers explained the study to the potential participants, including the risks and the benefits of joining the study. Two researchers implemented eportfolios for four topics in which one topic was conducted for two weeks (the first week include discussing the linguistic features, practicing the speaking skills, recording the videos, uploading them on YouTube; the second week comprised giving comments to friends' videos on YouTube, doing self-reflections on Google Docs, and performing in front of the class).

After the eight weeks of eportfolios implementation, a total of twenty students were recruited using a purposive sampling technique, ensuring rich emotional experiences (Palinkas et al., 2015) with the following criteria: (1) fully joined in eportfolios activities and (2) experienced positive/negative emotions during the implementation, (3) were willing to share the experiences. 40 students were invited to join the interviews, however only 20 students who gave their consents to participate in the study were finally recruited. The participants were divided into two groups: ten students who had mostly positive emotional experiences and ten who had negative emotional experiences related to eportfolio use. The context of this research comprised the positive and negative emotions experienced by the EEFL students during the implementation of eportfolios in speaking courses. The researchers had no personal or institutional connection with participants. Interestingly, this distance made students feel comfortable and open, even when discussing sensitive or emotional issues. All students gave informed consent, and ethical procedures were followed throughout the study. All students' identities were anonymized, and they were assigned pseudonyms. Participants P1-P10 (five males; five females) took part in interviews focusing on positive emotions, while P11-P20 (five males; five females) participated in interviews exploring negative emotions.

Data Collection and Instruments

The data were collected through semi-structured interviews. To guide the interviews, the researchers developed an interview protocol which was sent to two experts for validation. The interview guide only received minor revisions, such as language corrections. Then, individual interviews were conducted in two sessions by two researchers. The interviews allowed students to talk freely about their experiences and emotions when using eportfolios. Each interview lasted around 45-60 minutes and was conducted in Bahasa Indonesia to ensure the data clarity and comfortability of participants. The interviews were conducted from January to March. All interviews were recorded, transcribed and translated into Indonesian. The researchers also took notes on some important information shared by the participants.

Data Analysis

The transcribed interviews were analyzed using inductive thematic analysis, following the six-phase process by Braun and Clarke (2006):

1. Familiarization: two researchers carefully read the interview transcripts.
2. Generating initial codes: two researchers began the coding process by identifying meaningful segments of text related to emotional experiences.
3. Searching for themes: these codes were grouped into early sub-themes that described different patterns of emotions.
4. Reviewing themes: the sub-themes were discussed in a meeting with all four researchers. Together, they reviewed how the sub-themes connected and suggested possible main themes.
5. Defining and naming themes: the team refined and named the final themes based on shared argument.
6. Producing the report: the final themes were used to write the findings, supported by quotes and descriptions from both the interviews.

The analysis process was collaborative and iterative, with several rounds of discussion to ensure that the final themes truly reflected the students' experiences.

Trustworthiness and Credibility

By involving multiple researchers in the coding and analysis process, this collaborative approach mitigates individual biases and fosters a more nuanced understanding of the data. The process of collaborative coding allows for multiple perspectives to be considered during interpretation and reducing the influence of researcher's assumptions. The use of peer debriefing aligns with established criteria for ensuring credibility in qualitative research, as outlined by Lincoln and Guba (1985). Also, the open and trusting relationship with participants encouraged them to share honest and deep reflections, which strengthen the quality of the data.

FINDINGS & DISCUSSION

Findings

To answer the research questions, the inductive thematic analysis conducted in this study yielded 5 themes and 13 subthemes as summarized in Table 1 below.

Table 1. Themes and subthemes

Questions	Themes	Subthemes
What positive emotions were experienced by the English as a foreign language students during the implementation of eportfolios in speaking courses?	Affective Improvement in Students' Positive Emotional Well-being through Eportfolio activities	Enjoyment and Happiness in the Learning Process
		Sense of pride and boosting Self-Confidence
		Increased Motivation through Positive Feedback
		Sense of Relief and Calmness after completing the video assignment
What negative emotions were experienced by the English as a foreign language students during the implementation of eportfolios in speaking courses?	Anxiety and Nervousness About Speaking Performance	Fear of Speaking on Video
		Pronunciation Struggles and Fear of Making Mistakes
	Embarrassment and Lack of Confidence in making speaking task	Fear of Being Laughed at by Peers
		Self-Conscious About Mistakes
	Frustration with Technology	Difficulties in Uploading and Sharing Videos
		Struggles with Video Editing and Device Limitations
	Pressure to Perform Well Under Strict Expectations	Expectation to Produce High-Quality Content
		Feeling of Obligation Rather Than Enjoyment
	Stress and Mental Exhaustion from Overload	Feeling Overwhelmed by Assignments

Affective Improvement in Students' Positive Emotional Well-being through Eportfolio Activities

A total of nine students shared their experiences, indicating how engaging in eportfolio activities brought a strong sense of enjoyment and happiness during the learning process. These positive emotional responses demonstrate that beyond being a tool for academic assessment, the eportfolio also served as a source of motivation and emotional enrichment. Students not only engaged with the content but also found personal satisfaction and joy in tracking their development and receiving encouragement from peers. The participants often sensed happiness when they saw the tangible progress in their abilities and when they were acknowledged by others during the eporfolio activity. The participants claimed that eportfolios brought positive impact to learning and positive emotion. For instance, P7 said, "...Anyway, I'm also a bit happy because I can see the progress. From not understanding how to pronounce it to starting to become fluent after doing this eportfolios." The experience of moving from a place of uncertainty and difficulty to one of growing fluency gave the participants a sense of accomplishment. Students like P7 could witness the development first-hand by comparing the earlier works with the recent ones. This visible progress became a source of personal pride and encouragement for P7. P1, also highlighted the role of the peer comments that gave a sense of happiness during the eportfolio activity. P1 reported, "In terms of happiness, I feel happy. Yes, actually it was said [peer's comment] his voice was loud, his facial expression was good, as if telling the story was no longer stiff." Peer comments not only provided external validation but also helped build confidence in P1's presentation and speaking skills. Hearing specific feedback such as compliments on voice projection or expressive storytelling reassured P1 that the efforts were noticed and appreciated. P4 also reported the sense of happiness and enjoyment during the learning process because of receiving the constructive peer feedback. For students like P4, the learning process became more meaningful when it included not only teacher evaluations but also input from classmates. P4 said, "I feel happy when I get feedback from other friends. When they watch my own video, it turns out I have strengths like this, weaknesses like this. I feel happy because I know that I have my own strengths" P8 also stated a deep sense of happiness during the eportfolio activities as the result of positive feedback and completion of the assignment:

> *"Perhaps the positive feeling I felt was happiness and pride, because I had successfully completed the eportfolio assignment that had been given, like that. Especially when I see my videos and get positive comments from friends too. So that's what makes me feel like my efforts are appreciated...." (P8, Female, Individual Interview)*

A total of seven students experienced heightened self-confidence and a sense of pride as the result of their engagement in eportfolio activities, which encouraged regular practice, reflective processes, and the integration of peer feedback. These components worked together to provide a structured and supportive environment where students could track their progress, identify areas for improvement, and celebrate their achievements. Over time, repeated exposure to speaking tasks and opportunities to self-assess and receive feedback allowed students to build a stronger sense of capability. Many students felt more confident in their ability to speak English as a result of the combination of structured activities and honest peer responses. For example, P8 shared a breakthrough experience in gaining confidence and a sense of pride. P8 said, "Maybe my self-confidence and motivation emerged when I got feedback from friends. That's why I became confident to speak better using English...." Students like P8 started to trust their own abilities after realizing that their classmates appreciated and noticed their efforts. Receiving feedback was a pivotal event that strengthened the belief in one's own improvement by bringing external recognition into line with internal development. Similarly, P6 also reported on how self-assessment became a source of internal affirmation. P6 stated, "…I feel proud because before doing this eportfolios, my English was still poor. After doing the eportfolios, I know and can correct the words that are still wrong." P6 felt a sense of accomplishment after going over their own work and pinpointing areas where they could improve. Students could clearly see the results of their efforts when they were able to identify specific improvements, like improving word choice or correcting pronunciation. Moreover, P7 also reflected the emergence of self-confidence and a sense of pride after successfully delivering the eportfolio assignment as well as receiving positive comments from other friends. For example, P7 said, "…Then a sense of pride emerged when I was able to convey the story clearly and fluently even though there were still shortcomings in expression. And get praise from friends...." Praise from peers further reinforced this emotional response, helping students develop a more positive self-image regarding their English-speaking abilities. Additionally, P4 also expressed both a sense of pride and confidence through the role of peer comments and consistent practice when engaged in eportfolio activities:

> *"I felt a sense of pride. For example, oh I managed to record my own video smoothly. After several tries, oh this is still not enough, and after keep practicing, I was able to get better results. And besides that, when I read positive comments from my friends, I also feel happy. It turns out that my efforts are appreciated. Especially when they gave positive comments, I felt proud of myself." (P4, Female, Individual Interview)*

This student further added on how peer comments boosting self-confidence which supported to become better in improving speaking skills. This cycle of consistent practice, peer feedback, and improvement strengthens the student's emotional resilience and fostering sustainable confidence and pride:

"...I feel confident when my friends give positive comments that can support me to become more confident. For example, oh you are good, your vocabulary and pronunciation are better than before. It varies. This also really motivates me to continue practicing and improving my speaking skills again." (P4, Female, Individual Interview)

A total of four students noted that receiving positive peer comments from their friends played a crucial role in boosting their motivation throughout the eportfolio process. These supportive comments encouraged the students to keep trying and improving, especially when they saw that others noticed their progress. Knowing that their classmates were not only watching their videos but also offering kind and constructive responses gave students a sense of being seen and supported. This emotional reinforcement helped maintain their enthusiasm and determination, particularly during moments when they might have felt uncertain or tired. The experience of being acknowledged by peers became a motivational force that kept them engaged with the assignments. For many students, feedback from friends was not just helpful, but also made them motivated. It made the students want to do better and continue learning. The comments served as a reminder that their work had value, and that someone else was benefiting from or inspired by their performance. P9 reported the importance of interaction when receiving feedback from their friends. For example, P9 stated, "...Especially when they say [peer's comment], I am making continuous progress because I also practice many times, so I feel motivated to continue improving. And watching their videos also makes me more motivated to be better...." Similarly, P8 also shared an experience regarding the role of peer feedback in boosting motivation. P8 said, "... these friends gave positive comments, which made me more motivated." Even short, affirming phrases from classmates had a lasting effect, serving as emotional fuel during the task. It helped P8 gain energy and interest to keep improving their work. The effort put into the eportfolio was no longer just about fulfilling an assignment, it became about growth and positive interaction with peers. This sentiment was also echoed by P10 with the benefits of peer feedback, which contributed to their motivation. This kind of small encouragement was important because it helped students notice their progress. The recognition of improvement, even in specific areas such as pronunciation, brought a sense of validation. When someone else pointed out what they were doing well, it helped them feel like their hard work was being appreciated, which can lead them to boost their motivation. For example, P10 said, "...But because there are friends who give positive comments, like my pronunciation is good. I felt happy and motivated a little"

A total of five students expressed a sense of emotional relief and calmness after completing eportfolio assignments. For many of these students, finishing the task marked a moment of emotional closure, giving them the space to relax, reflect, and regain composure after completing the video assignments. The task's emotional intensity made its completion feel especially significant, especially for those who had dealt with anxiety, repeated attempts, or performance-related stress. Knowing that they had followed through, regardless of any previous challenges, gave students a sense of relief and satisfaction that allowed them to reset emotionally and mentally. For instance, P7 shared an experience after finishing the tasks that had been given. P7 said, "...Then there are times when I also feel relieved because I can complete the task on time and with somewhat satisfactory results." The participants also showed that the calmness after completing the assignment did not just come from meeting the deadline, but also from knowing that the work was done well. Students' assessments of their own performance served as the foundation for the ensuing sense of calm. They felt more at ease when they saw progress or felt like they had given it their all. For example, P1 reported the sense of relief and calmness after completing the eportfolio video assignment:

"I feel very relieved after making this eportfolio assignment. Because the assignment was sudden. This is my first time doing a task like that. But after doing that, after it was uploaded, after the assignment was accepted, I felt relieved. It turns out I can also make it, although not as much as possible." (P1, Male, Individual Interview)

Furthermore, P6 also emphasized similar experiences after finishing the speaking videos. For example, P6 said, "I am also relieved when I do my assignments because the task is finished. That's it." This sentiment was also echoed by P2, who highlighted the same emotional release after completing the assignment. This showed not only a sense of relief but also an increase in confidence moving forward. By completing one video successfully, the student experienced emotional calm and a sense of relief for the next task. The relief felt here also helped to lower stress for future assignments:

"After the video assignment was completed, I became very calm and relieved because one assignment was completed. After that there is another assignment, a video assignment. When I wanted to make the assignment, it was easier because I already knew that I could make video assignments beforehand." (P2, Male, Individual Interview)

Anxiety and Nervousness About Speaking Performance

One of the most dominant emotional barriers was the fear of speaking on video. A total of three students felt intimidated by the recording process and were afraid of making mistakes. This fear was particularly evident when students were asked to submit tasks in the form of video recordings, which required them not only to speak clearly in English but also to be mindful of how they looked and sounded. P11 expressed this clearly and stated, "I'm nervous when I speak in front of the camera, I'm afraid I'll make a mistake" P13 also showed similar concern as P13 said, "The first time I made a video I was nervous, afraid of saying the wrong thing" The novelty of the task seemed to be the source of this anxiety. Many students were nervous because it was their first time recording themselves speaking English in this strange format. They were more careful with their words because they knew that any mistake could be recorded forever. Concerns raised by P11 and P13 indicated a performance-related anxiety that frequently surfaced when students felt under pressure. They appeared to internalize the pressure to perform perfectly, which created a tension that interfered with their natural speech. Similarly, P18 admitted "I'm afraid that the assignment won't be good and I'll be asked to redo it again" P18 was afraid of being judged not only on how they used language but also on how well they presented, pronounced, and fluently used the language overall. This concern was about having to repeat the process, which was emotionally taxing in and of itself, rather than merely being incorrect.

A total of three students faced difficulties with pronunciation, which made them feel even more insecure about speaking. Pronunciation, as a skill, involves mastering the sounds and rhythms of a new language, and for many learners, this aspect can feel particularly foreign and difficult to control. For example, P11 admitted, "I don't know how to pronounce certain words…." P11 revealed a lack of confidence that made speaking in English daunting. This lack of confidence often left students unsure of the spoken output, leading them to doubt even the words they did know. Likewise, P17 also acknowledged his pronunciation difficulties and stated, "…The English pronunciation part. There are things I can't and difficult to pronounce it right" His statement confirms the challenge students faced with articulating English words accurately. The struggle to say words correctly can be discouraging, especially when students are not sure how the words are supposed to sound. Furthermore, P20 expressed additional concerns and mentioned, "How it will be if I do the task? Is this the right way to do it or not and I'm afraid of being asked to come to in front of the class" It suggested that these pronunciation struggles did not only affect performance during video recordings but also translated into anxiety in traditional classroom settings.

Embarrassment and Lack of Confidence in making speaking task

A total of three students experienced the fear of being laughed at by peers, which emerged as a particularly distressing emotional experience. Students' sensitivity to peer perception was the main cause of this fear. Students were greatly impacted by how they thought their peers would respond to their performance rather than just their own learning progress. They hesitated and felt emotionally uneasy because they were afraid of being judged negatively, especially when they had to talk in English on camera. P17 described this explicitly and mentioned, "Embarrassed... because my friends laughed at me…." It showed how negative peer responses left a lasting emotional impact. In addition to undermining their confidence, the experience of being made fun of made them dislike doing the same thing in the future. In addition, P11 expressed similar concerns and stated, "I'm afraid of being ridiculed because my English is still not good enough, I'm embarrassed" This demonstrated that emotional distress was triggered simply by the perception of being less proficient. Students were reluctant to perform because they were afraid of being viewed as not "good enough" in English. Their willingness to fully participate in video assignments was limited by their fear of ridicule, even before any actual ridicule took place. P15's concern further illustrated the digital dimension of the fear as P15 stated, "I'm afraid my friends will screenshot my video and make it a joke" Her statement reflected the worry that the video could become material for ridicule beyond the intended educational purpose.

Three students often expressed anxiety about making mistakes, particularly in pronunciation. It exacerbated their reluctance to speak, in addition to their fear of mockery. Anxiety was also brought on by the fear of making mistakes. P20 echoed this self-doubt and stated, "I'm not confident because I know my pronunciation is still bad" Her acknowledgment of weakness in a specific skill area showed how self-awareness could quickly shift into self-criticism, and it created a barrier to improvement. P13 added another dimension to this discomfort as P13 commented, "It's still embarrassing to be watched by so many people… Moreover, when I speak English, automatically has the mistakes." Her comments highlighted the dual strain of giving a performance in front of an audience and in a second language, where mistakes seemed more noticeable due to scrutiny. She felt uneasy not just because she was speaking English but also because she was doing it in front of people who could criticize or see every mistake. P12 also reflected the same uncertainty and mentioned, "I feel like I'm in doubt, is it good or not? The way I pronounce it is correct, or no…" This uncertainty showed the mental burden students carried when they were unsure about their abilities. They also had to perform under public or semi-public conditions.

Frustration with Technology

A total of four students expressed confusion and difficulty with the process of uploading videos to platforms like YouTube. This technical step presented a significant obstacle for some students. Some of them were unfamiliar with video-sharing platforms even though it was considered a simple procedure. P13 admitted, "I'm confused about how to upload videos to YouTube for the first time...." This misunderstanding demonstrated how the assignment called for digital literacy in addition to language proficiency, which not all students initially possessed. When P13 students encountered this new step, it increased the cognitive load because P13 had to concentrate on using new digital platforms in addition to the content and accuracy of the spoken English. P14 elaborated on the task's intricacy and noted, "The assignment is not only to record, but also to edit and upload. It's really complicated!" This statement underscored the multifaceted nature of video assignments and it required a combination of skills in recording, editing, and uploading. Each procedure presented its own set of challenges.

Regarding the uploading issues, a total of four students grappled with video editing challenges and device limitations that further complicated their ability to produce quality assignments. In some cases, technical limitations hindered their ability to carry it out smoothly even when students understood the task. P20 reported, "My handphone often freezes. When I start making a video, it turns off...." It indicated that older or less capable devices hindered the recording process. This type of disruption not only made the task harder but also discouraged repeated attempts, especially when progress was lost due to freezing or crashing. Students like P20 found themselves repeating the same steps over and over with no guarantee of success, and it led to frustration. P14 faced similar hardware issues and mentioned, "...network and camera. I was struggling with my camera. It's a bit blurry, there's fog...." This compromised the visual quality of the recordings. Blurry visuals or unclear images made students feel that their efforts might not meet expectations, even when their spoken English was good. These technical problems were not within the students' control. Even worse, the technical problems had a direct impact on the presentation of their work. In some cases, students may have hesitated to submit recordings that they felt looked unprofessional due to poor camera quality. P18's experience with device performance further illustrates these challenges as P18 stated, "The handphone was lagging and didn't support it when recording, so I had to restart it" Such technical interruptions not only disrupted the recording process but also heightened anxiety and diminished the quality of the final product.

Pressure to Perform Well Under Strict Expectations

Frustration and emotional exhaustion emerged as significant factors as four students grappled with the pressure to meet high standards in their video assignments during the implementation of eportfolios. For many, the effort to present themselves well on camera led to mental fatigue, particularly when things didn't go as expected. P12 described her struggle and stated, "The problem is that I repeat the video over and over again; sometimes it's already half the time, it's like I forget what I want to say, and then repeat it again, and it's really annoying" Her comment reveals the mental exhaustion that stems from repeatedly attempting to meet a personal standard of excellence. For some, even adequate preparation did not prevent frustration when things didn't go as planned. P15's experience of internal pressure when recording the speaking performance is evident in her words: "The emotions were like, 'Oh, it should be recalled. How could I forget this one? Even though it was already memorized" Her frustration revealed how memorization, often viewed as a safety net, could itself become a source of stress. The presence of a recording device or the pressure of performance disrupted the student's ability to recall information smoothly even when the material had been studied and rehearsed.

A total of four students perceived the assignments as compulsory and unenjoyable obligations, and it intensified their negative emotions. Students framed the task in the speaking courses as duties to be fulfilled. P14's reflected, "I feel this task is more of a burden than something enjoyable" P14 also showed that the video assignments, while educationally purposeful, were often experienced as emotionally draining. The weight of mandatory video assignments, combined with the anxiety of public speaking and performance expectations, made the learning process stressful rather than engaging. The interviews revealed that requirement to complete these tasks, regardless of whether students felt emotionally or mentally ready, added to their feelings of being overwhelmed. Even when students attempted to shift their mindset, stress remained a dominant factor. For example, P13 confessed, "Feeling stressed... just trying to enjoy doing the job..." While students made attempts to adjust their outlook and find some positivity in the process, the sense of obligation and fear of underperformance continued to affect their emotional state. This inner conflict between wanting to enjoy the task and feeling burdened by it contributed to a sense of exhaustion. P16, too, admitted to initial resistance, and said, "So at first I didn't seem to want to do the assignment," This P16's lack of motivation suggested that video assignments, rather than inspiring engagement, initially created resistance among students. The interviews also showed that the reluctance to begin the task revealed a deeper discomfort or dread surrounding the activity.

Stress and Mental Exhaustion From Overload

The overwhelming workload associated with eportfolio assignments added significant stress to four students, and it made the learning process mentally exhausting. These students were not only expected to complete speaking tasks but also manage multiple stages of preparation, performance, and submission. Often, they were in close succession with other academic responsibilities. The sheer volume of tasks, particularly video assignments, placed a heavy burden on them, and it left little room for relief. The steps, such as from brainstorming, writing scripts, rehearsing, recording, editing, uploading, and sometimes re-recording due to mistakes, accumulated into a time-consuming process that extended far beyond regular homework. P15 articulated this frustration, emphasizing how multiple assignments piled up, creating a cycle of continuous pressure:

> *"Lots of assignments, lots of thoughts, thoughts about this, about this, about this, but I'm tired of the assignments, I'm given more assignments. Especially the video assignments, if there are no video assignments yet what to do, not to mention searching, not to mention being written on video, not to mention the memory full" (P15, Female, Individual interview)*

P15's words indicated the overwhelming nature of these tasks, as students not only had to prepare and record their videos but also deal with technical limitations such as storage issues, and it made the process even more frustrating. The cognitive and emotional overload described here revealed how students were constantly shifting their focus from one assignment to the next without sufficient mental rest. The mention of "memory full" suggested that the technical side of things only added to the burden, especially when students had to delete files or repeat recordings due to lack of space. These overlapping academic and technical stressors compounded, creating a sense of nonstop pressure. The persistent workload led to a sense of mental fatigue, with students craving a break from academic stress. P18's stated, "I just want healing because there are always a lot of tasks that don't all fit in my brain…." For P18, the desire for "healing" points to a mental and emotional need for rest. The interviews revealed the continuous pace of tasks pushed students to the brink of burnout. The mental space needed for reflection and creativity was overtaken by the urgency to simply get through one assignment after another. P18's phrase "don't all fit in my brain" further revealed how overwhelmed the students felt. Similarly, P14 expressed frustration with the repetitive nature of video assignments and stated, "It's the first, second, third assignment, everyone is told to make another video, they're getting more and more tired" The interviews also showed that rather than becoming more comfortable with the process over time, the demand for multiple video assignments only contributed to increased fatigue and frustration. According to the interviews, repetition became a source of weariness. For students like P14, the expectation to keep producing video content again and again left little opportunity to experience growth or satisfaction, only growing tiredness. The accumulation of tasks without meaningful variation or rest periods further deepened students' sense of burnout, and it led them to associate eportfolio tasks with exhaustion rather than engagement.

DISCUSSION

The first research question explored how eportfolios influence students' positive emotional well-being. The findings of this study revealed that students experienced emotional growth in four primary aspects, such as (1) enjoyment and happiness in the learning process, (2) sense of pride and boosted self-confidence, (3) increased motivation through positive feedback, and (4) sense of relief and calmness after completing the video assignment. The findings of this study align with previous research that underscores the positive emotional impact of eportfolios in language learning contexts. For instance, several earlier studies (e.g., Cabrera-Solano, 2020; Cepik & Yastibas, 2013; Sun & Yang, 2015) reported that eportfolio activities improved students' motivation and self-confidence. It mirrors the present study's themes of increased motivation through positive feedback and boosted self-confidence. This similarity can be theoretically explained by Pekrun's (2006) Control-Value Theory, which posits that positive emotions emerge when students perceive a sense of control over their tasks and see value in their activities. In eportfolio-based learning, it seems that the students often feel empowered by the ability to plan, revise, and reflect on their work, which enhances both their emotional engagement and sense of achievement. This study found that feedback mechanisms, particularly peer and teacher comments, played a significant role in increasing learners' self-efficacy as also found by Kusuma and Waluyo (2023). This can be explained by Bandura's Social Cognitive Theory (Bandura, 1994, 1997), which highlights that self-efficacy is reinforced through positive social feedback and observational learning. When students receive affirmative feedback and see their peers succeed, they are more likely to develop confidence in their own abilities. Additionally, the constructivist approach, which underlies eportfolios (Yastibas & Yastibas, 2015), encourages reflective and autonomous learning, thereby supporting the development of a positive emotional climate.

This study also reveals nuanced differences. This study highlights momentary emotions, such as relief and calmness after completing video tasks, likely a result of the task-completion effect. They are unlike earlier research that emphasized long-term gains like confidence or reduced anxiety (e.g., Kusuma et al., 2022). Additionally, this study brings attention to enjoyment and happiness during the learning process, a less explored aspect in prior research. This may stem from the creative freedom afforded by modern eportfolio tools, such as video editing and multimedia integration.

The second research question examined the negative emotions experienced by the students in using eportfolios and the participants in this study experiences some negative emotions, such as (1) fear of speaking on video, (2) pronunciation struggles and fear of making mistakes, (3) fear of being laughed at by peers, (4) self-conscious about mistakes, (4) difficulties in uploading and sharing videos, (5) struggles with video editing and device limitations, (6) expectation to produce high-quality content, (7) feeling of obligation rather than enjoyment, and (8) feeling Overwhelmed by Assignments. One of the most common negative emotions reported was anxiety about speaking performance, particularly when students had to record themselves. Many expressed concerns over pronunciation and making mistakes, echoing findings by Sun and Yang (2015), where students initially felt stressed and some even considered dropping the course. Similarly, Kusuma, Mahayanti, Gunawan, et al. (2021) found that students became anxious when struggling to memorize lines during recording. Closely related was embarrassment and lack of confidence, especially regarding peer evaluation. Some feared being laughed at or judged by classmates. This mirrors Zheng and Barrot's (2022) findings, where students felt social pressure when their videos were shared online, making them more self-conscious and emotionally vulnerable. Technological frustration also emerged as a key issue. Students faced difficulties with editing, uploading, and device limitations, which hindered engagement. Similar technical challenges were reported in Kusuma, Mahayanti, Gunawan, et al. (2021), where time-consuming production processes reduced students' motivation. Additionally, many felt pressure to produce high-quality videos, which led to stress and perfectionism. As Sun and Yang (2015) noted, students often repeated recordings to meet high expectations, making the process feel more like a burden than a learning opportunity. Lastly, students reported stress and exhaustion due to workload overload. The need to repeatedly record, edit, and reflect caused burnout, echoing Kusuma, Mahayanti, Gunawan, et al. (2021), who found that the extensive demands of eportfolio tasks negatively affected student motivation. These findings highlight the importance of balancing task complexity with emotional support.

The negative emotions experienced during eportfolio implementation, such as anxiety, embarrassment, frustration, and mental exhaustion, can be explained. According to Pekrun's (2006) Control-Value Theory, anxiety emerges when students value a task but feel little control over it. This was evident when learners, concerned about pronunciation and mistakes, grew anxious while recording their speaking tasks. The permanent and shareable nature of eportfolio videos likely amplified this anxiety, particularly for students lacking speaking confidence. Similarly, embarrassment and self-consciousness, especially when sharing videos with peers, reflect emotional vulnerability in language learning contexts. As Dewaele et al. (2019) argue, emotions are central to learner engagement. Students' fear of being judged, as seen in Zheng and Barrot (2022), suggests that public exposure through eportfolios can intensify social anxiety and reduce participation. Technical difficulties, such as video editing, uploading, or device limitations, contributed to frustration, which can be linked to the constructivist model outlined by Yastibas and Yastibas (2015). While eportfolios aim to promote autonomy and reflection, technical barriers can disrupt this process, shifting focus from content to logistical struggles, and resulting in disengagement. The pressure to produce polished, high-quality videos also contributed to stress. As supported again by Pekrun (2006), when learners feel performance expectations exceed their abilities, perfectionism and demotivation can arise. Sun and Yang (2015) observed similar patterns, where students spent excessive time re-recording to meet perceived standards. Finally, the cumulative demands of eportfolio tasks, including planning, speaking, editing, and reflecting, led to mental exhaustion, disrupting students' emotional balance. Zhu and Fan (2018) define emotional well-being as a balance between positive and negative affect, which can be tipped by overload. As noted by Namaziandost et al. (2022), such stress reduces motivation and engagement, highlighting the need to manage workload and provide emotional support throughout the eportfolio process.

There are four implications drawn from the aforementioned findings and discussion. Firstly, the enjoyment and happiness experienced by students during eportfolio-based speaking tasks indicate the importance of integrating creative autonomy in language learning. Allowing students to personalize their video content through tools such as video editing, background music, and visual design can boost emotional engagement and intrinsic motivation. Secondly, the finding that positive feedback contributed to students' increased motivation and self-confidence highlights the need for structured, ongoing feedback mechanisms. Teachers should provide timely and affirming responses, while also encouraging peer feedback, to create a supportive learning climate. Third, the frequent reports of anxiety, fear of mistakes, and technical frustration underscore the importance of offering both emotional and technical support when implementing eportfolios. Educators should scaffold tasks by providing step-by-step tutorials, low-stakes practice opportunities, and clear rubrics to build students' confidence and reduce cognitive overload. Finally, students' experiences of stress and mental exhaustion due to the cumulative workload of recording, editing, and reflecting suggest that careful curriculum design is crucial. Instructors should avoid overloading students with too many high-stakes tasks and instead focus on quality over quantity, integrating meaningful but manageable assignments.

CONCLUSION

This study explored the positive and negative emotional experiences of Indonesian EFL students during the implementation of eportfolios in speaking courses. The findings revealed that while eportfolios supported emotional growth through enjoyment, pride, motivation, and relief, they also triggered various negative emotions, including anxiety, embarrassment, frustration, and exhaustion. These emotional experiences were shaped by both the learning tasks and the technological demands of the eportfolio process.

Despite its contributions, this study has several limitations. First, it focused on a relatively small group of 20 students from a single public senior high school in Indonesia, which may limit the generalizability of the findings to broader EFL contexts. Second, the emotional experiences were self-reported through semi-structured interviews, which, while rich in detail, may have been influenced by students' recall bias or willingness to disclose sensitive feelings. This study hopes that the researchers can conduct better future studies by addressing the above limitations.

REFERENCES

Bandura, A. (1994). Self-efficacy. In Ramachaudran, V. S. (Ed.), *Encyclopedia of human behavior* (4th ed., pp. 71–81). Academic Press.

Bandura, A. (1997). *Self-Efficacy: The exercise of control.* W.H. Freeman and Company.

Bensalem, E., & Hamed Alenazi, F. (2023). Using speaking eportfolios to enhance oral English language skills among English as a foreign language students. *Vivat Academia. Revista de Comunicación*, 25–47. https://doi.org/DOI: 10.15178/va

Bolliger, U., & Shepherd, E. (2010). Student perceptions of e-portfolio integration in online courses. *Distance Education, 31*(3), 295–314. DOI: 10.1080/01587919.2010.513955

Braun, V., & Clarke, V. (2006). Using thematic analysis in psychology. *Qualitative Research in Psychology, 3*(2), 77–101. DOI: 10.1191/1478088706qp063oa

Brown, H. D. (2004). *Language assessment : Principles and classroom practices* (Blanford, V., & Johnston, J., Eds.). Longman.

Cabrera-Solano, P. (2020). The use of digital portfolios to enhance English as a foreign language speaking skills in higher education. *International Journal of Emerging Technologies in Learning, 15*(24), 159–175. DOI: 10.3991/ijet.v15i24.15103

Cepik, S., & Yastibas, A. E. (2013). The use of e-portfolio to improve English speaking skill of Turkish EFL learners. *Anthropologist, 16*(1–2), 307–317. DOI: 10.1080/09720073.2013.11891358

Ching, C., & Chan, V. (2020). Positive emotions, positive feelings and health: A life philosophy. *Linguistics and Culture Review, 4*(1), 1–14. DOI: 10.21744/lingcure.v4n1.16

Cong-LEM, N. (2022). Emotion and its relation to cognition from Vygotsky's perspective. *European Journal of Psychology of Education, 38*(2), 865–880. Advance online publication. DOI: 10.1007/s10212-022-00624-x

Creswell, J. W., & Poth, C. N. (2018). *Qualitative inquiry & research design: Choosing among five approaches* (4th ed.). Sage Publications, Inc.

Dewaele, J., Chen, X., Padilla, A., & Lake, J. (2019). The flowering of positive psychology in foreign language teaching and acquisition research. *Frontiers in Psychology, 10*(SEP), 1–13. DOI: 10.3389/fpsyg.2019.02128 PMID: 31607981

Dougherty, E., & Coelho, D. (2017). Eportfolios in English language learning: perceptions of Arabic-speaking higher education students. *TESL-EJ, 21*(3), 1–21. http://www.tesl-ej.org/wordpress/issues/volume21/ej83/ej83int/

Efe, H. (2016). Reflective portfolio assessment in an EFL context. *Anthropologist, 24*(1), 157–163. DOI: 10.1080/09720073.2016.11892001

Erbas, Y., Ceulemans, E., Lee Pe, M., Koval, P., & Kuppens, P. (2014). Negative emotion differentiation: Its personality and well-being correlates and a comparison of different assessment methods. *Cognition and Emotion*, *28*(7), 1196–1213. DOI: 10.1080/02699931.2013.875890 PMID: 24410047

Georgi, D., & Crowe, J. (1998). Digital portfolios: A confluence of portfolio assessment and technology. *Teacher Education Quarterly*, *25*(1), 73–84.

Goh, C. C. M., & Burns, A. (2012). *Teaching speaking: A holistic approach*. Cambridge University Press.

Harmer, J. (2007). *The practice of English language teaching* (4th ed.). Longman.

Hsu, H. C. (2016). Voice blogging and L2 speaking performance. *Computer Assisted Language Learning*, *29*(5), 968–983. DOI: 10.1080/09588221.2015.1113185

Huang, H.-T. D., & Hung, S.-T. A. (2010). Implementing electronic speaking portfolios: Perceptions of EFL students: Colloquium. *British Journal of Educational Technology*, *41*(5). Advance online publication. DOI: 10.1111/j.1467-8535.2009.00996.x

Hung, S.-T. A. (2011). Pedagogical applications of Vlogs: An investigation into ESP learners perceptions. *British Journal of Educational Technology*, *42*(5), 736–746. DOI: 10.1111/j.1467-8535.2010.01086.x

Hung, S.-T. A., & Huang, H.-T. D. (2015). Video blogging and English presentation performance: A pilot study. *Psychological Reports*, *117*(2), 614–630. DOI: 10.2466/11.PR0.117c20z6 PMID: 26444840

Ilinawati, Sijono, & Elisa, H. (2021). The study of the students' speaking performance in demonstrating product advertising. Journal of English Education E-ISSN, 4(1), 59–70.

Jaber Rafidi, T., & Wagner, B. (2023). The impact of emotional scaffolding on online teaching and learning in Palestinian English language classrooms during COVID-19. *Innovation in Language Learning and Teaching*, 1–14. DOI: 10.1080/17501229.2023.2245810

Johnson, D. W., & Johnson, R. T. (2002). *Meaningful assessment: A manageable and cooperative process*. Allyn and Bacon.

Kusuma, I. P. I., Mahayanti, N. W. S., Adnyani, N. L. D., & Budiarta, L. G. R. (2021). Incorporating e-portfolios with flipped classrooms: An in-depth analysis of students' speaking performance and learning engagement. *The JALT CALL Journal, 17*(2), 93–111. DOI: 10.29140/jaltcall.v17n2.378

Kusuma, I. P. I., Mahayanti, N. W. S., Gunawan, M. H., Rachman, D., & Pratiwi, N. P. A. (2021). How well do e-portfolios facilitate students' learning engagement in speaking courses during the COVID-19 pandemic? *Indonesian Journal of Applied Linguistics, 11*(2), 351–363. DOI: 10.17509/ijal.v11i2.30583

Kusuma, I. P. I., Pratiwi, N., Paramartha, A., & Utami, I. (2022). Can e-portfolios reduce foreign language anxiety in online speaking courses during the COVID-19 pandemic? *Computer-Assisted Language Learning Electronic Journal (CALL-EJ), 23*(2), 38–55.

Kusuma, I. P. I., & Waluyo, B. (2023). Enacting e-portfolios in online English-speaking courses: Speaking performance and self-efficacy. *Iranian Journal of Language Teaching Research, 11*(1), 75–95. DOI: 10.30466/ijltr.2023.121273

Mihai, M., Albert, C. N., Mihai, V. C., & Dumitras, D. E. (2022). Emotional and social engagement in the English language classroom for higher education students in the COVID-19 online context. *Sustainability (Basel), 14*(8), 1–20. DOI: 10.3390/su14084527

Namaziandost, E., Alekasir, S., Hassan Mohammed Sawalmeh, M., & Miftah, M. Z. (2020). Investigating the Iranian EFL learners' attitudes towards the implementation of e-portfolios in English learning and assessment. *Cogent Education, 7*(1), 1856764. Advance online publication. DOI: 10.1080/2331186X.2020.1856764

Namaziandost, E., Razmi, M. H., Ahmad Tilwani, S., & Pourhosein Gilakjani, A. (2022). The impact of authentic materials on reading comprehension, motivation, and anxiety among Iranian male EFL learners. *Reading & Writing Quarterly, 38*(1), 1–18. DOI: 10.1080/10573569.2021.1892001

Ningrum, B., & Rokhuma, C. (2023). Positive and negative emotions of high school students in learning English through google classroom. *Journal of English Teaching and Learning Issues, 6*(1), 19–28. DOI: 10.21043/jetli.v6i1.19113

O'Malley, J. M., & Pierce, L. V. (1996). *Authentic assessment for English language learners: Practical approaches for teachers.* Addison-Wesley Pub. Co.

Pekrun, R. (2006). The control-value theory of achievement emotions: Assumptions, corollaries, and implications for educational research and practice. *Educational Psychology Review, 18*(4), 315–341. DOI: 10.1007/s10648-006-9029-9

Pratolo, B., Habibie, A., & Setiawan, A. (2019). Speaking Skill Challenges Encountered by Indonesian EFL Learners'. *Proceedings of the 2019 Ahmad Dahlan International Conference Series on Education & Learning, Social Science & Humanities (ADICS-ELSSH 2019)*. DOI: 10.2991/adics-elssh-19.2019.7

Rabbani Yekta, R., & Kana'ni, M. A. (2020). Using Google Drive as the e-portfolio for the self-assessment of speaking fluency. *International Journal of Research in English Education*, *5*(2), 49–60. DOI: 10.29252/ijree.5.2.49

Richards, J. C. (2020). Exploring emotions in language teaching. *RELC Journal*, *53*(1), 1–15. DOI: 10.1177/0033688220927531

Sun, Y. C., & Yang, F. Y. (2015). I help, therefore, I learn: Service learning on Web 2.0 in an EFL speaking class. *Computer Assisted Language Learning*, *28*(3), 202–219. DOI: 10.1080/09588221.2013.818555

Yastibas, A. E., & Cepik, S. (2015). Teachers' attitudes toward the use of e-portfolios in speaking classes in English language teaching and learning. *Procedia: Social and Behavioral Sciences*, *176*, 514–525. DOI: 10.1016/j.sbspro.2015.01.505

Zheng, Y., & Barrot, J. S. (2022). Social media as an e-portfolio platform: Effects on L2 learners' speaking performance. *Language Learning & Technology*, *26*(1), 1–19. https://hdl.handle.net/10125/73487. DOI: 10.64152/10125/73487

Zhu, J., & Fan, Y. (2018). Daily travel behavior and emotional well-being: Effects of trip mode, duration, purpose, and companionship. *Transportation Research Part A, Policy and Practice*, *118*, 360–373. DOI: 10.1016/j.tra.2018.09.019

Chapter 12

An In-Depth Study on Enhancing the Effectiveness of ICT in CLT Through Mixed Method Research From Teachers' Perspectives

Nafis Mahmud Khan
https://orcid.org/0000-0002-4430-0449
TIIAME National Research University, Uzbekistan

Nargiza Arifovna Kadirova
https://orcid.org/0000-0001-7913-5642
TIIAME National Research University, Uzbekistan

Jasur Umirzokov
Tashkent State University of Economics, Uzbekistan

ABSTRACT

The study aims to investigate English teachers' awareness of using different applications for personal and professional purposes at the secondary level, and explore the measures needed to adopt to enhance ICT's effectiveness in CLT. A mixed-method research design was employed to answer the research questions. In the first phase of the study, a random sampling technique was used to collect the data from the teacher participants of Bangladesh. In the second phase, a semi-structured interview was conducted to identify the dominant theme of this study. Results indicate that

DOI: 10.4018/979-8-3373-3366-3.ch012

majority of the teacher participants are aware of using ICT materials for personal and professional purposes. The findings of the study reveal teacher participants' expectation from the school authorities for incorporating ICT in their classroom language teaching. Thus, the study provides meaningful insights for policymakers, stakeholders, and authorities to take timely necessary initiatives for the successful implementation of ICT in classroom language teaching.

1. INTRODUCTION

Information and Communication Technology (ICT) has made a significant transformation in the process of teaching and learning. Language teachers have already understood the importance of digital literacy, and nowadays, integrating technology for language teaching has become a common phenomenon among EFL teachers, and stakeholders. The benefits of technology make teaching quality better and help learners learn the language effectively (Mudra, 2020). Gracemary, Dolly, and Njeri have also found that most teachers desire to use ICT in schools to enhance their instructional effectiveness in classroom teaching (Gracemary, Dolly and Njeri, 2014). In addition, it also invites an authentic environment for classroom teaching which helps the learner to enhance the four language skills: listening, reading, speaking and writing (Warschauer, 2004).

Similarly, Internet and computer use in the language classroom make classroom language teaching more effective. Recently, the usability of Mobile Assisted Language Learning has been increasing exponentially as it brings freedom of autonomous learning, especially for learners at their own time, pace and space. In addition, several studies indicate that devices of MALL such as smartphones, tablets, and iPods has the potential to enhance learners' capability to learn language easily (Dash and Kuddus, 2020; Kuddus, 2018; Chatterjee & Kuddus, 2014).

However, there are enormous benefits of integrating technology into language learning, the existing challenges impede users from implementing ICT into classroom language teaching. To mention a few, teachers' negative perceptions and beliefs about technology significantly affect classroom teaching. Moreover, limited ICT skills consistently abstain teachers from using technology effectively in classroom teaching (Gracemary, Dolly and Njeri, 2014). Factors, such as limited technical support, limited Internet accessibility, and limited class duration discourage teachers from integrating ICT in language classrooms (Salehi and Salehi, 2012; Khan and Kuddus, 2020).

Realizing the importance of ICT in education, almost all the education ministries have already consented to integrate ICT into their education system. Bangladesh has also taken several initiatives to develop the quality of education and integrate

ICT into its education system (Khan et al., 2012). As a part of the initiatives, 'National Education Policy 2010' (NEP) was rephrased with some modifications based on 'National ICT Policy 2009', particularly in the secondary school English language teaching curriculum (Talukder & Saba, 2016). Apart from these, to make the initiatives successful, various massive programs have been initiated by the government, such as projects on implementing multimedia/smart classrooms for the classroom, providing funding facilities to increase ICT use in classroom teaching, free laptop distributions, ensuring the availability of ICT devices, incentives and rewards, high-speed internet facilities, training on using ICT tools in the classroom setting for the secondary school teachers. In addition, the government emphasizes ICT-based English language teaching in secondary education rather than only ICT literacy (A2I; PMO 2009).

But as discussed above, the enormous significance of ICT in language learning fades away due to the challenges faced by the teachers in the real implementation of ICT in classroom language teaching. Hence, in most cases, planning and strategies taken by the authority are not implemented rightly due to the challenges teachers experience while implementing ICT in classroom language teaching. Understanding the hindrance in the way of ICT implementation, working on the initiatives and measures to overcome the existing and predominant challenges in the actual implementation of ICT in language learning becomes significant. Although, the literature review highlights information about the benefits of ICT in language learning and the challenges faced by the teachers and students in its ground level implementation, there is a paucity of studies on the recommendations, measures, and initiatives to overcome the inhabiting factors in implementing ICT in classroom language teaching, particularly in secondary school level in Bangladesh. Hence, the suggestions for adopting several measures and initiatives by the teachers, authorities, policymakers and other stakeholders in order to enhance the effectiveness of ICT in Communicative Language Teaching makes the present study significant.

2. LITERATURE REVIEW

2.1 Emergence of ICT in Bangladesh

Bangladesh has the vision to become 'Digital Bangladesh' and is determined to ensure equal education (Khan, Hasan, and Clement, 2012; dhakatribune.com). To accomplish the goal, ICT division announced 'National ICT Policy 2009', and, in light of that, 'National Education Policy' was rephrased in 2010, emphasizing the use of ICT in improving quality education. As a part of initiatives, UNESCO, the international organization, has extended their utmost support to introduce technology-based education systems in Bangladesh. In addition, different national and international private organizations and NGOs from a different levels, including Bangladesh Rural Advancement Committee (BRAC) and Grameen Bank, particularly have taken massive initiatives to ICT for education.

2.2 ICT in EFL Context

Implementing ICT in classroom language teaching successfully requires organized planning and proper monitoring. Most teachers understand the importance of technology integration in classroom teaching, but they cannot integrate it in their classroom due to their limited technology use (Gracemary, Dolly and Njeri, 2014). However, in this respect, to accommodate the limitations, many researchers suggest possible solutions through their studies. To mention a few, a study in secondary schools in Iran, conducted by Haider, indicates the necessity to introduce sufficient technical support, especially for teachers, accessibility to the internet, and ensure the availability of ICT tools in classroom teaching (Haider, 2012). At the same time, teachers' assertiveness and positivity toward technology use in classroom teaching play a significant role in implementing technology in classroom teaching; but, unfortunately, teachers attending various professional ICT training courses are not willing to learn about how to implement ICT in classroom teaching properly (Khan & Kuddus, 2020). Therefore, teachers' positivity toward technology integration must be ensured. Apart from this, assurance should be there in teachers' commitment to implementing ICT in classroom teaching.

In this relation, some common factors, such as lack of software, limited funding facilities, limited time, and lack of technology use skills, hinder accommodating technology in classroom settings (Ihmeideh, 2009). However, the conclusion of the study suggests a clear vision that needs to be set up by the school management. In the same vein, Anderson, Ronakd.e comes up with a similar recommendation that it is to be ensured the availability of hardware and software in institutions to use technology effectively (Anderson & Dexter, 2000).

2.3 Mobile-Assisted Language Learning (MALL) in Communicative Language Teaching (CLT): Current Insights:

Mobile Assisted Language learning serves as a driving force within communicative language teaching (CLT), enhancing learners' communicative competence through authentic, context-rich, and interactive language practice. Several studies highlight that MALL integrated CLT offers learners to develop meaningful communication skills, strengthen learners' motivation and provide opportunities for real time feedback and collaboration. Additionally, it offers learners to access to authentic materials and opportunities for continuous language learning beyond traditional classroom settings (Alda, 2023; Chua & Soon, 2021; Walsh, 2022).

Through WeChat, MALL facilitates instant texting, enhancing students' Englis proficiency by promoting multifunctionality, accessibility, and interactive engagement (Shi, Luo, & He, 2017). A large number of research highlight the advantages of Mobile Assisted Language Learning Program (MALL) to enhance vocabulary acquisition among EFL learners. Mobile applications, digital flashcards, and messaging platforms such as WhatsApp offer substantially enhancement to learners' vocabulary acquisition, motivation, and overall engagement in the learning process (Zakian et al., 2022; Benlaghrissi & Ouahidi, 2023; Hassan Taj et al., 2017).

Several studies highlight the advantages of mobile applications to enhance English Language Teaching (ELT) and teachers' professional development in Bangladesh. In recent years, the use of WhatsApp among language teachers has notably increased, as it offers flexible learning environments, assists teachers' speaking and writing skills development, and contributes to creating more interactive and learner-centered classrooms (Ibna Seraj, Prodhan, & Klimova, 2025).At the same time the integration of such technology in pedagogical pedagogical practices has brough about a significant change, contributing meaningfully to the enhancement of reading, writing, listening, and speaking skills (Hasan et al., 2022; Oh et al., 2022) .

Studies demonstrate that the use of Duolingo for language proficiency development, especially for vocabulary acquisition significantly enhance among EFL learners (Fakhrurriana, Nisa, & Noni, 2024; Ajisoko, 2020; de Sousa, Cardoso, & Toassi, 2018,). Learners completed Duolingo English courses consistently performed better score, Intermediate High, in both their reading and listening (Jiang et al., 2024; Jiang & Pajak, 2022,). Albeit, Duolingo offers significant improvements in developing learners speaking and writing skills, the progress is less compared to the progress observed in reading and listening. Study shows that due to the limited chance of real-life practice and extended writing activities learners are not able to perform significantly in speaking and writing (Lo et al., 2025; Firdausia & Hamdani, 2025, Irzawati, 2023). Therefore, Duolingo is considered to be used as a supplementary learning tool alongside with other learning methods considering active communication and feedback (Firdausia & Hamdani, 2025).

Concerning to use Duolingo by Bangladeshi EFL learners, result reveals that Duolingo contributes significantly in enhancing their vocabulary, Grammer, and overall proficiency skills. Learners show their positive responses and consider Duolingo as their self-assessment tools assisting them with measuring their learning progress but many of them are found not fully aware of the app's diversified features and potential use. The findings suggest to arrange the training facilities for the teachers to integrate Duolingo effectively into their classroom teaching practice (Sadaf, 2023).

2.4 ICT in Communicative Language Teaching

Information and Communication Technology introduce numerous tools for leaners to enhance English Language proficiency and support Communicative Language Teaching by making it engaging and easily accessible through both online and offline platforms ((Mashhura, 2024). Various ICT tools such as, video conferencing, email correspondence, social media interaction facilitate students developing speaking performances in real life situations. In addition, integration of technology in communicative language teaching enhances learners speaking ability to build rapport, developing fluency and accuracy, reduce stress level and foster better confidence among students in verbal communication (Sosas, 2021). Nowadays, different online platforms have emerged and are functioning as leading collaboration tools by promoting teamwork, easy interaction and knowledge building in English language education. For example, google docs supports collaborative work and develop their writing skills, Padlet- a dynamic online forum offers online assignment submission for multiple learners simultaneously, Trello- a structured project management platform, where multiple learners can engage in project works together, shared their pair feedback, and able to track their project progress visually (Mashhura, 2024)

The integration of ICT into task-based language teaching (TBLT) through CLT approach significantly enhance students' communicative competence level and prepare them using English in authentic context. Mulyadi et al. (2021) conducted a study on ESP students majoring in nursing and found that using technology within TBLT significantly improved learners' comprehension and performance in listening and speaking skills. Likewise, Nasrulloh et al. (2024) reported that the purposeful use technology into structured learning increases students' motivation and broadens their perspectives, preparing them to handle authentic, real-world communication challenges effectively.

ICT integrated CLT has emerged into a leading area in Communicative Language Teaching, demonstrating how technology helps learners to enhance their communication, interaction and learners' participation in classroom settings. Wardoyo et al. (2025) conducted a comprehensive literature review and investigate how successfully ICT tools can integrate into CLT to improve learners' language proficiency. Findings indicate that both hardware technologies, such as mobile phone and iPad, different software application online learning platform, educational websites play significant role in promoting communicative competence by offering accessibility, interaction and motivation.

2.5 ICT in Education: Bangladesh

Technology has brought a major shift in education, especially in English Language Teaching (ELT) in Bangladesh (Talukder & Saba, 2016). To cope with the changes concerning education policy by other developing countries, Bangladesh government has taken massive initiatives focusing on integrating ICT in English Language teaching in education rather than only ICT literacy. At the same time, NEP 2010 picked out strategies for implementing ICT-based education in all academic levels; primary, secondary, and higher secondary.

2.6 ICT in Education During the Covid - 19 Pandemic: Bangladesh Context

The outbreak of Covid-19 pandemic has led online teaching and learning popular and accelerated the adaptation of digital technology in Bangladesh Education. During Covid-19 pandemic, digital tools, such as Zoom, Google Meet, mobile learning application, and social media platform became popular digital tools for remote education. During this transition, both the learners and teachers had to adjust themselves to this sudden pedagogical shift. However, this transition brough several challenges to both learners and teachers, such as unstable internet connectivity, limited digital literacy, and insufficient institutional technological support that hinder them into smooth implementation of digital education (Shrestha et al., 2022). As a consequence of covid-19, traditional teaching methods have undergone a significant transformation as both the learners and educators had to adopt with the emerging educational and digital tools (Gamage, Ayres, & Behrend, 2022).

However, to ensure sustainable educational transformation, studies suggested several recommendations for policymakers and stakeholders. Several studies also highlighted leading factors motivate learners and educators to keep engaged with e-learning platform in the future. Maisha and Shetu (2023), conducted a study investigating the reason of keep using e-learning platform after the pandemic by the Bangladesh students. The findings of the study indicates that factors, such as performance expectancy, social influence, and behavioral intention played significant role in shaping students' willingness to keep using e-learning systems after the COVID-19 pandemic. Adopting the extended framework of UTAUT3 model, the study provides the valuable insights into the underlying motivation that leads to sustainable use of e-learning platform in the post pandemic context.

2.7 ICT in CLT Context: Bangladesh

Since integrating ICT into the rephrased national education policy, our education has been facing huge challenges. Just after the declaration of rephrased NEP 2010 policy, Parvin Shahnaj carried out a study where the researcher pinpointed various challenges in implementing ICT in education, including strategic challenges: ICT support infrastructure, lack of funding, lack of vision; social-cultural and political difficulties; other challenges, as such: teachers' negative attitude to ICT, Lack of ICT skills, limited classroom time duration. The study offers some possible solutions with few recommendations (Parvin Shahnaj, 2013). Likewise, Hasan and Khan are not different from their findings on the challenges faced by the teachers in the teachers' training program (Hasan et al., 2013). In the same line, Talukder discussed some difficulties faced by secondary school teachers while using ICT in a language teaching classroom (Talukder & Saba, 2016). The study reflects that most of the teachers do not have ICT knowledge and the usability of ICT tools in classroom teaching is unsatisfactory. At the same time, having mere knowledge of English is another hindrance to integrating ICT in language teaching fruitfully; and the Lack of qualified subject-based teachers is another cause for not using ICT devices properly in the classroom (Talukder & Saba, 2016). However, the provided recommendations of this study are not elaborated enough (Talukder & Saba, 2016). Another study exhibits various challenges faced by language teachers in implementing ICT in communicative language teaching, as such underdeveloped CLT curriculum, limited class duration, lack of proper ICT skills, insufficient technical support, lack of training facilities, lack of technical support and maintenance, poor internet connection and lack of ICT infrastructure (Khan and Kuddus 2020). Though Khan and Kuddus pointed out the challenges in implementing ICT in a communicative language teaching context in their study, it does not provide necessary suggestions to overcome the challenges faced by the stakeholders. (Khan and Kuddus 2020).

As much attention has been given to the investment of technology for classroom teaching by government and non-government organizations, the present study attempts to investigate the awareness of using different applications and ICT devices by the language teachers of secondary schools of Bangladesh for personal and professional purposes. Further, as the existing literature shows that a few studies have been conducted on the challenges in implementing ICT in classroom language teaching, particularly at the secondary school level in Bangladesh; no studies have been conducted dealing with the suggestions and recommendations as measures, microscopically, by the secondary school language teachers to overcome the challenges, and enhance the effectiveness of ICT in language teaching.

Therefore, the current study aims to address the possible measures that need to be adopted by the government and other relevant authorities to enhance the effectiveness of ICT in communicative language teaching using the Mixed Method research design. The specific objectives are raised as follows:

I. to know the availability of IT devices and other peripherals at institutions
II. to explore the awareness of using different applications and ICT devices by the language teachers for personal and professional purposes; and
III. to investigate the possible measures to enhance the effectiveness of ICT in CLT

3. RESEARCH METHODOLOGY

3.1 Background Information of Teachers' Participants

Table 1 exhibits teacher participants' demographic characteristics. Among 100 participants, 67 were male, and 33 were female. With respect to Board wise participants, it is found that 35% out of 100% of participants were from the Dhaka board, whereas 31% and 34%, respectively from Jessore and Comilla Board. Among 100% of teacher participants, a total number of 35% of participants were from urban areas, whereas 32% were from semi-urban areas, and 33% were from rural areas. All the teacher participants were assistant English teachers by designation. Moreover, 33% of participants were from Government schools, 31% and 36% were from semi-government and private schools, respectively.

Table 1. Demographic information of teacher participants

Variables	Category	Percentage (%)
Gender	Male	67%
	Female	33%
Board	Dhaka Board	35%
	Jessore Board	31%
	Comilla Board	34%
Designation	Assistant Teacher	100%
Locality	Urban	35%
	Semi-urban	32%
	Rural	33%
Type of School	Government	33%
	Semi-Government	31%
	Private	36%

For the purpose of the study, a closed- ended questionnaire was administered to 100 secondary-level English teachers of Bangladesh, and 100 responses were obtained. The participants were assured that the provided information would be kept confidential and would be strictly used for research purposes.

Among the 100 English language teachers who completed the questionnaire, the researcher invited 20 teachers to participate in the follow-up semi-structured interviews from government, semi-government, and private schools to check the validity of the study and to get further suggestions to enhance the effectiveness of ICT on CLT.

3.2 Research Design and Instruments

For the study, the researcher employed the mixed-method approach: Explanatory Sequential Mixed-Method research design. In this first phase, the close-ended questionnaire provided the analysis of questionnaire data and in the second phase; subsequently, the qualitative analysis of semi-structured interview was carried out to check the validity and explore the research objectives (Creswell & Creswell, 2017). The English language teachers were the participants selected randomly from various secondary level schools viz; government, semi-government, and private schools of urban, semi-urban, and rural areas of Bangladesh.

A four-section close-ended questionnaire was designed with a total of 18 items: section A describes participants' background information, section B describes participants' awareness of using ICT devices for personal and professional purposes; section C deals with computers and peripherals at institutions, and section D consists of 6 items in the close-ended format on a 5-point Likert scale from 1 strongly disagree to 5 strongly agree to investigate the measure to enhance the effectiveness of ICT on CLT. Three experts from academia were consulted to assess the content validity of the questionnaire. Some items were adjusted, and some were deleted to maintain overall reliability. The Alpha value of the questionnaire was .707, which indicates that the internal consistency of the questionnaire is acceptable (Taber Keith S., 2018).

Following the data collection and analysis of the questionnaires, twenty English language teachers were selected from three different types of schools to be interviewed. They were chosen randomly from different schools from where the data was collected. A semi-structured interview was conducted to know teacher participants' opinions about using technology for English langue teaching in the classroom settings, the availability of ICT tools in the institutions for classroom teaching, and their suggestions to enhance the effectiveness of ICT on CLT. To fulfil the research objectives, the researcher asked the following questions:

I. What kind of technology do you think can be used in the classroom?
II. What are the ICT tools available in your institution? Do you think existing ICT materials are sufficient to integrate into classroom teaching?
III. What steps could be taken to make language teachers skilled in integrating ICT in communicative language teaching?
IV. What measure could be taken to increase the number of ICT materials in the classroom?
V. Please provide your valuable suggestions to enhance the effectiveness of ICT in communicative language teaching.

The above semi-structured interview questions were developed based on the close-ended questionnaire. The interview was recorded and transcribed in English. Interviews were conducted individually, and each lasted approximately 20-30 minutes.

3.3 Procedure

Quantitative data were analyzed using the statistical analysis software (SPSS 20) program. Data calculated with the SPSS program were expected to determine frequency, percentages, the mean, and standard deviation to interpret the data. In addition, obtained data from the questionnaire were analyzed to get measures to enhance the effectiveness of ICT on CLT.

Interview data have been used to provide detailed descriptions of the possible suggestions that need to be taken. The recorded interview has been transcribed, coded, and categorized manually to investigate the possible dominant theme. The interview data were also used to supplement the quantitative findings, and the interpretation exhibits a detailed description of suggestions made by English language teachers to enhance the effectiveness of ICT on CLT.

4. ANALYSIS AND INTERPRETATION OF DATA (QUANTITATIVE RESULT)

For data analysis, a questionnaire was designed to know the following answer from the teachers. Firstly, their awareness of ICT, then information about computers and peripherals at institutions, and finally, measures for the successful integration of ICT in communicative language learning were examined.

4.1 Teachers' Awareness of ICT Devices and its Applications

This section of the questionnaire aims to explore teacher participants' awareness of ICT tools and ICT skills and their frequency of using ICT tools and the internet. It also examines the types of applications and software they use in their daily lives.

4.1.1 Teachers' Computer Usage:

Table 2. Teachers' computer use and having computer

Teachers' using computer		Teachers' having a computer at home	
Response	**Percent (%)**	**Response**	**Percent (%)**
Yes	94%	Yess	72%
No	6%	No	28%

Table 2 exhibits teachers' computer usage and having computers at home, dealing with the awareness of ICT devices and how frequently teachers use computers for personal and professional purposes. The table shows that 94% of teachers use computers frequently, whereas 72% of teachers have computers at home for personal and professional goals.

4.1.2 Teachers Using Different Software

Figure1 shows teacher participants using various types of software for personal and professional purposes. It reveals that majority of the teacher participants used MS Word (89%) and PowerPoint (68%) for personal and professional purposes. In comparison, 69% of teachers used PowerPoint, whereas very few teacher participants used other software, such as Paint, Flesh, Photoshop, Publishers, Movie makers, and Excel.

Figure 1. Teachers' using different software

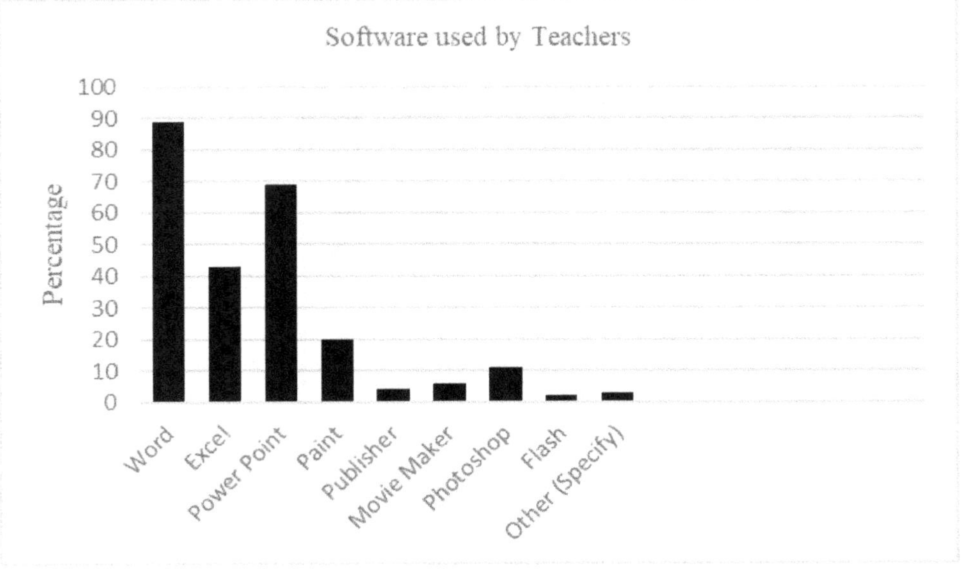

4.1.3 Using Computer Applications by Teachers for Teaching Purpose

Figure 2 shows teachers' using different applications for language teaching purposes'. The figure indicates that the majority of language teachers (70%) used internet for teaching and learning purposes. PowerPoint (65%) and MS Word (52%) were teachers' most commonly used applications for teaching purposes. However, other ICT applications were not popular among teachers.

Figure 2. Teachers using applications for language teaching

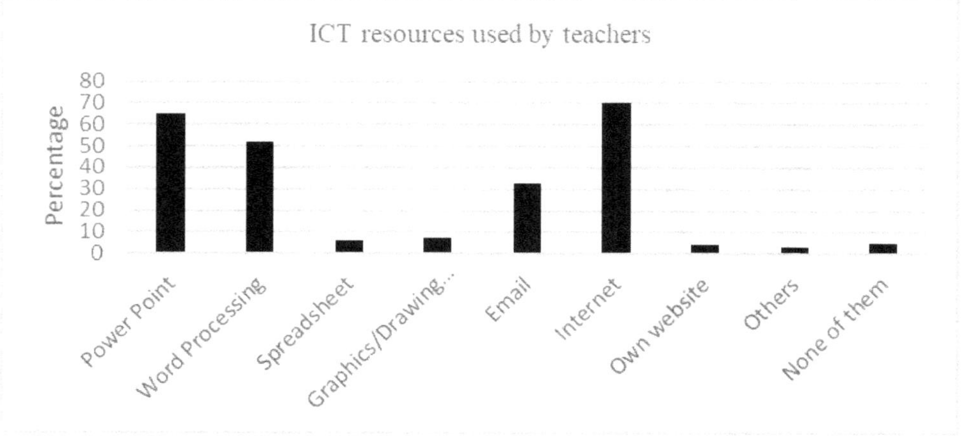

4.1.4 Frequency of Internet Usage

Regarding the frequency of the internet usage by the language teachers, the result indicates that the percentage of using internet (55%) on daily basis is not significant compare to the percentage of regular computer users, which is 94% (see Table 2). Only 30% of teachers were found using internet frequently, whereas 2% of the teacher participants never or hardly ever use it.

Figure 3. Periodicity of using internet

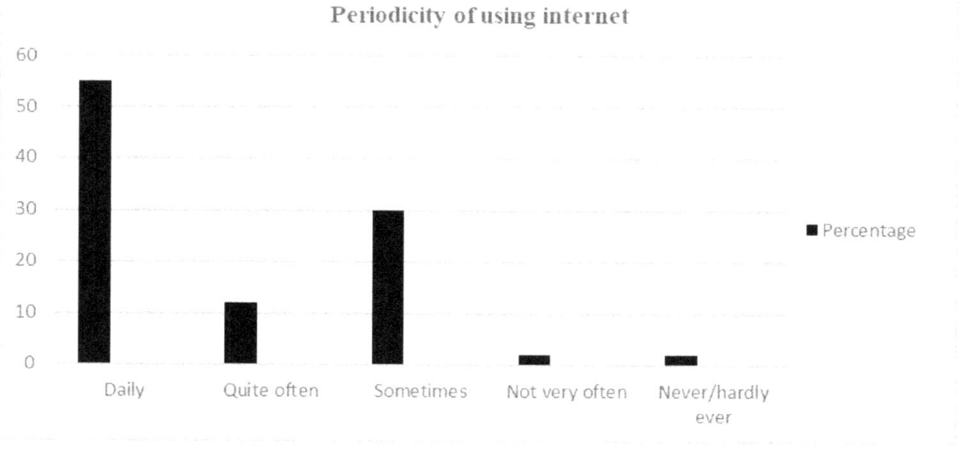

4.1.5 Member of Social Networking Sites

Figure 4 indicates the percentage of teachers being the members of social networking sites. The figure shows that a significant number of teacher participants used Facebook i.e.96 percent but only 36% of teachers used WhatsApp, and 25% used Instagram. However, hardly a few teachers use other social networking sites such as LinkedIn, Twitter etc. The figure also indicates the popularity of Facebook among English language teachers in Bangladesh. Hence, the result shows that the English teachers of secondary schools of Bangladesh are neither ignorant of ICT devices and their applications nor of social media platforms and their usages.

Figure 4. Members of social networking site

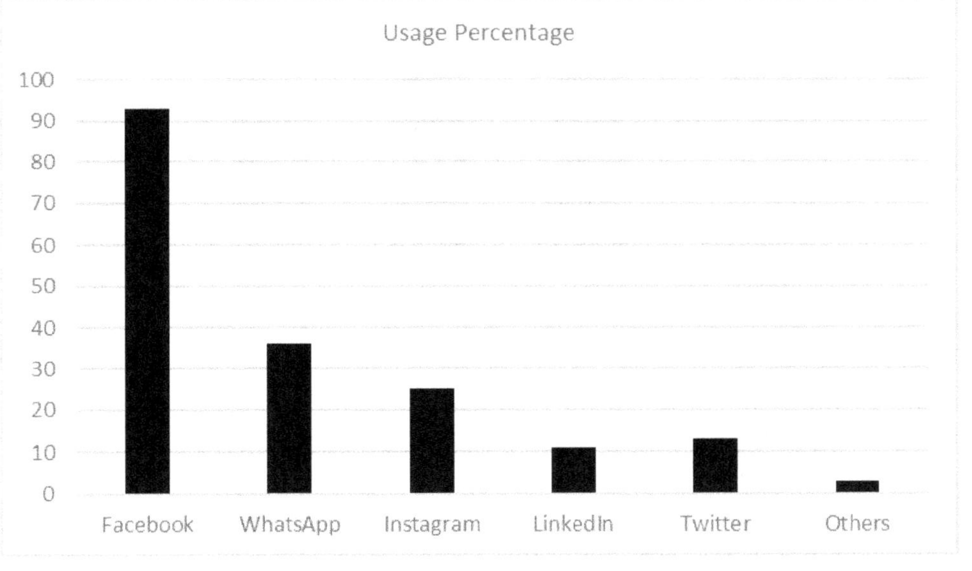

4.1.6 Teachers' Different Means of Communication with Friends and Relatives

Data also illustrates that all the teachers used mobile phones as their means of communication, unlike telephones, which is just 10%, indicating that the telephone is no longer popular among the teacher respondents as all of them have become the active users of mobile phones. Other significant means of communication which the teachers are using to communicate are SMS, Chat and Email. Data delineates that majority of the teacher participants (64%) used SMS, and 53% were chat users for communication (see Figure 5). However, interestingly, 43% of the teacher participants are also found using emails for communicating with their friends and relatives, which indicates the familiarity of emails address to the teachers for communication purposes.

Figure 5. Teachers' means of communications with friends and relatives

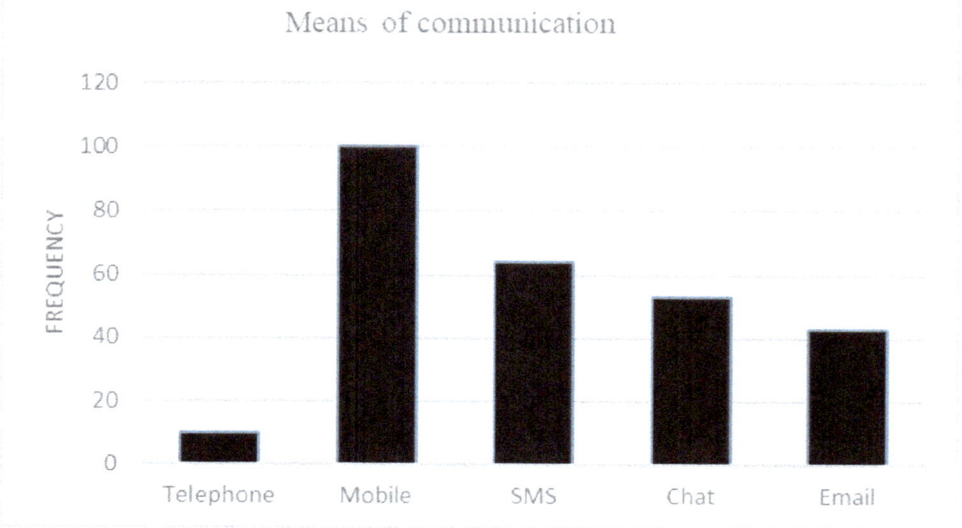

4.2 Computers and Peripherals at Institutions

This section of the teachers' questionnaire aims to elicit information about the availability of ICT tools and peripherals; the actual status of internet facilities provided to the teachers for teaching and learning purposes at secondary schools in Bangladesh. This section also attempts to explore the types of ICT applications used by the secondary school teachers of Bangladesh for language teaching and learning

4.2.1 Internet Connection Facility

Figure 6 shows that the authorities prioritize providing internet connection facilities more in the office rooms rather than classrooms, labs, and other sections. The finding reveals that more than 80% of the teachers responded positively to internet connection facilities in office rooms, whereas classrooms were limited to internet accessibility. However, other sections, such as the computer lab, library, and staffroom were rarely found with internet access.

Figure 6. Internet connection facility

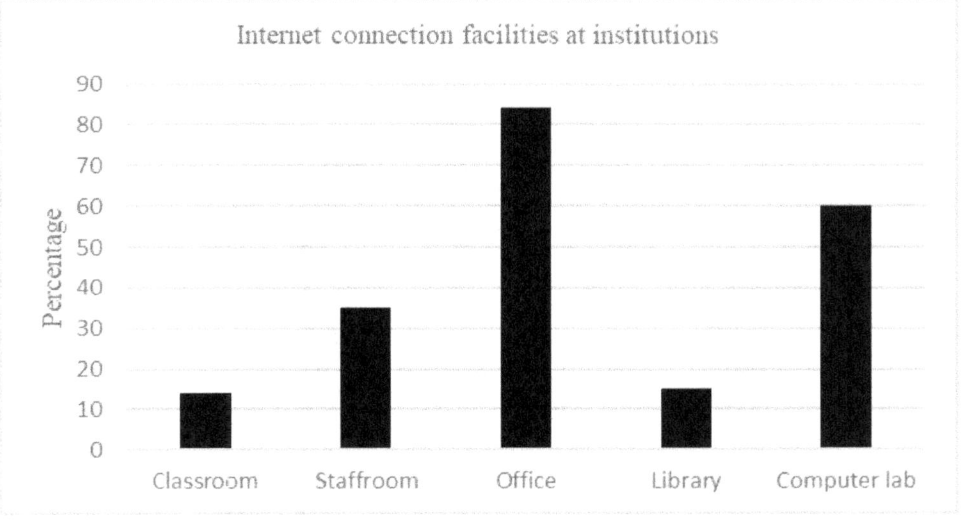

4.2.2 ICT Tools and Peripherals at Institutions

Regarding the availability of ICT peripherals at institutions, Figure 7 delineates that the majority of the teacher participants (80%) responded to have a digital whiteboard in their institutions. But the status of the availability of digital cameras is insignificant as less than 40% of language teachers consented to the availability of digital cameras at institutions. Likewise, the response to other ICT tools, such as video conferencing devices and Digital whiteboards, was also very poor, indicating the unavailability of those technological devices).

Figure 7. ICT and peripherals at institutions

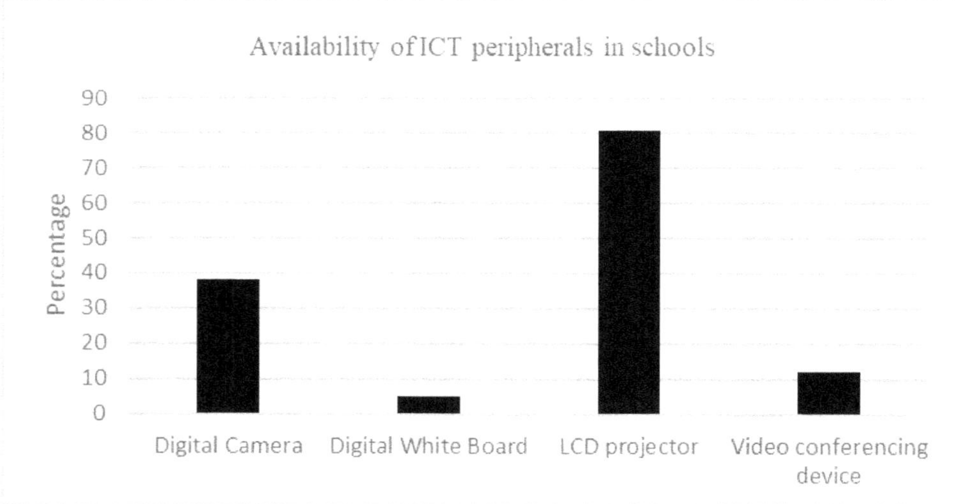

4.3 Measure to Enhance ICT in CLT

Table 3. Measures to enhance the effectiveness of ICT in CLT

Measures to enhance the effectiveness of ICT on CLT								
Sl. No.	**Statements**	**Strongly Agree**	**Agree**	**Neutral**	**Disagree**	**Strongly Disagree**	**Mean**	**Standard Deviation**
1	Teachers should be encouraged and motivated by the school management for incorporating ICT in teaching.	62%	38%	0%	0%	0%	4.62	.488
2	ICT integration in teaching should have an influence on teachers' annual performance assessment.	35%	56%	8%	1%	0%	4.25	.642
3	ICT certificate courses / training should be made a mandatory condition for the promotion of teachers.	34%	57%	7%	2%	0%	4.23	.664

continued on following page

Table 3. Continued

Measures to enhance the effectiveness of ICT on CLT								
Sl. No.	**Statements**	**Strongly Agree**	**Agree**	**Neutral**	**Disagree**	**Strongly Disagree**	**Mean**	**Standard Deviation**
4	Regular training on ICT should be organised for classroom teaching by the school authority.	55%	43%	1%	1%	0%	4.52	.577
5	Introducing free short term ICT courses conducted by the school management in collaboration with government bodies.	49%	48%	2%	1%	0%	4.45	.592
6	Classrooms should be developed enough to use ICT in classroom teaching.	59%	39%	2%	0%	0%	4.57	.573

Table 3 exhibits the measures that need to be adopted to enhance the effectiveness of ICT on CLT. In response to statement 1, as for measures to enhance the effectiveness of ICT on CLT, data reveals that all the teacher participants (100%) demanded that they should be encouraged and motivated by the school management to incorporate ICT in teaching. The mean for this statement is 4.62, which is very high, and the standard deviation is .488, respectively, indicating teachers are highly in need of motivation from the authority to use ICT in classroom settings.

The teachers also responded positively with the condition that ICT integration in teaching should influence teachers' annual performance. They also feel the necessity of ICT certificate course/training as mandatory and agreed to have it as a condition of getting promotion in their academic career. For this statement, the mean score is 4.23, and the standard deviation is .664, which indicates the importance of ICT certification training for the teachers to use ICT in classroom teaching.

Regarding having regular training organized by the school authority on ICT integrated classroom teaching, the response rate is above the midpoint, and the mean score is 4.52, indicating their interest to learn using ICT devices in language teaching. Apart from this, teachers are found indifferent regarding organizing short-term ICT and regular ICT courses, which should be conducted by the school management. The mean value for this statement is 4.45, and the standard deviation is .592. To enhance the effectiveness of ICT on CLT, the teachers strongly urge for classroom development. The mean score for this statement is 4.57, reflecting teachers' high recommendation to get required ICT materials and other tools sufficiently so that they can integrate ICT materials effectively in classroom settings. Results also show that all the mean values for the items are above the midpoint (3.00), and the standard deviation values of the items are above 0.26, indicating respondents' high positiveness towards all the statements.

4.4 Analysis and Interpretation of Data (Qualitative Result)

The qualitative data analysis of the semi-structured interview discusses the dominant theme underlying this research study. To analyze the theme, the researcher followed the guideline and outline designed by Braun and Clarke (Braun and Clarke, 2006). Though Braun and Clarke's thematic analysis particularly guides the researchers working in psychology as there are neither any hard and fast rules nor any clear agreement about thematic analysis and how to move further with it; therefore, the researcher follows step by step guide to analysis of the study (Stirling & Jennifer 2001). The data initially was transcribed, and then coded by writing notes on the texts to analyse. The highlighters and colored pens were used to indicate potential patterns to identify the data segments.

4.4.1 Teachers' Awareness About ICT Devices and Their Applications in Classroom Teaching

Regarding the integration of technological tools effectively in the language classrooms, all twenty teachers agreed about the usability of the projector, digital whiteboard, internet, and computer. One of the teachers- T2 (T refers to Teachers), also talked about using mobile phones and tablets to make lessons more interesting for the students.

"When a teacher gives information using mobile and tabs, the students feel much interested. Iff we can give lessons using a mobile phone, or tabs, it will be more interesting for students. Projector is also useful for teaching."(T: 2)

According to some teachers:

"Computer, Laptop, Interactive Board, Printer, Scanner, Modem, Graphic Tablet, Surround sound system (Multimedia Speakers), could be used so that the student can enjoy the class without the boredom which ensures effective learning."(T: 4)

".... projectors, slides or videos, related to the subjects can enhance the students learning capacity." (T: 20)

4.4.2 ICT Tools and Peripherals at Institutions

In response to the question of the availability of ICT in the institutions, all twenty teachers reported that they have computer lab facilities but limited access. According to T3,

"Existing electronic equipment was provided for the computer lab but not for each classroom for all courses."(T: 3)

T8 also confirmed the availability of Internet facilities at his institution:

"Yes, I use the internet in my English language classroom. Wi-fi net is available in our academy. When the wi-fi fails, I do net surfing through my mobile phone network". (T: 8)

4.4.3 Measure to Enhance ICT in CLT

From the interview data on measures to be adopted to enhance the effectiveness of ICT in CLT, the researcher identified the sub-them, which led to developing the dominant theme. Overall, the researcher identified four dominating themes with their sub-theme as follows:

I. Suggestions for using technology in classroom teaching
II. Suggestions for curriculum development
III. Recommendations for increasing ICT materials usage
IV. Suggestions to make language teachers skilled in using ICT in CLT

I. Suggestions for Using Technology in Classroom Teaching

a) Providing Sufficient ICT Materials by the School Authority to the Classroom

While the participants were asked about the measures to be taken for effective ICT use in language teaching and learning settings, most teachers suggested ensuring sufficient ICT materials for classroom teaching. Though they were enthusiastic about ICT use in their classroom settings, they could not show practical subject matter through PowerPoint presentation slides due to the lack of ICT materials. All twenty teachers strongly emphasized maintaining classroom standards equipped with ICT materials so that teachers can effectively integrate ICT into classroom teaching. Modern technological tools must be introduced in the classroom. According to T4, teachers must be skilled in basic computer operation and using digital classroom equipment. Using a projector and electronic whiteboard connected with broadband internet must be ensured.

One of the teacher participants, T4, talked about the class load and the increasing number of English Teachers.

"Providing ICT materials facility by school authority for classroom teaching must be ensured." (T: 4)

According to T: 16,

"Availability of ICT materials should be ensured by the school authority." (T: 16)

b) Ensuring Organized Classroom

In order to ensure organized classroom, new sections should be introduced to balance the number of students with adequate ICT use facilities in classroom settings. Majority of the teachers suggested increasing classroom sections for effective ICT use. T: 1 talked about classroom development to integrate ICT into classroom teaching and reported that:

"According to the government policy, a classroom will have 60 students maximum. But we cannot maintain it as we have 700 students total in our school. So, every class should increase new sections to balance the number and make it organized."(T: 1)

II. Suggestions for Curriculum Development

a) Improving Teachers' Teaching Methodology

Though the textbook is designed based on the CLT approach, it was found from the discussion with the teachers that they still follow traditional methods in their classroom teaching.

Therefore, it is also necessary to change teachers' teaching methodology and adopt the updated methodologies. T:1 encapsulated:

"Teaching methodology should be such that helps teachers engage the students and provide the students with the opportunity to speak." (T: 1)

T: 10 emphasizes on the curriculum oriented towards the important language skills like listening and speaking apart from reading and writing.

"Textbook favours practicing Listening and speaking whereas exam offers only reading and writing test." (T: 10)

b) Usability of ICT Devices for Listening and Speaking Practice

Besides adding more lessons on speaking and listening, teachers also suggested that the authority must ensure the availability of ICT devices for practicing speaking and listening. Teachers also reported that the government should provide CDs with textbooks for more practice. According to T4:

"If CD is added with the book for practicing listening and speaking skills by the government authority.'' (T: 4)

III. Suggestions for Increasing ICT Materials

a) Distribution of Laptops for the Teachers

The second interview question discusses the number of ICT materials in the classroom. Among the interviewees, strongly agreed that they do not have sufficient materials to use in the school. T1 reported the demand for laptop facilities for the teachers, and emphasized the accessibility of internet and computer lab classes for the students.

> *"As I told you, we have ICT tools for language teaching, but the classrooms are not organized enough to use computers and laptops, and sometimes teachers carry their laptops personally, but it's not enough because all the teachers are incapable of buying a laptop. So the gov't should provide laptop facilities for the teachers."(T: 1)*

b) Ensuring Computer lab Facilities and Access to the Internet

Most teachers feel interested in integrating ICT materials into their teaching, but some reported that they could not get computer lab facilities. Therefore, providing computer lab facilities by the authority should be ensured to increase ICT materials usability by the teachers. T1 recommended:

> *"Access to the internet and computer lab class functioning should be ensured." (T: 1)*

c) Training on Material maintenance and Budget facilities

T: 4's excerpt also encapsulated the demand of allocating the budget facilities and ensuring accountability in maintaining the ICT equipment by the school authority. According to T4:

> *"Accountability should be ensured in maintaining equipment. Enough budget should be allocated in this regard."(T: 4)*

IV. Suggestions to Make Language Teachers Skilled in Using ICT in CLT

a) Ensuring Proper Training Facilities

All the teachers highly reported ensuring free certification courses and training facilities, particularly for English language teachers to make them skilled in using ICT devices in classroom teaching. One of them (T1) reported:

> *"Teachers should be trained in language skills and ICT skills on both sides. Gov't should organize training facilities on using ICT devices in classroom teaching."(T: 1)*

b) Continuous Motivation for ICT use in Classroom Teaching

To answer the question of taking necessary steps to make language teachers skilled in integrating ICT in Classroom language teaching, three teachers in their interview sessions replied that they need motivation to use ICT devices in classroom settings. One of the teachers named T: 2 opined that a teacher should be positive and mentally motivated to integrate ICT into his subject. According to T: 2

> *"Teacher must be positive and mentally motivated to integrate ICT with his subject."(T: 2)*

5. DISCUSSION AND RECOMMENDATIONS

The findings of the study recommend various steps to employ to enhance ICT's effectiveness in communicative language teaching. The study reveals that language teachers are familiar with ICT devices and their usability for personal and professional purposes. Concerning teachers' awareness of ICT devices and its' applications, most teachers use MS word and PowerPoint presentations. Results are consistent with the findings of (Lau & Sim 2008; Slaouti & Baton 2007). Findings also reveal that using different applications for teaching purposes, PowerPoint Presentation is significantly more popular than MS Word. However, the percentages of PowerPoint and MS word users for teaching are not significant enough as most of the teachers use computers for different purposes. The Finding of the research is also true in many other studies (Gibson & Oberg, 2004; Gulbahar, 2008; Liu, 2012; Scheeler, 2008; Vanezky, 2004; Prasojo, et al., 2018). The study also indicates that language teachers are familiar with using mobile phones as the common means of communication, where they use internet daily, and the usability of text messages and chat come, respectively, by the teacher participants. Findings indicate that a significant percentage of language teachers use Facebook and sometimes WhatsApp for personal and professional contact.

Similarly, Atanda found that teachers used various applications, such as WhatsApp, Facebook, Wikipedia, Google+, Tube, Ask.com, Instagram, Twitter, and others, to support learning and teaching effectively (Atanda 2018). Connolly makes a similar point that teachers can use different apps such as Facebook for learning through sharing and for continuous student-teacher involvement in and outside the classroom (Connolly et al. 2019). In the same vein, Seraj also reported how teachers used a variety of applications, including WhatsApp and voice recorders, to tackle the problems of large classroom size, providing authentic materials, enhancing learners' activeness, as well as oral practice environment inside and outside the classroom (Seraj, 2021). The results also suggested several applications can be used to improve the setting for oral practice in classroom teaching, effectively manage large classrooms, offer authentic materials, and engage learners in interactive verbal tasks.

The outcomes of qualitative analysis corroborate the quantitative analysis result and add more to the understanding of the awareness of using ICT devices in classroom teaching. The researcher interviewed language teachers to determine their awareness of using different applications and ICT devices for personal and professional needs. The interview indicates teachers' positive attitude toward projectors, digital whiteboards, and computers can make classroom language teaching successful.

Regarding having ICT and peripherals at institutions, the study indicates that almost all of the schools have a common computer lab but with limited accessibility, which is insufficient. For Multimedia -based teaching facilities, data shows that only a few schools have LCD projection facilities in every classroom. About 38% of participants consented to have a common digital camera, whereas the number of other devices is fewer. A broadly similar point was made by Afshari, Bakar, Luan, Samah, and Fooi, who also came up with the unavailability of ICT tools in educational institutions through the study (Afshari, et. al., 2009). However, to accommodate the limitations suggested by the language teachers, the government and stakeholders should ensure free laptop distributions, especially for teachers. They also emphasize classroom infrastructure development to use ICT materials in language teaching and ensure standard classrooms are equipped with ICT materials with the responsibility of maintaining the ICT devices by the school authority.

Regarding the availability of computers and peripherals at institutions for classroom teaching, the interview report reveals that existing ICT materials for classroom language teaching were insufficient. In the same line, Afshari, Bakar, Luan, Samah, and Fooi reported a similar result that language teachers could not integrate ICT devices into classroom teaching due to the shortage of technological tools such as computers, audio aids, multimedia, and language labs in their institutions (Afshari et al. 2009). Talukder came up with a similar result in their qualitative research study that inadequate ICT-based infrastructure, absence of resources, and insufficient funds are the barriers to implementing ICT in ELT classroom settings (Talukder & Saba, 2016).

5.1 Integrating ICT in CLT: Teachers' Views

Concerning the measures to enhance the effectiveness of ICT in CLT, the quantitative result indicates most teachers' agreement with the point that the authorities should promote a welcoming environment so that English teachers feel motivated to incorporate ICT into language instructions. The findings are consistent with the result of a study done by Prasojo, Mukminin, Habibi, Marzulina, Sirozi, and Harto, which also recommended ensuring all necessary facilities and positive culture in the school to enhance ICT's successful integration into education (Prasojo, L. D. et al., 2018).

In addition, the study also reveals English teachers' strong recommendation for ensuring frequent ICT- based training facilities and short-term certificate courses offered by school authorities. They also reported to be provided regular training programs, which will offer them a series of experiences to keep learners/teachers up to date with rapidly changing technology (Fullan, 1992).

They also advised that ICT training should be made a prerequisite for promotion. Government, many non-government, and international organizations have already introduced programs to train teachers using ICT tools (Khan & Kuddus, 2020). However, respondents with extensive experience in the programs are still strongly reported to organize ICT certificate courses and regular training program facilities with broad descriptions of digital pedagogy.

Berge and Warschauer & Healey, in their study, also advocated some points for teachers to implement digital pedagogy: using discursive resources to facilitate learning, encouraging human relations among members of the group, establishing general procedures for discussion and development of activities, and ensuring the transparency of technology for a good relationship between the system, the software, and the interface selected (Berge, 1996; and Warschauer & Healey 1998). In the same loop, Yuen, Law and Wong also stated some conditions for implementing pedagogical practice using ICT devices that largely depend on stakeholders' mission, vision, and objectives and the role of ICT in the curriculum (Yuen et al., 2003). The language teachers also appreciated the free short-term certificate courses on using ICT tools effectively in classroom settings. It also helps them maintain ongoing competency and acquire lifelong professional skills (Queiroz, 2003).

In addition, Bairagi, Rajon, and Roy addressed some recommendations for developing ICT-based academic education and digital, including establishing lab facilities and internet access, the inclusion of basic ICT courses, the appointment of candidates with basic ICT knowledge in all forms of educational institutions, developing institutional websites, and creating a digital society in Bangladesh (Bairagi et al., 2011).

Qualitative study emphasizes to provide sufficient ICT materials in the classroom by the authorities. A study by Khan and Kuddus recommended that stakeholders and school authorities must ensure adequate IT resources for successful ICT-based classroom teaching (Khan and Kuddus, 2012). Bairagi, Rajon, and Roy come with the similar result that teachers reported to provide them with a standard classroom with technological facilities (Bairagi et al., 2011).

Regarding curriculum development, most teachers reported incorporating more lessons on listening and speaking rather than writing and reading. In addition, for listening practising, the availability of ICT gadgets, CDs, or web links must be ensured by the government, as teachers reported. A similar interpretation also came from a previous study by Haider, stating that students must be provided with ICT materials such as CDs' for listening practice at home (Haider, 2012).

To enhance the usability of ICT materials, language teachers suggested to include them into free laptop distribution project as it motivates teachers ICT to make teaching and learning situations effective (Nasser, 2008). At the same time, accessibility to the computer lab and usability of ICT-based materials by the teachers must be ensured. Furthermore, teachers also reported paying close attention to budget facilities and accountability in maintaining the ICT equipment by the school authority.

Teachers' knowing digital literacy is crucial to successfully implementing ICT in language classroom teaching. But language teachers in our country have a lack of knowledge of using ICT applications in classroom settings (Khan & Kuddus, 2020; Pravin, 2013). Most of the teachers reported to be provided with free certification courses and training facilities, notably for English language teachers. They also suggested ensuring training assessment before designing such ICT's training program.

5.2 Mobile-Assisted Language Learning (MALL) as a Solution for CLT Barriers

Mobile Assisted Language Learning offers functional remedy to several challenges including, limited classroom interaction, lack of authentic materials, learners' low motivation in communicative language teaching (CLT) settings.

MALL provides authentic settings for communication, shared assignment, and access to authentic language resources, which are often lacking in traditional CLT classrooms (Alda, 2023; Chua & Soon, 2021; Walsh, 2022). Mobile applications, such as WeChat, WhatsApp provides, interactive, and context-rich language use both in-class and out-of-class communicative practice (Shi, Luo, & He, 2017). Digital flashcards offer substantially enhancement to learners' vocabulary acquisition, motivation, and overall interactive engagement in the learning process (Zakian et al., 2022; Benlaghrissi & Ouahidi, 2023; Hassan Taj et al., 2017). In addition to that Duolingo can be an effective platform for enhancing vocabulary, developing pronunciation skills, beyond the classroom settings, providing learners with the opportunity to become autonomous and motivated which are two essential keys of communicative learning (Fakhrurriana, Nisa, & Noni, 2024; Ajisoko, 2020; de Sousa, Cardoso, & Toassi, 2018).

5.3 Measure Undertaken by the Policy makers

Khan and Kuddus highlighted some other measures (see table-4) that have already been undertaken by the Bangladesh government and different national and international organizations for successful ICT implementation in classroom teaching (Khan & Kuddus, 2020).

Table 4. Initiatives undertaken by the Bangladesh government and NGOs to reduce the challenges in implementing ICT in education and CLT (Khan & Kuddus, 2020)

Initiatives adapted	Full Form	Funding Agencies	Functions
(a2i) programme	Access to Information	Funded by United Nation Development Programme (UNDP) and the United States Agencies for International Development (USAID) and run by the PMO	**Shikkhok Batayon:** It develops a huge amount of e-content with the help of teachers and teacher educators.
			Muktopaath: It provides several professional development courses for teachers.
			It has produced talking books for students with disabilities, specifically for visually impaired children.
			It also works on infrastructure and human resource development.
BANBEIS	Bangladesh Bureau of Educational Information and Statistics	Government of Bangladesh	It produces training modules and conducts training programs on ICT.
			At the same time, BANBEIS is also working on the development of the first e-library in Bangladesh.
BOU	Bangladesh Open University		It is a national resource which produces text, audio, and video programmers for learners.
EIA program	English in Action programme	UK government	It focuses on the professional development of teachers using authentic audio-visual resources.
ARBAN	Activity of Reformation of Basic Needs	It is a non government development organisation.	**ICT and English Club** (It is supported by the British Council and Dnet.) The members of the club learn English through ICT.

continued on following page

Table 4. Continued

Initiatives adapted	Full Form	Funding Agencies	Functions
BNNRC	Bangladesh NGOs Network for Radio and Communication	It is a national networking body.	It provides training in basic English using the community radio.
			It produces episodes for radio broadcasting to engage both teachers and students.
			It develops handbooks on English language learning for teachers.
			It also publishes English language learning materials through local newspapers so that the common people can access English learning resources.
Dnet: ICT-based Education Initiatives		Dnet is a non-profitable social enterprise.	It provides training in ICT by effectively deploying multimedia content.

6. LIMITATIONS AND SCOPE FOR FURTHER STUDY

The present study is limited only to secondary-level English language teachers. Results may be different for the other secondary level teachers teaching other subjects. The sample size consisted of 67male teachers and 33 female teachers, which may come with different results if the participants are equal with respect to gender. The researcher collected the data from government, semi-government and private schools in urban, semi-urban, and rural areas under three different boards- Dhaka, Jessore, and Comilla of Bangladesh. Further research could be done considering the Boards, locality, and different types of schools individually. The study mainly focused on teachers' awareness of using ICT devices for personal and professional purposes and the measure that needs to be taken to enhance the effectiveness of ICT in CLT based on teachers' opinions.

Further extensive study could be done on students' awareness of using Mobile technology devices for personal and academic purposes, and the steps could be taken to enhance the effectiveness of MALL in communicative language learning based on students' opinions. The researcher used 100 samples for the research, with limited findings. Therefore, investigation with a large sample size may yield different significant results.

7. CONCLUSION

The findings of the study confirm teacher participants' awareness of using ICT devices for their personal and professional purpose, but the challenges they experience prohibit them from integrating ICT materials in language classroom teaching successfully. Although they are enthusiastic about integrating ICT in their classroom teaching, the lack of training facilities hinders teacher participants' integration of ICT in classroom language teaching. However, teacher participants, through this study, addressed the necessary measures to be adopted to reduce the challenges and enhance the effectiveness of ICT in communicative language classroom settings. They suggested several initiatives such as; providing sufficient ICT materials by the school authority to the classroom, distributing free laptops, computer lab facilities, and access to internet facilities for the teachers, improving teachers' teaching methodology, ensuring the availability of ICT devices for listening and speaking practice, hands-on training on material maintenance and ensuring budget facilities, and rendering continuous motivation for ICT use in classroom teaching, to be taken to increase the usability of technology in classroom teaching, develop the existing curriculum, enhance ICT materials in the classroom, and make language teachers skilled. To accommodate such limitations, initiatives have already been undertaken by many other national and international organizations along with the government, but the constructive outcome is not visible yet. However, some more initiatives, such as policy making, planning executions, training organization, and others, should be implemented based on collaborative discussion among the teachers' stakeholders, school management, and policymakers, as teachers suggested. Moreover, feedback and suggestions coming from teacher stakeholders must be counted and ensured, and execution should be done accordingly with proper monitoring.

REFERENCES

A2i Project, Office of the Prime Minister, Government of the People's Republic of Bangladesh. (2009).

Afshari, M., Bakar, K. A., Luan, W. S., Samah, B. A., & Fooi, F. S. (2009). Factors affecting teachers' use of information and communication technology. *Online Submission*, *2*(1), 77–104.

Ajisoko, P. (2020). The use of Duolingo apps to improve English vocabulary learning. [iJET]. *International Journal of Emerging Technologies in Learning*, *15*(7), 149–155. DOI: 10.3991/ijet.v15i07.13229

Alda, R. C. (2023). Mobile-assisted language learning (MALL) in senior high school English classes. *World Journal of English Language*, *13*(6), 211. DOI: 10.5430/wjel.v13n6p211

Anderson, R. E., & Dexter, S. L. (2000). *School technology leadership: Incidence and impact*. http://www.crito.uci.edu/tlc/html/findings.html

Atanda, A. I. (2018). Teacher trainees' readiness to adopt social media in teaching-learning process: Case study of Faculty of Education undergraduates, University of Ibadan. *The International Journal of the Arts in Society*, *11*(1), 499–511. http://www.universitypublications.net/ijas/1101/pdf/U8K696.pdf

Bairagi, A. K., Rajon, S. A. A., & Roy, T. (2011). Status and role of ICT in educational institutions to build a digital society in Bangladesh: Perspective of a divisional city, Khulna. *International Journal of Advances in Engineering and Technology*, *1*(4), 374.

Benlaghrissi, H., & Ouahidi, L. M. (2023). The impact of mobile-assisted language learning on developing EFL learners' vocabulary knowledge. *International Journal of Interactive Mobile Technologies*, *17*(22), 38–51. Advance online publication. DOI: 10.3991/ijim.v17i22.41665

Berge, Z. L. (1996). *The role of the online instructor*. Facilitator.

Braun, V., & Clarke, V. (2006). Using thematic analysis in psychology. *Qualitative Research in Psychology*, *3*(2), 77–101. DOI: 10.1191/1478088706qp063oa

. Chatterjee, B., & Kuddus, K. (2014). Second language acquisition through technology: A need for underdeveloped regions like Jharkhand. *Research Scholar: An International Referred e-Journal of Literary Explorations*, *2*(2), 252–259.

Chua, N. A., & Soon, G. Y. (2021). Performing communicative language teaching in Mandarin mobile learning. *International Journal of Interactive Mobile Technologies, 15*(5), 87. Advance online publication. DOI: 10.3991/ijim.v15i05.20899

Connolly, T., Willis, J., & Lloyd, M. (2019). Evaluating teacher and learner readiness to use Facebook in an Australian vocational setting. *Studies in Continuing Education, 41*(1), 61–75. DOI: 10.1080/0158037X.2018.1461619

Creswell, J. W., & Creswell, J. D. (2017). *Research design: Qualitative, quantitative, and mixed methods approaches* (5th ed.). Sage.

Dash, A., & Kuddus, K. (2020). Leveraging the benefits of ICT usage in teaching of English language and literature. In *Smart intelligent computing and applications* (pp. 225–232). Springer. DOI: 10.1007/978-981-32-9690-9_22

de Sousa, G. M. B., Cardoso, L. A., & Toassi, P. F. P. (2018). Duolingo as a tool to improve vocabulary writing in English as a foreign language. *Letras em Revista, 9*(1), 119–134.

Dhaka Tribune. (2018, December 7). *Digital Bangladesh a reality now*. https://archive.dhakatribune.com/bangladesh/2018/07/11/digital-bangladesh-a-reality-now

Fakhrurriana, R., Nisa, A., & Noni, N. (2024). The perceptions of using Duolingo application in learning English for students' vocabulary mastery. *BATARA DIDI: English Language Journal, 3*(1), 45–54. DOI: 10.56209/badi.v3i1.82

Firdausia, W., & Hamdani, B. (2025). Integrating artificial intelligence in English learning: Implementation of Duolingo application to improve English skills. *Jurnal Kependidikan, 13*(1), 177–188. DOI: 10.24090/jk.v13i1.13339

Fullan, M. (1992). *Successful school improvement: The implementation perspective and beyond*. McGraw-Hill Education.

Gamage, S. H. P. W., Ayres, J. R., & Behrend, M. B. (2022). A systematic review on trends in using Moodle for teaching and learning. *International Journal of STEM Education, 9*(1), 9. DOI: 10.1186/s40594-021-00323-x PMID: 35096513

Gibson, S., & Oberg, D. (2004). Visions and realities of Internet use in schools: Canadian perspectives. *British Journal of Educational Technology, 35*(5), 569–585. DOI: 10.1111/j.0007-1013.2004.00414.x

Haider, M. (2012). Promoting CLT within a computer assisted learning environment. *English Language Teaching, 5*(8), 91–102. DOI: 10.5539/elt.v5n8p91

Hasan, M. K., Fakih, A. H., Ibna Seraj, P. M., & Hasmirati, . (2022). The effect of technology-assisted language programme on vocabulary learning among EFL students at the tertiary level. *Heliyon*, *8*(8), e10313. DOI: 10.1016/j.heliyon.2022.e10313 PMID: 36051275

Hasan, M. S. H. K., Mahbub, & Khan, M. S. H. (2013). Introducing ICT into teacher-training programs: Problems in Bangladesh. *Journal of Education and Practice*, *4*(14), 79–85.

Hassan Taj, I., Ali, F., Aslam Sipra, M., & Ahmad, W. (2017). Effect of technology-enhanced language learning on vocabulary acquisition of EFL learners. *International Journal of Applied Linguistics and English Literature*, *6*(3), 262. DOI: 10.7575/aiac.ijalel.v.6n.3p.262

. Ibna Seraj, P. M., Mahbub, P., & Klimova, B. (2025). Smartphone apps-based teaching method to develop oral English communication skills at the tertiary level in an EFL context.

Ihmeideh, F. M. (2009). Barriers to the use of technology in Jordanian pre-school settings. *Technology, Pedagogy and Education*, *18*(3), 325–341. DOI: 10.1080/14759390903255619

Irzawati, I. (2023). The integration of Duolingo into EFL learning. *Esteem Journal of English Education Study Programme*, *6*(2), 328–337. DOI: 10.31851/esteem.v6i2.12317

Jiang, X., & Pajak, B. (2022). *Reading and listening outcomes of learners in the Duolingo English course for Spanish speakers* (Duolingo Research Report DRR-22-08).

Jiang, X., Peters, R., Plonsky, L., & Pajak, B. (2024). The effectiveness of Duolingo English courses in developing reading and listening proficiency. *CALICO Journal*, *41*(3), 249–272. DOI: 10.1558/cj.26704

Khan, M. S. H., Hasan, M., & Clement, C. K. (2012). Barriers to the introduction of ICT into education in developing countries: The example of Bangladesh. *International Journal of Instruction*, *5*(2), 61–80.

Khan, N. M., & Kuddus, K. (2020). Integrating ICT in English language teaching in Bangladesh. *Rupkatha Journal on Interdisciplinary Studies in Humanities*, *12*(5), 1–10. DOI: 10.21659/rupkatha.v12n5.rioc1s23n1

Kuddus, K. (2018). Emerging technologies and the evolving roles of language teachers. *Language in India*, *18*(6), 81–86.

Lau, B. T., & Sim, C. H. (2008). Exploring the extent of ICT adoption among secondary school teachers in Malaysia. *International Journal of Computing and ICT Research*, *2*(2), 19–36.

Liu, S.-H. (2012). A multivariate model of factors influencing technology use by preservice teachers. *Journal of Educational Technology & Society*, *15*(4), 137–149.

Lo, Y. Y., Shah, S. S. A., Hasim, Z., Senom, F., Saad, M. R. M., & Thienpermpool, P. (2025). Thai EFL university students' views on the effectiveness of Duolingo. *Studies in English Language and Education*, *12*(1), 154–171. DOI: 10.24815/siele.v12i1.40517

Maisha, K., & Shetu, S. N. (2023). Influencing factors of e-learning adoption. *Future Business Journal*, *9*(1), 37. DOI: 10.1186/s43093-023-00214-3

Mudra, H. (2020). Digital literacy among young learners. *Teaching English with Technology*, *20*(3), 3–24.

Nasser, R. (2008). A formative assessment of ICT in Lebanese schools. *International Journal of Education and Development Using ICT*, *4*(3), 63–77.

Oh, C. M., Krish, P., & Hamat, A. (2022). Reading on smartphones. *CALL-EJ*, *23*(1), 259–277.

Parvin, S. (2013). Integration of ICT in education sector. *International Journal of Computer Science and Information Technologies*, *5*(4), 81–92. DOI: 10.5121/ijcsit.2013.5406

Prasojo, L. D.. (2018). Learning to teach in a digital age. *Teaching English with Technology*, *18*(3), 18–32.

. Queiroz, V. (2003). Roles and competencies of online teachers. *The Internet TESL Journal, 9*(7).

Sadaf, M. (2023). Influences of Duolingo English test. *Open Journal of Modern Linguistics*, *13*(3), 392–405. DOI: 10.4236/ojml.2023.133024

Salehi, H., & Salehi, Z. (2012). Integration of ICT in language teaching. In *Proceedings of the 3rd International Conference on e-Education, e-Business, e-Management and e-Learning* (pp. 215–219).

Scheeler, M. C. (2008). Generalising effective teaching skills. *Journal of Behavioral Education*, *17*(2), 145–159. DOI: 10.1007/s10864-007-9051-0

. Seraj, P. M. I., et al. (2021). Exploring EFL teachers' perception on readiness to use smartphones. *MEXTESOL Journal, 45*(4).

Shi, Z., Luo, G., & He, L. (2017). Mobile-assisted language learning using WeChat. *International Journal of Emerging Technologies in Learning, 12*(2), 16. DOI: 10.3991/ijet.v12i02.6681

Shrestha, S.. (2022). Preparations for online education during Covid-19. *Education and Information Technologies, 27*(1), 243–265. DOI: 10.1007/s10639-021-10659-0 PMID: 34341654

Slaouti, D., & Barton, A. (2007). Opportunities for practice and development. *Journal of In-service Education, 33*(4), 405–424. DOI: 10.1080/13674580701687807

Taber, K. S. (2018). The use of Cronbach's alpha. *Research in Science Education, 48*(6), 1273–1296. DOI: 10.1007/s11165-016-9602-2

. Talukder, M. D., & Saba. (2016). ICT-based ELT at secondary level education in Bangladesh. *The NEHU Journal, 15*(1), 117–129.

Walsh, R. (2022). Teaching communicatively with mobile-sourced materials. *Advances in Language and Literary Studies, 12*(5), 23–29. DOI: 10.7575/aiac.alls.v.12n.5.p.23

Wardoyo, J., Iyati, W., & Citraningrum, A. (2025). *UAV-Based Thermal Mapping for Urban Heat Island Intensity Assessment in Malang.*

Warschauer, M. (2004). Technological change and the future of CALL. In *New perspectives on CALL for second language classrooms* (pp. 15–26).

Warschauer, M., & Healey, D. (1998). Computers and language learning. *Language Teaching, 31*(2), 57–71. DOI: 10.1017/S0261444800012970

Yuen, A. H. K., Law, N., & Wong, K. C. (2003). ICT implementation and school leadership. *Journal of Educational Administration, 41*(2), 158–170. DOI: 10.1108/09578230310464666

Zakian, M., Xodabande, I., Valizadeh, M., & Yousefvand, M. (2022). Out-of-the-classroom learning of English vocabulary. *Asian-Pacific Journal of Second and Foreign Language Education, 7*(1), 16. DOI: 10.1186/s40862-022-00143-8

Chapter 13
The Potential of AI, Blockchain, and IoT in Language Education

S. Ida Evangeline
https://orcid.org/0000-0003-2997-7897
Government College of Engineering, Tirunelveli, India

ABSTRACT

This chapter explores the transformative potential of artificial intelligence (AI), blockchain, and the Internet of Things (IoT) in reshaping language education through mobile technology integration. As digital learning environments continue to evolve, these emerging technologies offer new pathways for personalized instruction, secure credentialing, real-time assessment, and immersive learning experiences. The chapter begins by examining how AI enhances language learning through natural language processing, adaptive learning platforms, and speech recognition tools that personalize content and provide real-time feedback. It then investigates the role of blockchain in ensuring academic integrity, supporting decentralized learning networks, and enabling secure, tamper-proof credentialing for language proficiency. Finally, the discussion is extended to IoT, providing insight into how it is used in smart classrooms, wearable devices and context-aware environment for improving interactive and situated learning.

1. INTRODUCTION

Language education is undergoing a paradigm shift, driven by the rapid advancement and integration of emerging technologies into mobile learning environments. Among the most transformative of these technologies are Artificial Intelligence,

DOI: 10.4018/979-8-3373-3366-3.ch013

Blockchain, and the Internet of Things. They each have their own unique set of capabilities which, when synergized with mobile platforms, have potential to revolutionize how languages are taught, learned, assessed and certified (Zheng & Yang, 2024).

This chapter explores the convergence of these technologies and their transformative impact on language education. It delves into how AI can personalize and automate language learning through natural language processing, real-time feedback, and adaptive content delivery. It explores how blockchain can be used to establish secure, transparent, and distributed systems for credentialing and academic record management. It also explores how IoT technologies including wearable devices, smart classrooms can facilitate immersive and context aware learning experiences..

The urgency of this exploration lies in the increasing demand for scalable, inclusive, and reliable mobile learning solutions. With language skills increasingly important in global education, business and migration, traditional educational models are being forced to adapt to the needs of diverse learners in an efficient way. AI, blockchain and IoT solve, in a revolutionary manner, these gaps by making learning more interactive, trustworthy and accessible.

But there are challenges to the integration of these technologies. Issues of data privacy, digital ethics, infrastructure inequality, and algorithmic transparency need to be tackled in advance. This chapter therefore also critically explores the ethical, regulatory and logistical considerations that should be associated with technological adoption.

This chapter is intended to serve as a roadmap for those educators, technologists, policymakers, and researchers in focusing on case studies, practical implementations, and visionary trends. This will assist them to steer the promising but complicated terrain of tech empowered language learning and add to the advancement of innovative, sustainable and learner focused mobile education ecosystems.

2. AI IN LANGUAGE EDUCATION

The integration of Artificial Intelligence into language education is reshaping the way learners interact with new languages, instructors deliver content, and institutions assess proficiency. The combination of mobile technology and AI brings about personalized, adaptive, and scalable solutions to language learning in various contexts (Law, 2024).

2.1 Natural Language Processing (NLP) and Chatbots

Natural Language Processing enables machines to understand, interpret, and generate human language. NLP-powered chatbots in language education simulate realistic conversational practice, allowing learners to engage in dialogue, receive instant feedback, and build confidence in using the target language. These chatbots can fix grammar, propose vocabulary improvements, and adjust conversation difficulty to the level of the learner's proficiency. Chatbots provide 24/7 access, are nonjudgmental, and provide continuous practice to reduce the anxiety that comes with speaking a new language, as well as support continued skill development (Kökver et al. 2025).

2.2 Speech Recognition and Automated Assessment

AI-driven speech recognition systems analyze learners' pronunciation, intonation, and fluency in real time. Mobile language learning apps utilize this technology to offer immediate feedback on spoken responses, helping learners refine their accents and speaking abilities. AI algorithms used in automated assessment tools objectively assess oral and written performance and scale up, and reduce teacher workload. Additionally, these systems can find patterns in errors and customize further exercises in addressing the students' weaknesses (Kurbanazarova et al. 2024).

2.3 Personalized and Adaptive Learning Platforms

AI enables personalized learning by dynamically adapting content to suit each learner's unique needs and pace. AI algorithms can pick up learning gaps and suggest customized drills using the collection and analysis of behavioral data like time spent on activities, quiz results, and engagement metrics. Material is presented on adaptive platforms at the right level of difficulty to keep learners challenged but motivated, while the platforms adjust in real time according to user performance. By this approach, learning is more effective and self directed than traditional one size fits all methods (Sajja et al. 2024).

2.4 AI Tutors and Translation Tools

AI-powered virtual tutors provide on-demand guidance, explanations, and motivation. These tutors can answer questions, explain grammar rules, and facilitate practice exercises, simulating elements of human instruction. Moreover, advanced AI translation tools, like neural machine translation, provide almost instant translation among several languages. These help learners to access resources and overcome language barriers in communicating. Moreover, some platforms utilize AI for sentiment analysis to measure the user's emotion and adjust the content delivery to increase the learner's engagement and help (Yefymenko et al. 2024).

3. BLOCKCHAIN IN LANGUAGE EDUCATION

Blockchain technology is increasingly recognized as a transformative force in education due to its ability to ensure data integrity, transparency, and decentralization. For language education, blockchain provides new ways to handle credentials, assess their validity, and create decentralized learning ecosystems. It enables secure, peer driven learning experiences and credential validation in global contexts through integration with mobile technologies (Peng & Yang, 2024).

3.1 Secure Credentialing and Academic Records

One of the most impactful uses of blockchain in language education is secure credentialing. Blockchain allows for the issuance of tamper-proof digital certificates and academic records, eliminating the risks of forgery, misrepresentation, or loss of credentials. Once the language test score, course completion or proficiency certification is recorded on the blockchain, it becomes immutable and verifiable, and can be easily shared with an employer or educational institution via a mobile wallet or QR code. In global mobility, for instance, language skills need to be verified (Rustemi et al. 2024), which makes this particularly relevant.

Learners are beginning to pilot blockchain based diplomas on educational platforms and language testing services that allow them to control and share their academic achievements from their devices directly without the need of intermediaries (Nazari et al. 2024).

3.2 Decentralized Learning Ecosystems

Blockchain fosters the creation of decentralized learning platforms where learners and instructors interact without a central authority (Any et al. 2024). Language learners in such ecosystems can i) participate in peer to peer language exchanges, ii) access decentralized course libraries, iii) govern using blockchain tokens.

Automate learning incentives using smart contracts, self executing contracts on the blockchain, that can reward users with tokens for completing lessons, contributing content, and mentoring others. The consistent participation and the self sustaining learning community (Duy et al. 2025) are motivated by this tokenization model.

In underserved regions, these decentralized models provide mobile accessible learning outside of expensive infrastructure and the need for bureaucratic oversight.

3.3 Micro-Credentials and Lifelong Learning

In a world where language skills are developed continuously across careers and geographies, micro-credentialing is becoming essential. Blockchain facilitates the issue of verifiable skill badges for smaller, modular learning outcomes, such as mastery of a new grammar module, a language immersion session, or completion of a translation task (Shanmughan et al. 2024).

Stacked, these micro credentials represent broader competencies and can be integrated into lifelong mobile learning profiles that learners create and update throughout their lives. These credentials can be trusted by employers and institutions, making the recruitment and admissions processes streamlined.

4. IoT AND SMART LANGUAGE LEARNING ENVIRONMENTS

The Internet of Things represents a significant shift in how technology interacts with the physical world, and its role in language education is becoming increasingly profound. IoT enables mobile, context-aware and immersive learning experiences that go beyond the walls of the traditional classroom by connecting devices and environments in real time. These connected systems are evolving and they provide new opportunities for learners to interact with languages in authentic, meaningful, and personalized ways (Dong et al. 2024).

One of the most prominent applications of IoT in language education is the development of smart classrooms. In these environments, devices such as interactive whiteboards, tablets, microphones, and cameras work together to support collaborative and multimodal instruction. For instance, microphones can be embedded in learning stations that analyze pronunciation in real time, and provide immediate, individualized feedback. Interactive displays and classroom analytics platforms also track participation and engagement, and allow teachers to respond to students' progress and interaction patterns simultaneously. In terms of providing an active and responsive learning environment for mobile integration and synchronous learning, the smart classroom model is the one that is promoted.

Wearable devices also offer promising contributions to mobile language learning. Smartwatches and augmented reality (AR) glasses can deliver real-time prompts, vocabulary tips, or contextual instructions during physical activities such as guided tours or cultural excursions. These wearables can be synced with mobile applications in order to give learners microlearning, providing language input in short, focused bursts throughout the day. For instance, a learner could be presented with vocabulary questions as he or she walks through a city or responds to voice commands while walking, enabling learning to happen in the course of everyday activities (Tabuenca et al. 2024).

Beyond the classroom, IoT enables powerful context-aware learning through the use of sensors and environmental triggers. Sensors embedded in learning spaces can initiate specific language tasks based on user movement or location. For example, in a language lab, an approaching student would automatically activate audio exercises that are related to what they are studying in the current unit. Gesture recognition systems based on cameras and sensors can also be used in sign language instruction such that the hand movements can be analyzed and corrected in real time to facilitate proper learning. These systems foster kinesthetic engagement and help bridge the gap between knowing abstract knowledge and executing physical action to reinforce learning through embodiment (Rashid et al. 2025).

Voice-enabled learning assistants further illustrate the potential of IoT in language education. Devices such as Amazon Alexa and Google Assistant can conduct conversational drills, administer vocabulary quizzes, and deliver pronunciation feedback through natural voice interaction. More and more, these assistants are embedded in mobile ecosystems and users are able to interact with language exercises through their mobile devices and smart speakers. Its voice commands can be used to schedule regular practice sessions, receive progress summaries, or even participate in interactive storytelling. Its seamless, multisensory experience helps reinforce habit formation and language acquisition to be more organic part of the daily life (Patty, 2024).

In addition, AI powered avatars and virtual conversation partners on smart speakers or wearables enable learners to practice speaking on demand in a spontaneous manner. They provide learners with the ability to learn in a safe environment, simulating real world dialogues. They help sustain engagement by responding contextually to learner input and help deepen conversational competence.

The practice of applying IoT in language education is already bearing fruit. Smart classrooms, which dynamically adapt to student behaviour and performance, are being experimented in educational institutions in technologically advanced regions, such as South Korea and Finland. Such pilot programs demonstrate the possibility to support language pedagogy through data driven, learner centered integration of IoT (Kot & Nykyporets, 2024).

Although it has benefits, IoT integration in education also has major challenges. Device accessibility, infrastructure costs, interoperability between platforms, and the protection of learner data are all important issues that must be dealt with. Given that data privacy regulations and ethical guidelines must be observed in smart environments, where sensors and voice data are constantly collected, it is crucial for educators and developers to ensure compliance (Zhang et al. 2024).

In general, the integration of IoT in language education makes mobile learning from a passive, screen–bound activity into an active, physical, and deeply immersive experience. With the line between the digital and physical blurring, and with the increasing role of IoT, this technology will continue to be at the center of designing smart learning environments that are responsive to learners' needs, are adaptable to their contexts, and are optimized for language acquisition.

5. INTEGRATION SYNERGIES: AI + BLOCKCHAIN + IoT

While artificial intelligence, blockchain, and the Internet of Things each offer distinct advantages in language education, their true potential emerges when these technologies are integrated into a unified ecosystem. AI, blockchain and IoT are converging to create a technological triad that makes language learning better in every dimension—from personalizing learning content to authenticating credentials, to providing contextual immersions and real time responsiveness. The synergy of these technologies not only enhances each other's strengths but also forms a dynamic and interoperable framework for mobile, intelligent and secure language learning experiences (Zuo et al. 2024).

The fusion of AI and IoT, for instance, allows language learning platforms to leverage real-time data collected from IoT devices—such as microphones, cameras, and environmental sensors—and analyze it using AI algorithms to adapt instruction instantaneously. For instance, imagine a learner who puts on a smart headset that records how he pronounces and speaks. This input can be processed by AI in real time, errors can be detected and feedback can be given directly to the user through the same device. In a smart classroom or home environment, this system can be embedded to adjust lighting, sound or visual aids according to detected engagement levels, emotional cues or activity patterns for better learning conditions (Tyagi et al. 2024).

Meanwhile, blockchain adds a crucial layer of security, traceability, and trust to the AI-IoT system. AI-generated assessments—such as automated writing evaluations or speaking scores—can be recorded and verified through blockchain to ensure that the results are authentic, unaltered, and transparently linked to learner performance data. Such a digital record of language proficiency reduces the risks of score manipulation or bias, and gives educators, institutions, and employers a trusted digital record of their learners' proficiency. In addition, learners are in control of their credentials, which are stored in a decentralized and portable digital ledger, that they can access, or share from mobile devices, at any time (Alrashdi & Alqazzaz, 2024).

A compelling illustration of this synergy can be found in the concept of a "smart certification" system. Here, AI evaluates a learner's written and spoken language tasks in real time using NLP and speech recognition tools. Then the results are then certified by a blockchain ledger and made available in a mobile accessible profile. Contextual learning is tracked by IoT devices, and the behavioral data is fed back into the AI system for improved personalization, such as the amount of time spent practicing in multiple locations or the voice-based tasks completed in natural environments. This forms a feedback loop that makes learning more efficient and increases the reliability of certification and the learner's confidence in the process (Aloun, 2024).

In decentralized language learning ecosystems, smart contracts on the blockchain can facilitate automatic credential issuance or trigger micro-rewards when learners complete specific AI-driven tasks or reach milestones recorded by IoT sensors. For instance, a learner can earn verified badges for meeting a pronunciation benchmark tracked on a smart device, or for taking part in peer-to peer practice sessions tracked on mobile apps. These badges recorded on the blockchain are part of a lifelong learning record that can evolve with the individual and is available globally (Bhumichai et al. 2024).

Additionally, these technologies can be seamlessly integrated into mobile learning platforms to create holistic environments that are able to provide multiple modes of instruction, assessment and certification. For instance, an AI based vocabulary exercise could be done by a language learner using a mobile app in commuting hours, IoT enabled pronunciation tools for real time feedback at home and finally a blockchain authenticated certificate of fluency presented during a job interview, all within a single ecosystem (Ifedayo et al. 2025).

With this triadic model growing, it shifts the paradigm in education from institution driven, centralized models to learner centric, distributed systems. The days of passive recipients of content are over and learners are becoming active co-creators of their educational pathway. The mobile technology is the gateway to a global, connected learning network of experiences that are tailored, credentials trusted and progress continuous (Zainuddin et al. 2024).

It also provides new opportunities for research, data analytics, and policy development. Anonymized, aggregated data collected across AI, blockchain, and IoT systems can be used by educators and developers to study learner behaviors, test new methodologies, and design interventions for promoting equity, motivation, and effectiveness in language acquisition (Gouiza et al. 2024).

However, the success of this integration relies on strong infrastructure, data governance frameworks, and the cooperation of technologists, educators, ethicists, and policy makers. However, the complexity of integration without thoughtful implementation could make accessibility more difficult or worsen the digital divide (Omidian, 2024).

Therefore, the synergy between AI, blockchain and IoT redefines what mobile language learning can be. It provides a path to more personalized, transparent and immersive education built on technology that grows with the learner. This integration also leads to critical ethical and infrastructural challenges that must be addressed for the full potential of this integration to be realized in an inclusive and sustainable manner, as the next sections will show.

6. CHALLENGES AND ETHICAL CONSIDERATIONS

The integration of AI, blockchain, and IoT in language education, while promising in its potential to revolutionize mobile learning, also raises a host of challenges that must be critically examined. This includes data privacy, algorithmic ethics, infrastructural limitations, and global equity, to name a few. Because these technologies will be more and more embedded in educational ecosystems, their impact must be evaluated not only in terms of innovation but also in terms of fairness, accessibility, and sustainability (Mohamed, 2024).

One of the most immediate concerns is data privacy and security. All three technologies—particularly when used in conjunction—collect, process, and store vast amounts of learner data. AI systems rely on detailed user profiles, interaction patterns, voice recordings, and textual input to adapt and personalize instruction. In real time, IoT devices track movements, gestures and environmental context. Although tamper proof data storage is offered by blockchain systems, these still expose personal identifiers if not encrypted properly. Therefore, such aggregation of sensitive data, especially when transmitted across platforms and geographies, is vulnerable to breaches, surveillance, and unauthorized use. The issue of collecting data without consent and the loss of digital agency is of ethical concern, especially since learners, especially minors, may not be aware of the consequences of constant data collection (Kovalenko & Baranivska, 2024).

Legislation such as the General Data Protection Regulation (GDPR) in the European Union and similar data protection laws globally are beginning to address these issues by requiring explicit consent, transparency in data use, and the right to be forgotten. However, compliance is a difficult and ever evolving problem, especially for mobile applications that span many jurisdictions. To protect user data, educational institutions and edtech companies must put stringent safeguards in place and transparent policies to ensure that user data is anonymized where possible and never used beyond its educational use (Rane, 2024).

Another critical challenge lies in the ethical use of AI in assessment and decision-making. While AI offers valuable automation in evaluating writing, speaking, and comprehension, there is increasing concern over algorithmic bias and lack of transparency. In the case of biased or incomplete datasets, AI systems can reproduce stereotypes, fail to recognize non-native speech patterns, and even penalize learners for speaking with a regional accent or using cultural nuances. This can result in a bias towards unfair outcomes and demoralization of users who feel they have been judged unfairly by automated systems. Additionally, AI driven feedback lacks pedagogical nuance of human instructors and can lead to oversimplified or incorrect learning path (Mahmudulhassan et al. 2024).

To tackle this issue, organizations like UNESCO and the OECD have proposed ethical guidelines for AI in education and call for explainable AI, fairness audits, and inclusive training data. More and more, developers are being pushed to integrate educators, linguists, and cultural experts into the AI design process so that assessment systems are accurate and culturally responsive (Pack & Maloney, 2024).

Infrastructure disparity also poses a significant barrier to equitable implementation. Regions with reliable internet connectivity, access to smart devices and well resourced educational systems are the most accessible to the benefits of AI, blockchain and IoT in language education. However, learners in low income or rural areas may be disadvantaged by the lack of digital infrastructure, high cost of data, or no access to smartphones or wearables. As this digital divide persists, it is likely to reinforce existing educational inequalities in which technology becomes a privilege and not a right (Novawan et al. 2024).

Bridging this gap requires coordinated policy interventions, public private partnerships and open access models that emphasize inclusivity. International development programs that support the deployment of necessary infrastructure in underserved communities (Yan et al. 2024) can be supplemented with mobile first design strategies and lightweight, offline capable applications that can help expand reach.

Finally, interoperability and standardization are still technical and strategic challenges. The data formats, open protocols, and collaborative governance required to seamlessly integrate AI, blockchain, and IoT systems are shared. If language learning platforms are not interoperable, they will remain as silos and will not be able to scale. Additionally, cross border data sharing, certification recognition and mobile application deployment can be complicated by conflicting regulatory requirements in different countries (Al-khresheh, 2024).

In order to deal with these issues, the education sector requires flexible, modulable architectures and the development of international technology standards. Such technologies may also require ethical frameworks and oversight bodies to monitor their evolution and how they are used in combination to impact learners' rights, autonomy, and educational trajectories (Tiwari, 2024).

To conclude, the integration of AI, blockchain, and IoT in language education has the potential to transform language education, but it must be done with a critical eye and a responsible approach to innovation. To ensure that these tools serve all learners equitably and respectfully, we need ethical design, inclusive infrastructure, transparent data policies, and global collaboration. If we are to have a future where technology makes language education easier, not harder, then we need to address these challenges head on.

7. FUTURE TRENDS AND INNOVATIONS

As language education continues to evolve in response to technological advancements, the horizon reveals a landscape shaped by immersive experiences, hyper-personalization, and deeper human-machine interaction. The convergence of AI, blockchain and IoT is not just improving current practices, it is paving the way for entirely new paradigms. Environments in which mobile language learning is likely to be intelligent, adaptive, emotionally attuned, immersive and globally interconnected will define the future of mobile language learning (Ulutaş et al. 2025).

One of the most promising frontiers in this evolution is the metaverse-based virtual classroom. In this envisioned space, learners engage with language instruction through fully immersive 3D environments, often accessed via mobile-compatible virtual reality (VR) or augmented reality headsets. Learners practice language in simulated real world contexts (markets, cafes, airports, job interviews) where learners practice language in dynamic, interactive scenarios. The metaverse enables language learners to go beyond static lessons to experience language as a lived social phenomenon. Learners can engage in real-time dialogues with AI driven avatars that adjust complexity and tone according to the learner's fluency and objectives (Yadav et al. 2025).

The integration of AR/VR with AI is also enhancing mobile language education by enabling context-aware immersion. AR overlays can be used to project translated labels on real world objects in order to help learners build vocabulary in situ, and VR simulations can provide controlled yet realistic speaking practice in multilingual environments. When used together with speech recognition and NLP algorithms, these tools give learners feedback on pronunciation, grammar and intonation, helping them improve their skills in lifelike environments (Ramya & Ramamoorthy, 2025).

Another emerging trend is the incorporation of emotion-aware AI into language education. Future language learning platforms will be powered by affective computing and behavioral analytics that will be able to recognize users' emotional states through facial expressions, voice tone, or biometric signals gathered by wearable devices. The system might simplify content, encourage, or suggest a break if a learner seems frustrated or fatigued. On the contrary, if high engagement is detected, the platform could present more difficult material or continue a learning session. Thus, emotional adaptability makes learning more empathetic and human centered, which is crucial for keeping motivation in mobile, self-paced environments (Salloum et al. 2025).

Blockchain will continue to evolve as a foundation for decentralized and learner-owned educational ecosystems. We will see increased usage of portable learning wallets — secure digital profiles where learners keep their credentials, badges, and learning histories across different platforms and providers. These wallets will allow for more fluid and flexible lifelong language learning, where achievements will be recognized across institutions, employers, and borders without bureaucratic friction. Furthermore, the emergence of Decentralized Autonomous Organizations (DAOs) in education may bring community governed language learning platforms where learners vote on what course content to learn, who should be instructors, or what criteria for assessment is to be used using token based governance models (Javed & e Zehra, 2025).

Environment responsive learning will become a phenomenon at the intersection of IoT and language pedagogy. Wearable devices and smart homes will start to be language learning environments themselves. A smart kitchen might, for instance, read instructions in a second language as the user cooks, or a wearable headset might offer vocabulary suggestions depending on where a learner is, what they are doing. The embedding of language into daily routines at this deep level supports continuous, holistic, and deeply personalized learning that is both practical and learning (Ferguson-Sams et al. 2025).

Furthermore, they will aid in the development of new multilingual interaction forms using cross-linguistic AI models. Instead of studying languages by themselves, learners may interact with systems that allow for code switching, contrastive grammar instruction, and comparative linguistics. The learners of these models would be invaluable in multicultural societies, global workplaces, or multilingual curricula (Mizumoto et al. 2025).

This is all dependent on continued user experience design innovation, equitable access to new technologies, and commitment to ethical AI development. Looking ahead, educators, developers, and policymakers will need to work together to ensure that the next generation of language learning technologies are inclusive, culturally responsive and pedagogically grounded (Shirinova, 2025).

In short, the future of language education is toward an AI, blockchain, and IoT coexisting ecosystem that is intelligent, secure, immersive, emotional, and adaptive. These advances will change the way we acquire, teach and validate languages: the learning will be not only mobile but also meaningfully integrated into every dimension of everyday living.

8. CASE STUDIES AND REAL-WORLD APPLICATIONS

To move from conceptual vision to practical understanding, it is essential to examine how AI, blockchain, and IoT are currently being applied in real-world language education contexts. The pilot programs and case studies presented in this thesis provide a glimpse of the tangible impact of emerging technologies and provide important lessons learned on successes, challenges and future scalability of these innovations. They also show the increasing involvement of educational institutions, technology companies and international organizations in developing mobile and integrated technologies that can transform language learning.

One prominent example of AI-driven language education is Duolingo, a mobile application used by over 500 million learners worldwide. Duolingo uses advanced natural language processing (NLP), speech recognition, and adaptive learning algorithms to provide a highly individualized experience. In real time, the app analyzes user performance and adapts difficulty of tasks, timing of repetition, and feedback to match the users' strengths and weaknesses. Furthermore, the platform utilizes machine learning to sequence the lessons and retain the vocabulary, illustrating that AI can enable scalable and engaging mobile language learning for both casual learners and academic learners (Handini et al. 2022).

Another noteworthy example is ELSA Speak (English Language Speech Assistant), which focuses on spoken English proficiency through real-time pronunciation feedback. ELSAs are powered by deep learning models trained on thousands of speech samples to provide personalized guidance on stress patterns, intonation, and articulation. Because it is particularly good for those learning English as a second language and preparing for standardized language exams or international work environments. The success of ELSA shows how AI can help bridge speaking confidence gaps and provide speech coaching via mobile platforms (Karim et al. 2023).

In the realm of blockchain, MIT Media Lab's Digital Credentials Initiative provides a significant precedent. While not exclusive to language education, the initiative demonstrates how blockchain can be used to issue secure, tamper-proof academic certificates. These digital credentials are stored on the blockchain and can be accessed through a mobile app, which is well suited for students or global learners who need to provide proof of their language proficiency across institutions or employers. Following such efforts, smaller edtech startups like BitDegree and ODEM are creating blockchain based ecosystems for language learning. The micro-credentials earned on these platforms are shareable verified achievements, and token rewarded peer tutoring can happen on these platforms, demonstrating the possibilities of decentralized learning communities in language education (Jirgensons & Kapenieks, 2018).

On the IoT front, South Korea's Ministry of Education has piloted smart classrooms in selected public schools. These learning spaces feature interconnected devices including AI assistants, interactive displays, and real-time analytics systems. In language classrooms, students practice pronunciation with AI powered microphones, get personalized help through voice assistants and utilize gesture controlled boards to reinforce vocabulary. Engagement can be monitored and lesson flow can be adjusted dynamically by the teachers. Yet, as such implementations are still in the experimental phase, they hint at scalable models of IoT enhanced, data responsive language education (Lee & Jeong, 2023).

Another exciting innovation involves the use of AR/VR in immersive language learning. For example, the virtual reality based language simulations on the Mondly VR platform enable the learners to practice speaking with virtual characters in real world situations, such as ordering food in a restaurant or checking into a hotel. These scenarios promote understanding of context and reduce anxiety. Conversational fluency development experiences are powerful tools to develop an environment that is immersive, but not too stressful (Sánchez et al. 2023), when combined with AI and mobile accessibility.

In terms of integration, the Speak2Me project in Finland presents an example of AI, IoT, and mobile learning working together. The AI avatars incorporated in this initiative are available via mobile devices and smart speakers, and they can perform dynamic language conversations with learners, and adapt to learners' tone, pace, and grammar. The system pairs with IoT sensors that measure learning environments, like noise levels, posture, and time on task, to build a whole, responsive language learning ecosystem. Mobile alerts, visual dashboards, and automated feedback are provided to learners to support daily language goals in home or school settings (Martínez Durán, 2024).

Globally, there is increasing interest in using smart contracts for language learning verification, particularly in contexts such as migration and workforce development. In countries such as Estonia and the UAE, pilot programs are being tested to design blockchain backed platforms that automatically issue language proficiency certificates upon successful completion of AI verified language tasks. Learners' digital records can then be integrated into their mobile identities profiles and shared with employers or universities (Almakhour et al. 2020).

Each of these case studies reveals a facet of what is possible when technology is strategically deployed to support language education. However, as they demonstrate, the transition from theory to practice is happening, and despite the barriers (scalability, equity, and ethical oversight among them), there is mounting evidence that AI, blockchain, and IoT can all work together to create responsive, trustworthy, and hyperpersonalized language learning experiences.

These applications also serve as blueprints for future development. They highlight the significance of interdisciplinary collaboration, learner centered design and regulatory alignment. The early successes are providing more institutions and developers the opportunity to build on top of them, and the opportunity for transforming scaleable mobile language education is becoming a reality.

9. CONCLUSION

The convergence of artificial intelligence, blockchain, and the Internet of Things is ushering in a new era of transformation in language education—one that is more personalized, immersive, secure, and globally connected than ever before. As the mobile platform becomes the means to access a majority of educational resources, the combination of these three emerging technologies provides an unprecedented opportunity to reenvision how languages are learned, taught, assessed and certified.

Throughout this chapter, we have explored how AI is enhancing language learning through adaptive platforms, speech recognition, natural language processing, and intelligent tutoring systems. These tools are not only personalized to the individual learner profile but also offer continuous data driven feedback to keep the learner engaged and master the subject. At the same time, blockchain is reshaping the territory of educational trust by introducing tamperproof credentialing, decentralized learning networks, tokenized incentive models supporting peer to peer language exchange and lifelong learning. Meanwhile, IoT is extending the classroom beyond the classroom to everyday environments, making homes, workplaces, and public places responsive language learning spaces with sensor feedback, smart devices, and wearable technology.

The triadic integration that emerges here thus transforms the basic nature of language education from static, classroom bound instruction to dynamic, learner driven ecosystems. Language ceases to be studied as a subject, and becomes an experience lived in the context of a wider digital environment that adapts itself to learners' needs, behaviors and contexts. Educators and institutions shift from sole providers of knowledge to facilitators of the interactive, decentralized and learner owned digital ecosystems.

However, like any major change, these technologies promise to be as complicated to realize as they are to realize. Data privacy, algorithmic bias, the digital divide, and interoperability are concerns that must be addressed proactively through responsible design, transparent policy frameworks, and equitable access strategies. However, this will not be the future of language learning based solely on technological capability, but rather based on the values and governance structures that define how it is implemented.

Looking forward, the trajectory of innovation will continue to be even more sophisticated: emotionally intelligent learning agents, metaverse based language immersion and real time credentialing via smart contracts. However, the foundation needs to be laid today; it must be inclusive, it must be of quality, and the learners must be empowered. This implies that for educators, developers, researchers, and policymakers, this is an opportunity to work together to ensure that mobile language education is not only technologically advanced, but also pedagogically sound and ethically grounded.

The aim of this chapter was to offer an extensive and future oriented view of how AI, blockchain, and IoT could be utilized in mobile language education. It is a resource for those wanting to understand and influence the future of language learning in a digital, decentralized and data driven world. Thoughtful embracing of these innovations can contribute to a more accessible, engaging, and authentic language learning experience for everyone.

REFERENCES

Al-khresheh, M. H. (2024). The Future of Artificial Intelligence in English Language Teaching: Pros and Cons of ChatGPT Implementation through a Systematic Review. *Language Teaching Research Quarterly*, *43*, 54–80. DOI: 10.32038/ltrq.2024.43.04

Almakhour, M., Sliman, L., Samhat, A. E., & Mellouk, A. (2020). Verification of smart contracts: A survey. *Pervasive and Mobile Computing*, *67*, 101227. DOI: 10.1016/j.pmcj.2020.101227

Aloun, M. S. (2024). Synergistic Integration of Artificial Intelligence and Blockchain Technology: Advancements, Applications, and Future Directions. *Journal of Intelligent Systems and Applied Data Science*, *2*(2).

Alrashdi, I., & Alqazzaz, A. (2024). Synergizing AI, IoT, and Blockchain for diagnosing pandemic diseases in Smart Cities: Challenges and opportunities. *Sustainable Machine Intelligence Journal*, *7*, 6–1. DOI: 10.61356/SMIJ.2024.77106

Any, B., Ramadhan, T., & Nabila, E. A. (2024). Decentralized academic platforms: The future of education in the age of blockchain. *Blockchain Frontier Technology*, *3*(2), 112–124. DOI: 10.34306/bfront.v3i2.467

Bhumichai, D., Smiliotopoulos, C., Benton, R., Kambourakis, G., & Damopoulos, D. (2024). The convergence of artificial intelligence and blockchain: The state of play and the road ahead. *Information (Basel)*, *15*(5), 268. DOI: 10.3390/info15050268

Dong, W., Pan, D., & Kim, S. (2024). Exploring the integration of IoT and Generative AI in English language education: Smart tools for personalized learning experiences. *Journal of Computational Science*, *82*, 102397. DOI: 10.1016/j.jocs.2024.102397

Duy, P. T., Khoa, N. H., Quyen, N. H., Trinh, L. C., Kien, V. T., Hoang, T. M., & Pham, V. H. (2025). Vulnsense: Efficient vulnerability detection in ethereum smart contracts by multimodal learning with graph neural network and language model. *International Journal of Information Security*, *24*(1), 48. DOI: 10.1007/s10207-024-00965-2

Ferguson-Sams, N., Howell, E., Kaminski, R., Gazioglu, M., & Pennington, V. (2025). Multilingual Voices: Transforming Professional Development Based on Student Perceptions of Literacy Learning. tesol journal, 16(1), e70005.

Gouiza, N. I. S. S. R. I. N. E., Jebari, H. A. K. I. M., & Reklaoui, K. A. M. A. L. (2024). Integration Of Iot-Enabled Technologies and Artificial Intelligence in Diverse Domains: Recent Advancements And Future Trends. *Journal of Theoretical and Applied Information Technology*, *102*(5), 1975–2029.

Handini, B. S., Nurhasanah, N., & Panly, F. I. (2022). The effect of artificial intelligent technology used (Duolingo application) to enhance English learning. ELLITE. *Journal of English Language, Literature, and Teaching*, *7*(2), 86–94. DOI: 10.32528/ellite.v7i2.8354

Ifedayo, A. E., Olugbade, D., & Hamid, S. (2025). Integrating Artificial Intelligence with Blockchain: A Literature Review on Opportunities, Challenges, and Applications. *Blockchain, Artificial Intelligence, and Future Research*, *1*(1), 52–69. DOI: 10.70211/bafr.v1i1.179

Javed, S., Zehra, S. R., Ullah, H., & Naveed, M. (2025). How AI can detect Emotional cues in Students, Improving Virtual Learning Environments by providing Personalized support and Enhancing Social-emotional Learning. *Review of Applied Management and Social Sciences*, *8*(2), 605–622. DOI: 10.47067/ramss.v8i2.500

Jirgensons, M., & Kapenieks, J. (2018). Blockchain and the future of digital learning credential assessment and management. Journal of teacher education for sustainability, 20(1), 145-156.

Karim, S. A., Hamzah, A. Q. S., Anjani, N. M., Prianti, J., & Sihole, I. G. (2023). Promoting EFL students' speaking performance through ELSA Speak: An artificial intelligence in English language learning. *Journal of Languages and Language Teaching*, *11*(4), 655–668. DOI: 10.33394/jollt.v11i4.8958

Kökver, Y., Pektaş, H. M., & Çelik, H. (2025). Artificial intelligence applications in education: Natural language processing in detecting misconceptions. *Education and Information Technologies*, *30*(3), 3035–3066. DOI: 10.1007/s10639-024-12919-1

Kot, S. O., & Nykyporets, S. S. (2024). *Utilization of artificial intelligence in enhancing English language proficiency in tertiary education. Science and Education in the Third Millennium: Information Technology*. Education, Law, Psychology, Social Sphere, Management.

Kovalenko, I., & Baranivska, N. (2024). Integrating Artificial Intelligence in English Language Teaching: Exploring the potential and challenges of AI tools in enhancing language learning outcomes and personalized education. Європейські соціо-правові та гуманітарні студії, (1), 86-95.

Kurbanazarova, N., Shavkidinova, D., Khaydarov, M., Mukhitdinova, N., Khudoymurodova, K., Toshniyozova, D., Karimov, N., & Alimova, R. (2024). Development of Speech Recognition in Wireless Mobile Networks for An Intelligent Learning System in Language Education. *Journal of Wireless Mobile Networks, Ubiquitous Computing and Dependable Applications*, *15*(3), 298–311. DOI: 10.58346/JOWUA.2024.I3.020

Law, L. (2024). Application of generative artificial intelligence (GenAI) in language teaching and learning: A scoping literature review. *Computers and Education Open*, *6*, 100174. DOI: 10.1016/j.caeo.2024.100174

Lee, J., & Jeong, H. (2023). Keyword analysis of artificial intelligence education policy in South Korea. *IEEE Access : Practical Innovations, Open Solutions*, *11*, 102408–102417. DOI: 10.1109/ACCESS.2023.3317261

Mahmudulhassan, M., Muthoifin, M., & Begum, S. (2024). Artificial Intelligence in Multicultural Islamic Education: Opportunities, Challenges, and Ethical Considerations. *Solo Universal Journal of Islamic Education and Multiculturalism*, *2*(01), 19–26. DOI: 10.61455/sujiem.v2i01.114

Martínez Durán, L. M. (2024). Enhancing Speaking Skills through Small-talk2Me AI Tool in the English Language Learning Process among High School Students at a Public Institution in Riobamba City, the school year 2023-2024 (Master's thesis, Riobamba: Universidad Nacional de Chimborazo).

Mizumoto, M., Nguyen, D. T., Sytsma, J., Alfano, M., Izumi, Y., Fujita, K., & Minh, N. L. (2025). Cross-linguistic disagreement as a conflict of semantic alignment norms in multilingual AI~ Linguistic Diversity as a Problem for Philosophy, Cognitive Science, and AI~. arXiv preprint arXiv:2503.04792.

Mohamed, M. S. P. (2024). Exploring ethical dimensions of AI-enhanced language education: A literature perspective. *Technology in Language Teaching & Learning*, *6*(3), 1813–1813. DOI: 10.29140/tltl.v6n3.1813

Nazari, Z., Vahidi, A. R., & Musilek, P. (2024). Blockchain and Artificial Intelligence Non-Formal Education System (BANFES). *Education Sciences*, *14*(8), 881. DOI: 10.3390/educsci14080881

Novawan, A., Walker, S. A., & Ikeda, O. (2024). The new face of technology-enhanced language learning (TELL) with artificial intelligence (AI): Teacher perspectives, practices, and challenges. *Journal of English in Academic and Professional Communication*, *10*(1), 1–18. DOI: 10.25047/jeapco.v9i1.3754

Omidian, H. (2024). Synergizing blockchain and artificial intelligence to enhance healthcare. *Drug Discovery Today*, *29*(9), 104111. DOI: 10.1016/j.drudis.2024.104111 PMID: 39034026

Pack, A., & Maloney, J. (2024). Using artificial intelligence in TESOL: Some ethical and pedagogical considerations. *TESOL Quarterly*, *58*(2), 1007–1018. DOI: 10.1002/tesq.3320

Patty, J. (2024). The use of AI in language learning: What you need to know. Jurnal Review Pendidikan dan Pengajaran (JRPP), 7(1), 642-654.

Peng, W., & Yang, Y. (2024). Framework Design of Blockchain Intelligent Technology for Chinese Language Teaching Management System. *Procedia Computer Science*, *243*, 306–312. DOI: 10.1016/j.procs.2024.09.038

Ramya, R., & Ramamoorthy, S. (2025). AI in Communication Technology and Innovation of AR/VR in Communication. In Impact of Digitalization on Communication Dynamics (pp. 1-28). IGI Global Scientific Publishing.

Rane, N. (2024). Enhancing the quality of teaching and learning through Gemini, ChatGPT, and similar generative Artificial Intelligence: Challenges, future prospects, and ethical considerations in education. *TESOL and Technology Studies*, *5*(1), 1–6. DOI: 10.48185/tts.v5i1.1000

Rashid, A., Zafar, N., Mir, B., Asim, S., Sheikh, M. G., & Jahan, M. (2025). Artificial Intelligence and Smart Pedagogy: Machine Learning in Digital Media for Language Education. *Metallurgical and Materials Engineering*, *31*(2), 7–22. DOI: 10.63278/1324

Rustemi, A., Dalipi, F., Atanasovski, V., & Risteski, A. (2024, June). Enhancing Academic Credentials: The Synergy of Blockchain and Artificial Intelligence. In 2024 7th International Balkan Conference on Communications and Networking (BalkanCom) (pp. 206-211). IEEE.

Sajja, R., Sermet, Y., Cikmaz, M., Cwiertny, D., & Demir, I. (2024). Artificial intelligence-enabled intelligent assistant for personalized and adaptive learning in higher education. *Information (Basel)*, *15*(10), 596. DOI: 10.3390/info15100596

Salloum, S. A., Alomari, K. M., Alfaisal, A. M., Aljanada, R. A., & Basiouni, A. (2025). Emotion recognition for enhanced learning: Using AI to detect students' emotions and adjust teaching methods. *Smart Learning Environments*, *12*(1), 21. DOI: 10.1186/s40561-025-00374-5

Sánchez, E., Burgos, M., & Saborido, A. (2023). Virtual Reality in Linguistic Teaching to Immigrants. *Journal of Education For Sustainable Innovation*, *1*(2), 95–102. DOI: 10.56916/jesi.v1i2.561

Shanmughan, P., & Joseph, J. S, B. N., SM, A., CS, A., & Jose, J. (2024, November). The AI revolution in micro-credentialing: personalized learning paths. In Frontiers in Education (Vol. 9, p. 1445654). Frontiers Media SA.

Shirinova, S. (2025). Leveraging Artificial Intelligence in Linguistics: Innovations in Language Acquisition and Analysis. *EuroGlobal Journal of Linguistics and Language Education*, *2*(1), 50–57. DOI: 10.69760/egjlle.250006

Tabuenca, B., Uche-Soria, M., Greller, W., Hernández-Leo, D., Balcells-Falgueras, P., Gloor, P., & Garbajosa, J. (2024). Greening smart learning environments with Artificial Intelligence of Things. *Internet of Things : Engineering Cyber Physical Human Systems*, *25*, 101051. DOI: 10.1016/j.iot.2023.101051

Tiwari, H. P. (2024). Artificial intelligence in the classroom: Revolutionizing English language teaching. [JET Li]. *Journal of English Teaching and Linguistics Studies*, *6*(1), 42–59. DOI: 10.55215/jetli.v6i1.9757

Tong, Q., & Liu, S. (2025). Intelligent Classroom Environment: Application of Internet of Things and Artificial Intelligence in English Teaching. *International Journal of High Speed Electronics and Systems*, 2540364. DOI: 10.1142/S012915642540364X

Tyagi, P., & Shrivastava, N. Sakshi, & Jain, V. (2024). Synergizing Artificial Intelligence and Blockchain. In Next-Generation Cybersecurity: AI, ML, and Blockchain (pp. 83-97). Singapore: Springer Nature Singapore.

Ulutaş, N. K., Durmaz, M. E., Elbuğa, N., Sezgin, S., Kaya, K., & İsmailoğulları, B. (2025). The Use of Augmented Reality (AR), Virtual Reality (VR), and Artificial Intelligence (AI) in Language Education: A Systematic Literature Review. Reimagining Intelligent Computer-Assisted Language Education, 401-430.

Yadav, R., Nghiem, X. H., & Yadav, M. (2025). Adaptive AI in Metaverse Language Education. In Harnessing Chatbots for Effective Language Teaching (pp. 49-74). IGI Global Scientific Publishing. DOI: 10.4018/979-8-3693-9974-3.ch003

Yan, L., Sha, L., Zhao, L., Li, Y., Martinez-Maldonado, R., Chen, G., Li, X., Jin, Y., & Gašević, D. (2024). Practical and ethical challenges of large language models in education: A systematic scoping review. *British Journal of Educational Technology*, *55*(1), 90–112. DOI: 10.1111/bjet.13370

Yefymenko, T., Bilous, T., Zhukovska, A., Sieriakova, I., & Moyseyenko, I. (2024). Technologies for using interactive artificial intelligence tools in the teaching of foreign languages and translation. *Revista Amazonia Investiga*, *13*(74), 299–307. DOI: 10.34069/AI/2024.74.02.25

Zainuddin, A. A., Handayani, D., Ridza, I. H. M., Rahman, S. H. A., Kamarudin, S. I., Ahmad, K. Z., . . . Dhuzuki, N. H. M. (2024, May). Converging for Security: Blockchain, Internet of Things, Artificial Intelligence-Why Not Together? In 2024 IEEE 14th Symposium on Computer Applications & Industrial Electronics (ISCAIE) (pp. 181-186). IEEE.

Zhang, C., Liu, Z., Aravind, B. R., & Hariharasudan, A. (2024). Synergizing language learning: SmallTalk AI In industry 4.0 and Education 4.0. *PeerJ. Computer Science*, *10*, e1843. DOI: 10.7717/peerj-cs.1843 PMID: 38435575

Zheng, L., & Yang, Y. (2024). Research perspectives and trends in Artificial Intelligence-enhanced language education: A review. *Heliyon*, *10*(19), e38617. DOI: 10.1016/j.heliyon.2024.e38617 PMID: 39416824

Zuo, Y. (2024). Exploring the Synergy: AI Enhancing Blockchain, Blockchain Empowering AI, and their Convergence across IoT Applications and Beyond. *IEEE Internet of Things Journal*.

Compilation of References

Abad-Segura, E., González-Zamar, M. D., Rosa, A. L. D. L., & Cevallos, M. B. M. (2020). Sustainability of educational technologies: An approach to augmented reality research. *Sustainability (Basel)*, *12*(10), 4091. DOI: 10.3390/su12104091

Adams, W. C. (2015). Conducting semi-structured interviews. In Newcomer, K. E., Hatry, H. P., & Wholey, J. S. (Eds.), *Handbook of practical program evaluation* (4th ed., pp. 492–505). Jossey-Bass. DOI: 10.1002/9781119171386.ch19

Adipat, S., Laksana, K., Busayanon, K., Asawasowan, A., & Adipat, B. (2021). Engaging students in the learning process with game-based learning: The fundamental concepts. *International Journal of Technology in Education*, *4*(3), 542–552. DOI: 10.46328/ijte.169

Adkins-Jablonsky, S. J., Shaffer, J. F., Morris, J. J., England, B., & Raut, S. (2021). A tale of two institutions: Analyzing the impact of gamified student response systems on student anxiety in two different introductory biology courses. *CBE Life Sciences Education*, *20*(2), ar19. DOI: 10.1187/cbe.20-08-0187 PMID: 33797283

Afshari, M., Bakar, K. A., Luan, W. S., Samah, B. A., & Fooi, F. S. (2009). Factors affecting teachers' use of information and communication technology. *Online Submission*, *2*(1), 77–104.

Agarwal, P. K., Nunes, L. D., & Blunt, J. R. (2021). Retrieval Practice Consistently Benefits Student Learning: A Systematic Review of Applied Research in Schools and Classrooms. *Educational Psychology Review*, *33*(4), 1409–1453. DOI: 10.1007/s10648-021-09595-9

Ahmad, M., & Wilkins, S. (2024). Purposive sampling in qualitative research: A framework for the entire journey. *Quality & Quantity*. Advance online publication. DOI: 10.1007/s11135-024-02022-5

Ahmed, S. K. (2024). The pillars of trustworthiness in qualitative research. *Journal of Medicine, Surgery, and Public Health*, *2*, 100051. DOI: 10.1016/j.glmedi.2024.100051

AirAlarms.in.ua. (2023). Statistics of air alarms in Ukraine, February 24, 2022–December 26, 2023. https://air-alarms.in.ua/?from=2022-02-24&to=2023-12-26#statistic

Ajisoko, P. (2020). The use of Duolingo apps to improve English vocabulary learning. [iJET]. *International Journal of Emerging Technologies in Learning*, *15*(7), 149–155. DOI: 10.3991/ijet.v15i07.13229

Akçayır, M., & Akçayır, G. (2017). Advantages and challenges associated with augmented reality for education: A systematic review of the literature. *Educational Research Review*, *20*, 1–11. DOI: 10.1016/j.edurev.2016.11.002

Al-Ansi, A. M., Jaboob, M., Garad, A., & Al-Ansi, A. (2023). Analyzing augmented reality (AR) and virtual reality (VR) recent development in education. In *Social Sciences and Humanities Open* (Vol. 8, Issue 1). DOI: 10.1016/j.ssaho.2023.100532

Albarqi, G. (2023). Padlet as a Formative Assessment Tool in the Online Language Classroom: Action Research. In Chong, S. W., & Reinders, H. (Eds.), *Innovation in Learning-Oriented Language Assessment*. New Language Learning and Teaching Environments. Palgrave Macmillan., DOI: 10.1007/978-3-031-18950-0_11

Alda, R. C. (2023). Mobile-assisted language learning (MALL) in senior high school English classes. *World Journal of English Language*, *13*(6), 211. DOI: 10.5430/wjel.v13n6p211

Aldemir, T., Celik, B., & Kaplan, G. (2018). A qualitative investigation of student perceptions of game elements in a gamified course. *Computers in Human Behavior*, *78*, 235–254. DOI: 10.1016/j.chb.2017.10.001

Aldosari, M. S., & Alsager, H. N. (2023). A step toward autonomy in education: Probing into the effects of practicing self-assessment, resilience, and creativity in task supported language learning. *BMC Psychology*, *11*(1), 434. DOI: 10.1186/s40359-023-01478-8 PMID: 38062504

Alexandra, T. (2024). Foreign Language Acquisition Difficulties: Diagnosing Individual Differences.

Algerafi, M., Zhou, Y., Oubibi, M., & Wijaya, T. (2023). Unlocking the potential: A comprehensive evaluation of augmented reality and virtual reality in education. *Electronics (Basel)*, *12*(18), 3953. DOI: 10.3390/electronics12183953

Alizadeh, M. (2019). Augmented/virtual reality promises for ELT practitioners. *Diversity and Inclusion*, 360–368.

Al-khresheh, M. H. (2024). The Future of Artificial Intelligence in English Language Teaching: Pros and Cons of ChatGPT Implementation through a Systematic Review. *Language Teaching Research Quarterly*, *43*, 54–80. DOI: 10.32038/ltrq.2024.43.04

Almakhour, M., Sliman, L., Samhat, A. E., & Mellouk, A. (2020). Verification of smart contracts: A survey. *Pervasive and Mobile Computing*, *67*, 101227. DOI: 10.1016/j.pmcj.2020.101227

Aloun, M. S. (2024). Synergistic Integration of Artificial Intelligence and Blockchain Technology: Advancements, Applications, and Future Directions. *Journal of Intelligent Systems and Applied Data Science*, *2*(2).

Alrashdi, I., & Alqazzaz, A. (2024). Synergizing AI, IoT, and Blockchain for diagnosing pandemic diseases in Smart Cities: Challenges and opportunities. *Sustainable Machine Intelligence Journal*, *7*, 6–1. DOI: 10.61356/SMIJ.2024.77106

Amaaz, A., Mouradi, A., & Erradi, M. (2024). Pre-service Elementary Teachers' Attitudes Toward the Use of Mobile Learning in Elementary Science Education. In Motahhir, S., & Bossoufi, B. (Eds.), *Digital Technologies and Applications. ICDTA 2024. Lecture Notes in Networks and Systems* (Vol. 1099). Springer., DOI: 10.1007/978-3-031-68653-5_2

Amalia, I. (2020). The application of mobile assisted language learning (MALL) in teaching pronunciation. *International Journal of Language Education and Cultural Review*, *6*(2), 194–203. DOI: 10.21009/IJLECR.062.20

Amineh, R. J., & Asl, H. D. (2015). Review of constructivism and social constructivism. *Journal of Social Sciences. Literature and Languages*, *1*(1), 9–16.

Amrate, M., & Tsai, P. (2024). Computer-assisted pronunciation training: A systematic review. *ReCALL*. Advance online publication. DOI: 10.1017/S0958344024000181

Anderson, R. E., & Dexter, S. L. (2000). *School technology leadership: Incidence and impact*. http://www.crito.uci.edu/tlc/html/findings.html

Andujar, A., & Hussein, S. A. (2019). Mobile-mediated communication and students' listening skills: A case study. *International Journal of Mobile Learning and Organisation*, *13*(3), 309. DOI: 10.1504/IJMLO.2019.100432

Anil, Ö., & Batdi, V. (2023). Use of augmented reality in science education: A mixed-methods research with the multi-complementary approach. *Education and Information Technologies*, *28*(5), 5147–5185. DOI: 10.1007/s10639-022-11398-6

Annamalai, N., Mangaleswaran, S., Ab Rashid, R., Harb Alqaryouti, M., & Sadeq, A. E. (2024). Learner experiences of community of inquiry mode (CoI) in WhatsApp for English language learning. *SAGE Open*, *14*(2), 21582440241251944. DOI: 10.1177/21582440241251944

Any, B., Ramadhan, T., & Nabila, E. A. (2024). Decentralized academic platforms: The future of education in the age of blockchain. *Blockchain Frontier Technology*, *3*(2), 112–124. DOI: 10.34306/bfront.v3i2.467

Arashnia, M., & Shahrokhi, M. (2016). Mobile assisted language learning: English pronunciation among Iranian pre-intermediate EFL learners. *Journal of Applied Linguistics and Language Research*, *3*(4), 149–162. http://www.jallr.com/index.php/JALLR/article/view/433

Arce, N. H., & Valdivia, A. C. (2020). Adapting competitiveness and gamification to a digital platform for foreign language learning. [iJET]. *International Journal of Emerging Technologies in Learning*, *15*(20), 194–209. DOI: 10.3991/ijet.v15i20.16135

Arena, F., Collotta, M., Pau, G., & Termine, F. (2022). An Overview of Augmented Reality. In *Computers* (Vol. 11, Issue 2). DOI: 10.3390/computers11020028

Arias, O. E. (2021). Learning through Transitions: Exploring the Relationships between Emotional Intelligence Competencies and Learning Agility (Doctoral dissertation, University of Pennsylvania).

Aros, M., Arnold, H., & Chaparro, B. (2025). Exploring Device Usability: A Heuristic Evaluation of the Apple Vision Pro and Meta Quest 3. *Communications in Computer and Information Science, 2522 CCIS*, 3–12. DOI: 10.1007/978-3-031-94150-4_1

Asrifan, A., Kaddas, B., Mulyadi, M., Pratiwi, W. R., Supriadi, S., & Jabu, B. (2025). Cross-Cultural Gamification: Understanding Design and User Experience in Different Cultural Contexts. In Enhancing Engagement With Gamification: Education, Business, and Healthcare Perspectives (pp. 1-28). IGI Global Scientific Publishing.

Atanda, A. I. (2018). Teacher trainees' readiness to adopt social media in teaching-learning process: Case study of Faculty of Education undergraduates, University of Ibadan. *The International Journal of the Arts in Society*, *11*(1), 499–511. http://www.universitypublications.net/ijas/1101/pdf/U8K696.pdf

Atibuni, D. Z., Manyiraho, D., & Nabitula, A. N. (2022). A Fourth Industrial Revolution Paradigm Shift in Teacher Education? *International Journal of African Higher Education*, *9*(2), 1–21. DOI: 10.6017/ijahe.v9i2.15365

Azabdaftari, B., & Mozaheb, M. A. (2012). Comparing vocabulary learning of EFL learners by using two different strategies: Mobile learning vs. flashcards. *The EuroCALL Review*, *20*(2), 47. DOI: 10.4995/eurocall.2012.11377

Azem Qashou, A. M., Bahar, N., & Mohamed, H. (2025). Qualitative Exploration of Data Security Risks in Mobile Cloud Computing for Higher Education. *Security and Privacy*, *8*(2), e70001. DOI: 10.1002/spy2.70001

Bachmair, B., & Pachler, N. (2015). Framing Ubiquitous Mobility Educationally: Mobile Devices and Context-Aware Learning. In Wong, L. H., Milrad, M., & Specht, M. (Eds.), *Seamless Learning in the Age of Mobile Connectivity*. Springer., DOI: 10.1007/978-981-287-113-8_3

Bacigalupo, M., Bekh, O., Binasco, A., Israel, H., & WEIKERT, G. L. (2024). " Scaffold", a deck of cards to design competence-oriented learning experiences (No. JRC136622). Joint Research Centre.

Bailey, D., Almusharraf, N., & Hatcher, R. (2021). Finding satisfaction: Intrinsic motivation for synchronous and asynchronous communication in the online language learning context. *Education and Information Technologies*, *26*(3), 2563–2583. DOI: 10.1007/s10639-020-10369-z PMID: 33169066

Baines, A., Medina, D., & Healy, C. (2023). *Amplify student voices: Equitable practices to build confidence in the classroom*. ASCD.

Bairagi, A. K., Rajon, S. A. A., & Roy, T. (2011). Status and role of ICT in educational institutions to build a digital society in Bangladesh: Perspective of a divisional city, Khulna. *International Journal of Advances in Engineering and Technology*, *1*(4), 374.

Bajorek, J. (2017). L2 pronunciation in CALL: The unrealized potential of Rosetta Stone, Duolingo, Babbel, and Mango languages. *Issues and Trends in Educational Technology*, *5*(1), 60–87. DOI: 10.2458/azu_itet_v5i1_bajorek

Bajura, M., Fuchs, H., & Ohbuchi, R. (1992). Merging virtual objects with the real world: Seeing ultrasound imagery within the patient. *Computer Graphics*, *26*(2), 203–210. DOI: 10.1145/142920.134061

Bakar, S. (2021). Investigating the dynamics of contemporary pedagogical approaches in higher education through innovations, challenges, and paradigm shifts. *Social Science Chronicle*, *1*(1), 1–19. DOI: 10.56106/ssc.2021.009

Bakkali, I., Pérez-Torregrosa, A. B., Cebrián-Robles, V., & Cebrián-de-la-Serna, M. (2023). Digital Inclusion in the Moroccan Context: A Review of Research and Educational Initiatives in the Twenty-First Century. In Tomczyk, Ł., Guillén-Gámez, F. D., Ruiz-Palmero, J., & Habibi, A. (Eds.), *From Digital Divide to Digital Inclusion. Lecture Notes in Educational Technology*. Springer., DOI: 10.1007/978-981-99-7645-4_17

Baleghizadeh, S. (2015). The Procedural Syllabus and the Task Syllabus: How Similar, How Different?. *How, 22*(2), 104-113.

Bandura, A. (1994). Self-efficacy. In Ramachaudran, V. S. (Ed.), *Encyclopedia of human behavior* (4th ed., pp. 71–81). Academic Press.

Bandura, A. (1997). *Self-Efficacy: The exercise of control.* W.H. Freeman and Company.

Barik, S. (2023). Embracing the transformative dynamics of English teaching and technology in higher education: A posthumanist perspective. Smart Moves Journal IJELLH, 17-31.

Barnard, M., & Ferrier-Kerr, J. (2021). Moving to an Innovative Learning Environment: Exploring Teachers' Liminal Space. Pedagogy and Partnerships in Innovative Learning Environments: Case Studies from New Zealand Contexts, 141-164.

Basal, A., Yilmaz, S., Tanriverdi, A., & Sari, L. (2016). Effectiveness of Mobile Applications in Vocabulary Teaching. *Contemporary Educational Technology*, *7*(1), 47–59. DOI: 10.30935/cedtech/6162

Basaran, S., & Cabaroglu, N. (2014).Language Learning Podcasts and Learners' Belief Change.TESL-EJ, 17(4), n4 retrieved from http://eric.ed.gov/?id=EJ1024978 on 30.11.2015

Bashori, M., van Hout, R., Strik, H., & Cucchiarini, C. (2020). Web-based language learning and speaking anxiety. *Computer Assisted Language Learning*, •••, 1–32. DOI: 10.1080/09588221.2020.1770293

Bastawros, H. (2024). Enhancing Productive Vocabulary of ESL Learners: A Qualitative Case Study.

Baxter, P., & Jack, S. (2015). Qualitative case study Methodology: Study design and implementation for novice researchers. *The Qualitative Report*. Advance online publication. DOI: 10.46743/2160-3715/2008.1573

Belda-Medina, J. (2022). USING AUGMENTED REALITY (AR) AS AN AUTHORING TOOL IN EFL THROUGH MOBILE COMPUTER-SUPPORTED COLLABORATIVE LEARNING. *Teaching English with Technology*, *22*(2), 115–135.

Belhiah, H., & Abdelatif, A. (2016). English as a medium of instruction in Moroccan higher education. SSRN *Electronic Journal*. https://doi.org/DOI: 10.2139/ssrn.2895569

Benali, M., Kaddouri, M., & Azzimani, T. (2018). Digital competence of Moroccan teachers of English. *International Journal of Education and Development Using Information and Communication Technology*, *14*(2), 99–120. https://files.eric.ed.gov/fulltext/EJ1190022.pdf

Benlaghrissi, H., & Ouahidi, L. M. (2023). The impact of mobile-assisted language learning on developing EFL learners' vocabulary knowledge. *International Journal of Interactive Mobile Technologies*, *17*(22), 38–51. Advance online publication. DOI: 10.3991/ijim.v17i22.41665

Bensalem, E., & Hamed Alenazi, F. (2023). Using speaking eportfolios to enhance oral English language skills among English as a foreign language students. *Vivat Academia. Revista de Comunicación*, 25–47. https://doi.org/DOI: 10.15178/va

Bensalem, E. (2018). The impact of WhatsApp on EFL students' vocabulary learning. *Arab World English Journal*, *9*(1), 23–38. Advance online publication. DOI: 10.24093/awej/vol9no1.2

Berge, Z. L. (1996). *The role of the online instructor*. Facilitator.

Bertrand, M. G., Sezer, H. B., & Namukasa, I. K. (2024). Exploring AR and VR tools in mathematics education through culturally responsive pedagogies. *Digital Experiences in Mathematics Education*, *10*(3), 462–486. DOI: 10.1007/s40751-024-00152-x PMID: 39497730

Bhumichai, D., Smiliotopoulos, C., Benton, R., Kambourakis, G., & Damopoulos, D. (2024). The convergence of artificial intelligence and blockchain: The state of play and the road ahead. *Information (Basel)*, *15*(5), 268. DOI: 10.3390/info15050268

Bin Dahmash, N. (2021). *Synchronous and asynchronous English writing classes in the EFL context: Students' practices and benefits. Arab World English Journal* (Vol. 12). AWEJ.

Boller, S., & Kapp, K. (2017). *Play to learn: Everything you need to know about designing effective learning games*. ATD Press.

Bolliger, U., & Shepherd, E. (2010). Student perceptions of e-portfolio integration in online courses. *Distance Education, 31*(3), 295–314. DOI: 10.1080/01587919.2010.513955

Bos, D., Miller, S., & Bull, E. (2022). Using virtual reality (VR) for teaching and learning in geography: Fieldwork, analytical skills, and employability. *Journal of Geography in Higher Education, 46*(3), 479–488. DOI: 10.1080/03098265.2021.1901867

Bouzidi, H. E. I. (2024). Advancing an educational framework to enhance students' learning experiences. *E-Learning, 2*(11). ISCAE. DOI: 10.48423/IMIST.PRSM/rmere-v2i11.46258

Bowen, G. A. (2009). Document analysis as a qualitative research method. *Qualitative Research Journal, 9*(2), 27–40. DOI: 10.3316/QRJ0902027

Bower, M., Howe, C., McCredie, N., Robinson, A., & Grover, D. (2014). Augmented Reality in education–cases, places and potentials. *Educational Media International, 51*(1), 1–15. DOI: 10.1080/09523987.2014.889400

Braun, V., & Clarke, V. (2006). Using thematic analysis in psychology. *Qualitative Research in Psychology, 3*(2), 77–101. DOI: 10.1191/1478088706qp063oa

Braun, V., & Clarke, V. (2019). Reflecting on reflexive thematic analysis. *Qualitative Research in Sport, Exercise and Health, 11*(4), 589–597. DOI: 10.1080/2159676X.2019.1628806

Brown, H. D. (2015). *Teaching by principles: An interactive approach to language pedagogy* (4th ed., eText). Pearson Education.

Brown, H. D. (2004). *Language assessment : Principles and classroom practices* (Blanford, V., & Johnston, J., Eds.). Longman.

Bryant, L., Bailey, B., & Hemsley, B. (2024). Developing and evaluating an immersive augmented reality application for children with developmental communication disability: InterPlay for language learning. *International Journal of Speech-Language Pathology*. Advance online publication. DOI: 10.1080/17549507.2024.2361734 PMID: 39135408

Buchner, J., & Kerres, M. (2023). Media comparison studies dominate comparative research on augmented reality in education. *Computers & Education, 195*, 104711. Advance online publication. DOI: 10.1016/j.compedu.2022.104711

Bueno-Alastuey, M. C., & Nemeth, K. (2020). Quizlet and podcasts: Effects on vocabulary acquisition. *Computer Assisted Language Learning, 35*(7), 1407–1436. DOI: 10.1080/09588221.2020.1802601

Bülbül, A. E., & Özelçi, S. Y. (2025). The mediating role of mobile learning motivation in the relationship between pre-service teachers' digital literacy and pre-service teacher identity. *Education and Information Technologies*, *30*(2), 1729–1743. DOI: 10.1007/s10639-024-12868-9

Burden, P. R. (2025). *Classroom management: Creating a successful K-12 learning community*. John Wiley & Sons.

Burston, J. (2014). MALL: The pedagogical challenges. *Computer Assisted Language Learning*, *27*(4), 344–357. DOI: 10.1080/09588221.2014.914539

Burston, J. (2015). Twenty years of MALL project implementation: A meta-analysis of learning outcomes. *ReCALL*, *27*(1), 4–20. DOI: 10.1017/S0958344014000159

Cabrera-Solano, P. (2020). The use of digital portfolios to enhance English as a foreign language speaking skills in higher education. *International Journal of Emerging Technologies in Learning*, *15*(24), 159–175. DOI: 10.3991/ijet.v15i24.15103

Cai, S., Liu, E., Shen, Y., Liu, C., Li, S., & Shen, Y. (2020). Probability learning in mathematics using augmented reality: Impact on student's learning gains and attitudes. *Interactive Learning Environments*, *28*(5), 560–573. Advance online publication. DOI: 10.1080/10494820.2019.1696839

Cai, Y., Pan, Z., & Liu, M. (2022). Augmented reality technology in language learning: A meta-analysis. *Journal of Computer Assisted Learning*, *38*(4), 929–945. DOI: 10.1111/jcal.12661

Cakmakci, G., Aydeniz, M., Brown, A., & Makokha, J. M. (2025). *Situated Cognition and Cognitive Apprenticeship Learning*. 293–311. DOI: 10.1007/978-3-031-81351-1_17

Caldwell-Harris, C. L., & MacWhinney, B. (2023). Age effects in second language acquisition: Expanding the emergentist account. *Brain and Language*, *241*, 105269. DOI: 10.1016/j.bandl.2023.105269 PMID: 37150139

Canagarajah, A. (1999). *Resisting linguistic imperialism in English lan guage teaching*. Oxford University Press.

Cappelli, M. A., & Akkari, A. (2025). Academic ICTs training in South Africa, Cameroon, and Nigeria. Strategies for ICT training course design. *Education and Information Technologies*, *30*(8), 10463–10510. DOI: 10.1007/s10639-024-13200-1

Carmona-Galindo, V. D., Velado-Cano, M. A., & Groat-Carmona, A. M. (2025). The ecology of climate change: Using virtual reality to share, experience, and cultivate local and global perspectives. *Education Sciences*, *15*(3), 290. DOI: 10.3390/educsci15030290

Caudell, T. P., & Mizell, D. W. (1992). Augmented reality: An application of heads-up display technology to manual manufacturing processes. In Nunamaker, J. F., & Sprague, R. H. (Eds.), *Proceedings of the Twenty-Fifth Hawaii International Conference on System Sciences* (Vol. 2, pp. 659–669). IEEE Computer Society Press. DOI: 10.1109/HICSS.1992.183317

Çeken, B., & Taşkın, N. (2025). Examination of multimedia learning principles in augmented reality and virtual reality learning environments. *Journal of Computer Assisted Learning*, *41*(1), e13097. DOI: 10.1111/jcal.13097

Celce-Murcia, M., Brinton, D. M., & Goodwin, J. M. (2010). *Teaching pronunciation: A course book and reference guide*. Cambridge University Press.

Çelik, F., & Baturay, M. H. (2024). The effect of metaverse on L2 vocabulary learning, retention, student engagement, presence, and community feeling. *BMC Psychology*, *12*(1), 58. DOI: 10.1186/s40359-024-01549-4 PMID: 38308350

Cepik, S., & Yastibas, A. E. (2013). The use of e-portfolio to improve English speaking skill of Turkish EFL learners. *Anthropologist*, *16*(1–2), 307–317. DOI: 10.1080/09720073.2013.11891358

Chakrani, B. (2020). Languages in Conflict: EXAMINING THE STATUS OF STANDARD ARABIC AND FRENCH IN MOROCCO. *Al-'Arabiyya, 53*, 37–56. https://www.jstor.org/stable/48659135

Chang, B., Sheldon, L., Si, M., & Hand, A. (2012). Foreign language learning in immersive virtual environments. In *Proceedings of SPIE 8289, The Engineering Reality of Virtual Reality 2012* (p. 82890E). SPIE. https://doi.org/DOI: 10.1117/12.909835

Chang, H. Y., Binali, T., Liang, J. C., Chiou, G. L., Cheng, K. H., Lee, S. W. Y., & Tsai, C. C. (2022). Ten years of augmented reality in education: A meta-analysis of (quasi-) experimental studies to investigate the impact. *Computers & Education*, *191*, 104641. Advance online publication. DOI: 10.1016/j.compedu.2022.104641

Chang, S. C., Hsu, T. C., Chen, Y. N., & Jong, M. S. Y. (2020). The effects of spherical video-based virtual reality implementation on students' natural science learning effectiveness. *Interactive Learning Environments*, *28*(7), 915–929. DOI: 10.1080/10494820.2018.1548490

Chang, Y.-S., Chen, C.-N., & Liao, C.-L. (2020). Enhancing English-Learning Performance through a Simulation Classroom for EFL Students Using Augmented Reality—A Junior High School Case Study. *Applied Sciences (Basel, Switzerland)*, *10*(21), 7854. DOI: 10.3390/app10217854

Chapelle, C. A. (2001). *Computer applications in second language acquisition*. Cambridge University Press., DOI: 10.1017/CBO9781139524681

Chapelle, C. A. (2007). Technology and second language acquisition. *Annual Review of Applied Linguistics*, *27*, 98–114. DOI: 10.1017/S0267190508070050

Chen, C., Wang, X. Y., Chen, J., Liu, Q., & Peng, S. L. (2022). An Active Security System Based on AR Smart Classes and Face Recognition Technology. *Wangji Wanglu Jishu Xuekan*, *23*(2), 245–253. Advance online publication. DOI: 10.53106/160792642022032302005

Chen, Y. M. (2022). Understanding foreign language learners' perceptions of teachers' practice with educational technology with specific reference to Kahoot! and Padlet: A case from China. *Education and Information Technologies*, *27*(2), 1439–1465. DOI: 10.1007/s10639-021-10649-2

Chetry, K. K. (2024). Transforming education: Paradigm shifts in 21st century learning. *International Journal of Multidisciplinary Research*, *2*(5), 469–479.

Ching, C., & Chan, V. (2020). Positive emotions, positive feelings and health: A life philosophy. *Linguistics and Culture Review*, *4*(1), 1–14. DOI: 10.21744/lingcure.v4n1.16

Chou, Y.-K. (2021). *The Octalysis framework for gamification & behavioral design*. https://yukaichou.com/gamification-examples/octalysis-complete-gamification-framework/

Christopoulos, A., & Mystakidis, S. (2023). Gamification in education. *Encyclopedia*, *3*(4), 1223–1243. DOI: 10.3390/encyclopedia3040089

Chua, N. A., & Soon, G. Y. (2021). Performing communicative language teaching in Mandarin mobile learning. *International Journal of Interactive Mobile Technologies*, *15*(5), 87. Advance online publication. DOI: 10.3991/ijim.v15i05.20899

Chun, D. M., & Jiang, Y. (2022). Using technology to explore L2 pronunciation. In Levis, J., Derwing, T., & Sonsaat-Hegelheimer, S. (Eds.), *Second Language Pronunciation: Bridging the Gap Between Research and Teaching* (pp. 129–150). Wiley-Blackwell., DOI: 10.1002/9781394259663.ch7

Cong-LEM, N. (2022). Emotion and its relation to cognition from Vygotsky's perspective. *European Journal of Psychology of Education, 38*(2), 865–880. Advance online publication. DOI: 10.1007/s10212-022-00624-x

Connolly, T., Willis, J., & Lloyd, M. (2019). Evaluating teacher and learner readiness to use Facebook in an Australian vocational setting. *Studies in Continuing Education, 41*(1), 61–75. DOI: 10.1080/0158037X.2018.1461619

Couper, G. (2021). Pronunciation teaching issues: Answering teachers' questions. *RELC Journal, 52*(1), 128–143. DOI: 10.1177/0033688220964041

Creely, E. (2025). Innovating language learning in higher education: Creativity, creative pedagogies and technologies in the 21st century. In *2024 International Graduate Research Symposium.*

Creswell, J. W., & Creswell, J. D. (2017). *Research design: Qualitative, quantitative, and mixed methods approaches* (5th ed.). Sage.

Creswell, J. W., & Creswell, J. D. (2018). Research design: Qualitative, quantitative, and mixed methods approaches. *Sage (Atlanta, Ga.).*

Creswell, J. W., & Plano Clark, V. L. (2017). Designing and conducting mixed methods research. *Sage (Atlanta, Ga.).*

Creswell, J. W., & Poth, C. N. (2018). *Qualitative inquiry & research design: Choosing among five approaches* (4th ed.). Sage Publications, Inc.

Creswell, J. W., & Poth, C. N. (2018). *Qualitative Inquiry and Research Design Choosing among Five Approaches* (4th ed.). SAGE Publications, Inc.

Dai, Y., & Wu, Z. (2023). Mobile-assisted pronunciation learning with feedback from peers and/or automatic speech recognition: A mixed-methods study. *Computer Assisted Language Learning, 36*(5-6), 861–884. DOI: 10.1080/09588221.2021.1952272

Danilov, I. V., & Mihailova, S. (2021). Neuronal coherence agent for shared intentionality: A hypothesis of neurobiological processes occurring during social interaction. *OBM Neurobiology, 5*(4), 1–28. DOI: 10.21926/obm.neurobiol.2104113

Dash, A., & Kuddus, K. (2020). Leveraging the benefits of ICT usage in teaching of English language and literature. In *Smart intelligent computing and applications* (pp. 225–232). Springer. DOI: 10.1007/978-981-32-9690-9_22

Das, M. K. (2024). *Teacher education with ICT tools and techniques.* Academic Guru Publishing House.

De Costa, P. I. (2007). J. P. Lantolf and S. L. Thorne: Sociocultural Theory and the Genesis of Second Language Development. Oxford University Press, 2006. *Applied Linguistics*, *28*(3), 477–480. DOI: 10.1093/applin/amm027

de Sousa, G. M. B., Cardoso, L. A., & Toassi, P. F. P. (2018). Duolingo as a tool to improve vocabulary writing in English as a foreign language. *Letras em Revista*, *9*(1), 119–134.

De Souza, R., Parveen, R., Chupradit, S., Velasco, L. G., Arcinas, M., Tabuena, A. C., Pentang, J., & Ventayen, R. J. M. (2021). Language teachers' pedagogical orientations in integrating technology in the online classroom: Its effect on students' motivation and engagement. *Turkish Journal of Computer and Mathematics Education*, *12*. Advance online publication. DOI: 10.2139/ssrn.3844678

Dehganzadeh, H., & Dehganzadeh, H. (2020). Investigating effects of digital gamification-based language learning: A systematic review. *Two Quarterly Journal of English Language Teaching and Learning University of Tabriz*, *12*, 53–93. DOI: 10.22034/elt.2020.10676

Dehghanzadeh, H., Fardanesh, H., Hatami, J., Talaee, E., & Noroozi, O. (2021). Using gamification to support learning English as a second language: A systematic review. *Computer Assisted Language Learning*, *34*(6), 934–957. DOI: 10.1080/09588221.2019.1648298

Demirbilek, M., Talan, T., & Alzouebi, K. (2022). An examination of the factors and challenges to adopting gamification in English foreign language teaching. *International Journal of Technology in Education*, *5*(4), 654–668. DOI: 10.46328/ijte.358

Denzin, N. K., & Lincoln, Y. S. (Eds.). (2017). *The SAGE handbook of qualitative research* (5th ed.). SAGE Publications.

Derwing, T. M., & Munro, M. J. (2022). Pronunciation learning and teaching. In Derwing, T. M., Munro, M. J., & Thomson, R. I. (Eds.), *The Routledge handbook of second language acquisition and speaking* (pp. 147–159). Routledge. DOI: 10.4324/9781003022497-14

Deterding, S., Khaled, R., Nacke, L., & Dixon, D. (2011). *Gamification: Toward a definition*. In *CHI 2011, Gamification Workshop Proceedings* (pp. 1–9). Vancouver.

Dewaele, J., Chen, X., Padilla, A., & Lake, J. (2019). The flowering of positive psychology in foreign language teaching and acquisition research. *Frontiers in Psychology*, *10*(SEP), 1–13. DOI: 10.3389/fpsyg.2019.02128 PMID: 31607981

Dhaka Tribune. (2018, December 7). *Digital Bangladesh a reality now*. https://archive.dhakatribune.com/bangladesh/2018/07/11/digital-bangladesh-a-reality-now

Dillon, T., & Wells, D. (2021). Student perceptions of mobile automated speech recognition for pronunciation study and testing. *English Teaching*, *76*(4), 101–122. DOI: 10.15858/engtea.76.4.202112.101

Ding, Y., & Zhu, J. (2025). Mobile based peer feedback in EFL speaking: Learners' motivation, behavioral engagement in feedback provision, and achievement. *Asian. J. Second. Foreign. Language and Education*, *10*(1), 10. DOI: 10.1186/s40862-024-00314-9

Dolezal, D., Motschnig, R., & Ambros, R. (2025). Pre-Service Teachers' Digital Competence: A Call for Action. *Education Sciences*, *15*(2), 160. DOI: 10.3390/educsci15020160

Dombo, E., & Sabatino, C. A. (2019). Trauma and its sequelae in children and adolescents. In *Creating trauma-informed schools: A guide to school social workers and educators*. Oxford University Press, Incorporated. DOI: 10.1093/oso/9780190873806.003.0002

Dong, W., Pan, D., & Kim, S. (2024). Exploring the integration of IoT and Generative AI in English language education: Smart tools for personalized learning experiences. *Journal of Computational Science*, *82*, 102397. DOI: 10.1016/j.jocs.2024.102397

Dörnyei, Z., & Henry, A. (2022). Accounting for long-term motivation and sustained motivated learning: Motivational currents, self-concordant vision, and persistence in language learning. In *Advances in motivation science* (Vol. 9, pp. 89–134). Elsevier.

Dougherty, E., & Coelho, D. (2017). Eportfolios in English language learning: perceptions of Arabic-speaking higher education students. *TESL-EJ, 21*(3), 1–21. http://www.tesl-ej.org/wordpress/issues/volume21/ej83/ej83int/

Doyle, T. (2023). *Helping students learn in a learner-centered environment: A guide to facilitating learning in higher education*. Taylor & Francis. DOI: 10.4324/9781003445067

Duman, G., Orhon, G., & Gedik, N. (2015). Research trends in mobile assisted language learning from 2000 to 2012. *ReCALL*, *27*(2), 197–216. DOI: 10.1017/S0958344014000287

Dunne, C. (2024). Design strategies and dyslexia: Improving the accessibility of course material for third-level students with dyslexia (Doctoral dissertation, Institute of Art, Design+ Technology).

Durgungoz, F. C., & Durgungoz, A. (2025). "Interactive lessons are great, but too much is too much": Hearing out neurodivergent students, Universal Design for Learning and the case for integrating more anonymous technology in higher education. *Higher Education*. Advance online publication. DOI: 10.1007/s10734-024-01389-6

Duy, P. T., Khoa, N. H., Quyen, N. H., Trinh, L. C., Kien, V. T., Hoang, T. M., & Pham, V. H. (2025). Vulnsense: Efficient vulnerability detection in ethereum smart contracts by multimodal learning with graph neural network and language model. *International Journal of Information Security*, *24*(1), 48. DOI: 10.1007/s10207-024-00965-2

Ebbinghaus, H. (2013). [image] Memory: A Contribution to Experimental Psychology. *Annals of Neurosciences*, *20*(4), 155. DOI: 10.5214/ans.0972.7531.200408 PMID: 25206041

Efe, H. (2016). Reflective portfolio assessment in an EFL context. *Anthropologist*, *24*(1), 157–163. DOI: 10.1080/09720073.2016.11892001

EFTHYMIA-AGAPI. K. A. R. A. T. H. A. N. O. U. (2024). UNLOCKING VOCABULARY ACQUISITION: THEORETICAL FOUNDATIONS AND PRACTICAL APPROACHES FOR AUTISTIC EFL STUDENTS (Doctoral dissertation, Aristotle University of Thessaloniki).

Ehrlich, S., & Avery, P. (2013). *Teaching American English pronunciation*. Oxford University Press.

El Zahraa, F. (2025). Leveraging Artificial Intelligence and Digital Technologies to Enhance Sociolinguistic Competence and Arabic Language Skills. In *Proceeding of the International Conference on Religious Education and Cross-Cultural Understanding* (Vol. 1, No. 1, pp. 33-49).

Ellis, R. (2009). Corrective feedback and teacher development. *L2 Journal: An electronic refereed journal for foreign and second language educators, 1*(1). https://doi.org/DOI: 10.5070/L2.V1I1.9054

Elola, I., & Oskoz, A. (2016). Supporting second language writing using multimodal feedback. *Foreign Language Annals*, *49*(1), 58–74. DOI: 10.1111/flan.12183

Elwarraki, O., Aammou, S., Jdidou, Y., & Lahiassi, J. (2023). Toward the future of personalized learning: Emerging trends and challenges in recommendation systems. ICERI2023 Proceedings, 7226-7235.

Erbas, Y., Ceulemans, E., Lee Pe, M., Koval, P., & Kuppens, P. (2014). Negative emotion differentiation: Its personality and well-being correlates and a comparison of different assessment methods. *Cognition and Emotion*, *28*(7), 1196–1213. DOI: 10.1080/02699931.2013.875890 PMID: 24410047

Evurulobi, C. I., Dagunduro, A. O., & Ajuwon, O. A.Chinasa Iroabughichi Evurulo-biAdebukola Olufunke DagunduroOlanike Abiola Ajuwon. (2024). Language learning technologies: A review of trends in the USA and globally. *World Journal of Advanced Research and Reviews*, *23*(03), 2697–2707. DOI: 10.30574/wjarr.2024.23.3.2851

Fadli, S., & El Mediouni, A. (2024). Digital Transformation in Higher Education: Moroccan University Students' Perceptions and Expectations on Learning Analytics Implementation for Learning Optimization. In Adoui, A. (Ed.), *Internationalization of Higher Education and Digital Transformation*. Palgrave Macmillan., DOI: 10.1007/978-3-031-76444-8_14

Fakhrurriana, R., Nisa, A., & Noni, N. (2024). The perceptions of using Duolingo application in learning English for students' vocabulary mastery. *BATARA DIDI: English Language Journal*, *3*(1), 45–54. DOI: 10.56209/badi.v3i1.82

Familoni, B., & Onyebuchi, N.Babajide Tolulope FamiloniNneamaka Chisom Onyebuchi. (2024). Augmented and virtual reality in U.S. education: A review: Analyzing the impact, effectiveness, and future prospects of AR/VR tools in enhancing learning experiences. *International Journal of Applied Research in Social Sciences*, *6*(4), 1043. Advance online publication. DOI: 10.51594/ijarss.v6i4.1043

Fan, M., Antle, A., & Warren, J. L. (2020). Augmented reality for early language learning: A systematic review of augmented reality application design, instructional strategies, and evaluation outcomes. *Journal of Educational Computing Research*, *58*(6), 1059–1100. DOI: 10.1177/0735633120927489

Fauzi, M. F., & Anindiati, I. (2021). Improving the motivation of students in Arabic language learning through learning management system. *Izdihar: Journal of Arabic Language Teaching, Linguistics, and Literature*, *4*(3), 257–274. DOI: 10.22219/jiz.v4i3.19110

Ferguson-Sams, N., Howell, E., Kaminski, R., Gazioglu, M., & Pennington, V. (2025). Multilingual Voices: Transforming Professional Development Based on Student Perceptions of Literacy Learning. tesol journal, 16(1), e70005.

Figueroa, R. B.Jr, & Jung, I. (2025). Integrating VR tours in online language learning: A design-based research study. *Australasian Journal of Educational Technology*, *41*(1), 70. DOI: 10.14742/ajet.9498

Fijačko, N., Masterson Creber, R., Gosak, L., Štiglic, G., Egan, D., Chaka, B., Debeljak, N., Strnad, M., & Skok, P. (2021). Evaluating quality, usability, evidence-based content, and gamification features in mobile learning apps designed to teach children basic life support: Systematic search in app stores and content analysis. *JMIR mHealth and uHealth*, *9*(7), e25437. DOI: 10.2196/25437 PMID: 34283034

Filiz, O., Kaya, M. H., & Adiguzel, T. (2025). Teachers and AI: Understanding the factors influencing AI integration in K-12 education. *Education and Information Technologies*, *30*(13), 1–37. DOI: 10.1007/s10639-025-13463-2

Firdausia, W., & Hamdani, B. (2025). Integrating artificial intelligence in English learning: Implementation of Duolingo application to improve English skills. *Jurnal Kependidikan*, *13*(1), 177–188. DOI: 10.24090/jk.v13i1.13339

Floor, N. (2023). *This is Learning Experience Design: What it is, how it works, and why it matters*. New Riders.

Flores, J., Huffman, L., Lozada, V., & Dávila, E. (2022). Adopting VR in the classroom: Perceptions of bilingual and ESL education pre-service teachers. *EDMETIC*, *11*(1), 13610. DOI: 10.21071/edmetic.v11i1.13610

Forbes, H., & Maki, D. (2020). *Classroom180: A framework for creating, sustaining, and assessing the trauma-informed classroom*. Beyond Consequences Institute.

Ford, D. R. (2023). Arrhythmanalysis and revolutionary rupture of détournement: Lefebvre and the pedagogical rhythms of post-fordism. *Postdigital Science and Education*, *5*(2), 440–454. DOI: 10.1007/s42438-021-00285-x

Fortunato, M., Moreira, A., & Simões, A. R. (2023). 'All hands-on deck': Octalysis framework in primary English language learning. *Sisyphus—. Journal of Education*, *10*, 225–243. DOI: 10.25749/sis.26985

Fraser, H. (2000). *Coordinating improvements in pronunciation teaching for adult learners of English as a second language*. DETYA (ANTA Innovative project).

Fullan, M. (1992). *Successful school improvement: The implementation perspective and beyond*. McGraw-Hill Education.

Fune, A. (2024). *A Study of Quizizz's User Experience (UX)*. Measuring the Impact of Gamification and Usability on Language Acquisition.

Gamage, S. H. P. W., Ayres, J. R., & Behrend, M. B. (2022). A systematic review on trends in using Moodle for teaching and learning. *International Journal of STEM Education*, *9*(1), 9. DOI: 10.1186/s40594-021-00323-x PMID: 35096513

Garzón, J. (2021). An overview of twenty-five years of augmented reality in education. In *Multimodal Technologies and Interaction* (Vol. 5, Issue 7). DOI: 10.3390/mti5070037

Geisen, M., Fox, A., & Klatt, S. (2023). VR as an Innovative Learning Tool in Sports Education. *Applied Sciences 2023, Vol. 13, Page 2239, 13*(4), 2239. DOI: 10.3390/app13042239

George, A. S. (2024). Technology Tension in Schools: Addressing the Complex Impacts of Digital Advances on Teaching, Learning, and Wellbeing. *Partners Universal Multidisciplinary Research Journal, 1*(3), 49–65.

Georgi, D., & Crowe, J. (1998). Digital portfolios: A confluence of portfolio assessment and technology. *Teacher Education Quarterly, 25*(1), 73–84.

Ghomi, M., & Redecker, C. (2019). Digital Competence of Educators (DigCompEdu): Development and Evaluation of a Self-assessment Instrument for Teachers' Digital Competence. *In Proceedings of the 11th International Conference on Computer Supported Education (CSEDU 2019), Pages* 541-548. DOI: 10.5220/0007679005410548

Ghosheh Wahbeh, D., Najjar, E. A., Sartawi, A. F., Abuzant, M., & Daher, W. (2021). The role of project-based language learning in developing students' life skills. *Sustainability (Basel), 13*(12), 6518. Advance online publication. DOI: 10.3390/su13126518

Gibson, S., & Oberg, D. (2004). Visions and realities of Internet use in schools: Canadian perspectives. *British Journal of Educational Technology, 35*(5), 569–585. DOI: 10.1111/j.0007-1013.2004.00414.x

Gigerenzer, G. (2022). *How to stay smart in a smart world: Why human intelligence still beats algorithms*. MIT Press.

Godwin-Jones, R. (2011). Mobile apps for language learning. *Language Learning & Technology, 15*(2), 2–11. https://scholarspace.manoa.hawaii.edu/bitstream/10125/44244/1/15_02_emerging.pdf. DOI: 10.64152/10125/44244

Godwin-Jones, R. (2016). Augmented reality and language learning: From annotated vocabulary to place-based mobile games. *Language Learning & Technology, 20*(3), 9–19. DOI: 10.64152/10125/44475

Goh, C. C. M., & Burns, A. (2012). *Teaching speaking: A holistic approach*. Cambridge University Press.

Goi, C. L. (2018). The use of business simulation games in teaching and learning. *Journal of Education for Business, 94*(5), 342–349. DOI: 10.1080/08832323.2018.1536028

Gómez-Carrasco, C. J., Monteagudo-Fernández, J., Moreno-Vera, J. R., & Sainz-Gómez, M. (2020). Evaluation of a gamification and flipped-classroom program used in teacher training: Perception of learning and outcome. *PLoS One*, *15*(7), e0236083. DOI: 10.1371/journal.pone.0236083 PMID: 32673373

Goodwin, A. L., Madalińska-Michalak, J., & Flores, M. A. (2023). Rethinking teacher education in/for challenging times: Reconciling enduring tensions, imagining new possibilities. *European Journal of Teacher Education*, *46*(5), 840–855. DOI: 10.1080/02619768.2023.2299926

Goodwin, B., Rouleau, K., Abla, C., Baptiste, K., Gibson, T., & Kimball, M. (2022). *The new classroom instruction that works: The best research-based strategies for increasing student achievement*. ASCD.

Gorbunova, N., Boginskaia, Y., & Khrulyova, A. (2025). Analysis of the Benefits of Digital Services in Inclusive Higher Education. In: Lapina, M., Prakasha, G.S., Grigoriev, S. (eds) International Conference on Innovative Approaches to the Application of Digital Technologies in Education and Research. SLET 2022. Lecture Notes in Networks and Systems, vol 1222. Springer, Cham. https://doi.org/DOI: 10.1007/978-3-031-78776-8_48

Gouiza, N. I. S. S. R. I. N. E., Jebari, H. A. K. I. M., & Reklaoui, K. A. M. A. L. (2024). Integration Of Iot-Enabled Technologies and Artificial Intelligence in Diverse Domains: Recent Advancements And Future Trends. *Journal of Theoretical and Applied Information Technology*, *102*(5), 1975–2029.

Graham, C. R., Woodfield, W., & Harrison, J. B. (2013). A framework for institutional adoption and implementation of blended learning in higher education. *The Internet and Higher Education*, *18*, 4–14. DOI: 10.1016/j.iheduc.2012.09.003

Grant, L., & Brinton, D. (2014). *Pronunciation myths: Applying second language research to classroom teaching*. University of Michigan Press. DOI: 10.3998/mpub.4584330

Greene, J. C., Caracelli, V. J., & Graham, W. F. (1989). Toward a conceptual framework for mixed-method evaluation designs. *Educational Evaluation and Policy Analysis*, *11*(3), 255–274. DOI: 10.3102/01623737011003255

Habib, S., Haider, A., Suleman, S. S. M., Akmal, S., & Khan, M. A. (2022). Mobile Assisted Language Learning: Evaluation of Accessibility, Adoption, and Perceived Outcome among Students of Higher Education. *Electronics (Basel)*, *11*(7), 1113. DOI: 10.3390/electronics11071113

Hadi Mogavi, R., Guo, B., Zhang, Y., Haq, E. U., Hui, P., & Ma, X. (2022, June). When gamification spoils your learning: A qualitative case study of gamification misuse in a language-learning app. In Proceedings of the Ninth ACM Conference on Learning@ Scale (pp. 175-188). DOI: 10.1145/3491140.3528274

Haider, M. (2012). Promoting CLT within a computer assisted learning environment. *English Language Teaching*, *5*(8), 91–102. DOI: 10.5539/elt.v5n8p91

Hamad, A., & Jia, B. (2022). How Virtual Reality Technology Has Changed Our Lives: An Overview of the Current and Potential Applications and Limitations. In *International Journal of Environmental Research and Public Health* (Vol. 19, Issue 18). DOI: 10.3390/ijerph191811278

Hamad, D., Dafaallah, A. S., & Alhaj, A. A. (2021). The challenges of online English language teaching from EFL instructors' point of view in KKU, Tehama campus during COVID-19 period. *Journal of Language Teaching and Research*, *12*(4), 543–556. DOI: 10.17507/jltr.1204.04

Hamlin, M. D. (2022). Developing self-directed learning skills for lifelong learning. In *Self-directed learning and the academic evolution from pedagogy to andragogy* (pp. 209–234). IGI Global Scientific Publishing. DOI: 10.4018/978-1-7998-7661-8.ch012

Hampel, R. (2009). Training teachers for the multimedia age: Developing teacher expertise to enhance online learner interaction and collaboration. *Innovation in Language Learning and Teaching*, *3*(1), 35–50. DOI: 10.1080/17501220802655425

Handini, B. S., Nurhasanah, N., & Panly, F. I. (2022). The effect of artificial intelligent technology used (Duolingo application) to enhance English learning. ELLITE. *Journal of English Language, Literature, and Teaching*, *7*(2), 86–94. DOI: 10.32528/ellite.v7i2.8354

Hannon, M. (2017). A solution to knowledge's threshold problem. *Philosophical Studies*, *174*(3), 607–629. DOI: 10.1007/s11098-016-0700-9

Hanson-Smith, E. (2016). Games, gaming, and gamification: Some aspects of motivation. *TESOL Journal*, *7*(2), 227–232. DOI: 10.1002/tesj.233

Haoming, L., & Wei, W. (2024). A systematic review on vocabulary learning in AR and VR gamification context. *Computers & Education: X Reality*, *4*, 100057. DOI: 10.1016/j.cexr.2024.100057

Harmer, J. (2007). *The practice of English language teaching* (4th ed.). Longman.

Hasan, M. K., Fakih, A. H., Ibna Seraj, P. M., & Hasmirati, . (2022). The effect of technology-assisted language programme on vocabulary learning among EFL students at the tertiary level. *Heliyon*, *8*(8), e10313. DOI: 10.1016/j.heliyon.2022.e10313 PMID: 36051275

Hasan, M. S. H. K., Mahbub, & Khan, M. S. H. (2013). Introducing ICT into teacher-training programs: Problems in Bangladesh. *Journal of Education and Practice*, *4*(14), 79–85.

Hase, S., & Blaschke, L. M. (2021). *Unleashing the power of learner agency*. Unleashing the Power of Learner Agency.

Hassan Taj, I., Ali, F., Aslam Sipra, M., & Ahmad, W. (2017). Effect of technology-enhanced language learning on vocabulary acquisition of EFL learners. *International Journal of Applied Linguistics and English Literature*, *6*(3), 262. DOI: 10.7575/aiac.ijalel.v.6n.3p.262

Hayes, A. T., Dhimolea, T. K., Meng, N., & Tesh, G. (2021). Levels of Immersion for Language Learning from 2D to Highly Immersive Interactive VR. In *Chinese Language Learning Sciences*. DOI: 10.1007/978-981-16-3416-1_4

Heidari-Shahreza, M. A. (2024). When language is play: Looking into linguistic creativity through language play. Creativity Journal: Theories. *Research & Applications*, *11*(1), 18–35.

Higher Council of Education, Training and Scientific Research. (2015). *For a school of equity, quality and promotion: A strategic vision of reform 2015–2030*. https://www.csefrs.ma/wp-content/uploads/2015/05/Re%CC%81sume%CC%81-vision-Anglais.pdf

Hirsh-Pasek, K., Zosh, J. M., Golinkoff, R. M., Gray, J. H., Robb, M. B., & Kaufman, J. (2015). Putting education in "educational" apps: Lessons from the science of learning. *Psychological Science in the Public Interest*, *16*(1), 3–34. DOI: 10.1177/1529100615569721 PMID: 25985468

Hockly, N. (2015). Developments in online language learning. *ELT Journal*, *69*(3), 308–313. DOI: 10.1093/elt/ccv020

Holovatenko, T. (2023). Implementation of the principles of a trauma-informed environment in English lessons in elementary grades: a meta-analysis of foreign practices. *Pedagogical Education: Theory and Practice. Psychology. Pedagogy*, *39*(1), 80–86. DOI: 10.28925/2311-2409.2023.3912

Holozsai, J., & József, H. (2024). Aspects of digital learning among foreign language learners in secondary education.

Hossain, K. I. (2024). Reviewing the role of culture in English language learning: Challenges and opportunities for educators. *Social Sciences & Humanities Open*, *9*, 100781. DOI: 10.1016/j.ssaho.2023.100781

Hrastinski, S. (2021). Digital tools to support teacher professional development in lesson studies: A systematic literature review. *International Journal for Lesson & Learning Studies*, *10*(2), 138–149. DOI: 10.1108/IJLLS-09-2020-0062

Hsu, H. C. (2016). Voice blogging and L2 speaking performance. *Computer Assisted Language Learning*, *29*(5), 968–983. DOI: 10.1080/09588221.2015.1113185

Huang, H.-T. D., & Hung, S.-T. A. (2010). Implementing electronic speaking portfolios: Perceptions of EFL students: Colloquium. *British Journal of Educational Technology*, *41*(5). Advance online publication. DOI: 10.1111/j.1467-8535.2009.00996.x

Huang, X., Zou, D., Cheng, G., & Xie, H. (2021). A systematic review of AR and VR enhanced language learning. *Sustainability (Basel)*, *13*(9), 4639. DOI: 10.3390/su13094639

Hung, S.-T. A. (2011). Pedagogical applications of Vlogs: An investigation into ESP learners perceptions. *British Journal of Educational Technology*, *42*(5), 736–746. DOI: 10.1111/j.1467-8535.2010.01086.x

Hung, S.-T. A., & Huang, H.-T. D. (2015). Video blogging and English presentation performance: A pilot study. *Psychological Reports*, *117*(2), 614–630. DOI: 10.2466/11.PR0.117c20z6 PMID: 26444840

Hwang, G. J., & Tsai, C. C. (2011). Research Trends in Mobile and Ubiquitous Learning: A Review of Publications in Selected Journals from 2001 to 2010. *British Journal of Educational Technology*, *42*(4), E65–E70. DOI: 10.1111/j.1467-8535.2011.01183.x

Ifedayo, A. E., Olugbade, D., & Hamid, S. (2025). Integrating Artificial Intelligence with Blockchain: A Literature Review on Opportunities, Challenges, and Applications. *Blockchain, Artificial Intelligence, and Future Research*, *1*(1), 52–69. DOI: 10.70211/bafr.v1i1.179

Iftikhar, N. (2025). MOBILE-ASSISTED LANGUAGE LEARNING (MALL): REVOLUTIONIZING SECOND LANGUAGE ACQUISITION. [JALT]. *Journal of Applied Linguistics and TESOL*, *8*(1), 1038–1045.

Ihmeideh, F. M. (2009). Barriers to the use of technology in Jordanian pre-school settings. *Technology, Pedagogy and Education*, *18*(3), 325–341. DOI: 10.1080/14759390903255619

Ihnatova, O., Poseletska, K., Matiiuk, D., Hapchuk, Y., & Borovska, O. (2021). The application of digital technologies in teaching a foreign language in a blended learning environment. Linguistics and Culture Review, 114-127.

Ikeda, S., Villagran, C. T., Fukuda, T., Okada, Y., Arai, F., Negoro, M., Hayakawa, M., & Takahashi, I. (2008). Patient-Specific IVR Endovascular Simulator with Augmented Reality for Medical Training and Robot Evaluation. *Journal of Robotics and Mechatronics*, *20*(3), 441–448. Advance online publication. DOI: 10.20965/jrm.2008.p0441

Ilinawati, Sijono, & Elisa, H. (2021). The study of the students' speaking performance in demonstrating product advertising. Journal of English Education E-ISSN, 4(1), 59–70.

Illés, É., & Akcan, S. (2017). Bringing real-life language use into EFL classrooms. *ELT Journal*, *71*(1), 3–12. Advance online publication. DOI: 10.1093/elt/ccw049

Inayati, N., & Waloyo, A. A. (2022). The influence of Quizziz-online gamification on learning engagement and outcomes in online English language teaching. *Journal of English as a Foreign Language*, *12*(2), 249–271. DOI: 10.23971/jefl.v12i2.3546

Irzawati, I. (2023). The integration of Duolingo into EFL learning. *Esteem Journal of English Education Study Programme*, *6*(2), 328–337. DOI: 10.31851/esteem.v6i2.12317

Ismaili, Y. (2024). AI Literacy and Adaptive Learning in Moroccan Education: Advancing Critical Thinking and Personalized Learning. *Research on Education and Media*, *16*(2), n2. DOI: 10.2478/rem-2024-0017

Issa, D. A. (2020). *Morocco - Financial and Digital Inclusion Development Policy Financing Project: Appraisal Program Information Document (PID) - Financial and Digital Inclusion Development Policy Financing - P171587 (English)*. World Bank Group. Retrieved from https://documents.worldbank.org/curated/en/690571587680057013

Jaber Rafidi, T., & Wagner, B. (2023). The impact of emotional scaffolding on online teaching and learning in Palestinian English language classrooms during COVID-19. *Innovation in Language Learning and Teaching*, 1–14. DOI: 10.1080/17501229.2023.2245810

Jang, B., Yunus, M., & Hashim, H. (2024). Virtual reality (VR) and its acceptance among English language teachers. *International Journal of Academic Research in Progressive Education and Development*, *13*(3), 21546. DOI: 10.6007/IJARPED/v13-i3/21546

Jantankul, W. (2024). Immersive reality in education: Transforming teaching and learning through AR, VR, and mixed reality technologies. *Journal of Education and Learning Reviews* . https://doi.org/DOI: 10.60027/jelr.2024.750

Javed, S., Zehra, S. R., Ullah, H., & Naveed, M. (2025). How AI can detect Emotional cues in Students, Improving Virtual Learning Environments by providing Personalized support and Enhancing Social-emotional Learning. *Review of Applied Management and Social Sciences*, *8*(2), 605–622. DOI: 10.47067/ramss.v8i2.500

Jenkins, J. (2000). *The phonology of English as an international language*. Oxford university press.

Jiang, X., & Pajak, B. (2022). *Reading and listening outcomes of learners in the Duolingo English course for Spanish speakers* (Duolingo Research Report DRR-22-08).

Jiang, X., Peters, R., Plonsky, L., & Pajak, B. (2024). The effectiveness of Duolingo English courses in developing reading and listening proficiency. *CALICO Journal*, *41*(3), 249–272. DOI: 10.1558/cj.26704

Jiménez-Valverde, G., Fabre-Mitjans, N., Heras-Paniagua, C., & Guimerà-Ballesta, G. (2025). Tailoring Gamification in a Science Course to Enhance Intrinsic Motivation in Preservice Primary Teachers. *Education Sciences*, *15*(3), 300. DOI: 10.3390/educsci15030300

Jirgensons, M., & Kapenieks, J. (2018). Blockchain and the future of digital learning credential assessment and management. Journal of teacher education for sustainability, 20(1), 145-156.

Johnson, D. W., & Johnson, R. T. (2002). *Meaningful assessment: A manageable and cooperative process*. Allyn and Bacon.

Johnson, L., Becker, S. A., Cummins, M., Estrada, V., Freeman, A., & Hall, C. (2016). *NMC horizon report: 2016 higher education edition*. The New Media Consortium.

Jones, H., & Sudlow, B. (2022). A contemporary history of Silicon Valley as global heterotopia: Silicon Valley metaphors in the French news media. *Globalizations*, *19*(7), 1122–1136. DOI: 10.1080/14747731.2022.2034308

Jones, T. (2018). Pronunciation with other areas of language. In Kang, O., Thomson, R. I., & Murphy, J. M. (Eds.), *The Routledge Handbook of Contemporary English Pronunciation* (pp. 370–384). Routledge.

Jung, I. (2024). Personalized education for all: The future of open universities. *Open Praxis*, *16*(1), 24–36. DOI: 10.55982/openpraxis.16.1.612

Kabel, A. (2021). 'The tide is coming in fast': Ideologies of English, global linguistic coloniality and decolonial pluriversalingualism. In Tupas, R., & Rubdy, R. (Eds.), *World Englishes: Ideologies* (pp. 43–63). Bloomsbury.

Kalisz, D. E. (2024). Beyond Education: Scenarios of the Re-Imagined Higher Education from the Future.

Kalyuga, S. (2009). *Cognitive load factors in instructional design for advanced learners*. Nova Science Publishers.

Kamińska, D., Zwoliński, G., Laska-Leśniewicz, A., Raposo, R., Vairinhos, M., Pereira, E., Urem, F., Ljubić Hinić, M., Haamer, R. E., & Anbarjafari, G. (2023). Augmented Reality: Current and New Trends in Education. In *Electronics (Switzerland)* (Vol. 12, Issue 16). DOI: 10.3390/electronics12163531

Kang, O., Thomson, R. I., & Murphy, J. (2018). *The Routledge handbook of contemporary English pronunciation*. Routledge.

Kaplan-Rakowski, R., & Wojdyński, T. (2018). Students' attitudes toward high-immersion virtual reality assisted language learning. In *Proceedings of the 26th International Conference on Computers in Education* (pp. 824–829). Asia-Pacific Society for Computers in Education. https://doi.org/DOI: 10.14705/rpnet.2018.26.824

Karakaya, K., & Bozkurt, A. (2022). Mobile-assisted language learning (MALL) research trends and patterns through bibliometric analysis: Empowering language learners through ubiquitous educational technologies. *System*, *110*, 102925. DOI: 10.1016/j.system.2022.102925

Karim, S. A., Hamzah, A. Q. S., Anjani, N. M., Prianti, J., & Sihole, I. G. (2023). Promoting EFL students' speaking performance through ELSA Speak: An artificial intelligence in English language learning. *Journal of Languages and Language Teaching*, *11*(4), 655–668. DOI: 10.33394/jollt.v11i4.8958

Katona, J., & Gyonyoru, K. I. K. (2025). Integrating AI-Based Adaptive Learning into the Flipped Classroom Model to Enhance Engagement and Learning Outcomes. *Computers and Education Artificial Intelligence*, *100392*, 100392. Advance online publication. DOI: 10.1016/j.caeai.2025.100392

Kaufmann, H., & Dünser, A. (2007). Summary of usability evaluations of an educational augmented reality application. In *International Conference on Virtual Reality* (pp. 660–669). Springer. DOI: 10.1007/978-3-540-73335-5_71

Kayyali, M. (2021). Positive impact of high technology on higher education.

Kayyali, M. (2024 a). Global Perspectives on Personalized Learning: Case Studies and Best Practices. In Transforming Education for Personalized Learning (pp. 66-84). IGI Global.

Kayyali, M. (2024 b). Immersive Technologies: Virtual and Augmented Reality in Higher Education. In Reshaping learning with next generation educational technologies (pp. 99-114). IGI Global.

Kayyali, M. (2024 d). Inclusive Practices in Higher Education and Workplaces Strategies for Empowering Women. In A Cross-Cultural Examination of Women in Higher Education and the Workplace (pp. 237-262). IGI Global. DOI: 10.4018/979-8-3693-0102-9.ch012

Kayyali, M. (2025 a). E-Learning Platforms and Their Influence on Lifelong Learning. In Impact of Digitalization on Education and Social Sustainability (pp. 155-176). IGI Global.

Kayyali, M. (2025 b). E-Learning Platforms and Their Influence on Lifelong Learning. In Impact of Digitalization on Education and Social Sustainability (pp. 155-176). IGI Global.

Kayyali, M. (2025 b). Transforming Education: The Future of Mastery-Based Learning and Its Impact. In Cultivating Flourishing Practices and Environments by Embracing Positive Education (pp. 119-148). IGI Global Scientific Publishing.

Kayyali, M. (2025 c). Transforming Education: The Future of Mastery-Based Learning and Its Impact. In Cultivating Flourishing Practices and Environments by Embracing Positive Education (pp. 119-148). IGI Global Scientific Publishing.

Kayyali, M. (2025 d). Classroom Management: Techniques for Creating a Positive, Friendly Learning Environment. In Exploration of K-12 Teaching and Learning for Teacher Educators (pp. 53-78). IGI Global Scientific Publishing.

Kayyali, M. (2025). Classroom Management: Techniques for Creating a Positive, Friendly Learning Environment. In Exploration of K-12 Teaching and Learning for Teacher Educators (pp. 53-78). IGI Global Scientific Publishing.

Kayyali, M. (2024 e). Future possibilities and challenges of AI in education. In *Transforming education with generative AI: Prompt engineering and synthetic content creation* (pp. 118–137). IGI Global Scientific Publishing. DOI: 10.4018/979-8-3693-1351-0.ch006

Kayyali, M. (2024). Future directions towards sustainable educational equity: Policies and practices. In *Exploring Educational Equity at the Intersection of Policy and Practice* (pp. 81–100). IGI Global. DOI: 10.4018/979-8-3693-1614-6.ch005

Kayyali, M. (2025 b). The Impact of Personalized Learning on Student Engagement and Achievement in STEAM. In *Integrating Personalized Learning Methods Into STEAM Education* (pp. 51–78). IGI Global Scientific Publishing. DOI: 10.4018/979-8-3693-7718-5.ch003

Kearney, M., & Maher, D. (2019). Mobile learning in pre-service teacher education: Examining the use of professional learning networks. *Australasian Journal of Educational Technology*, *35*(1). Advance online publication. DOI: 10.14742/ajet.4073

Kerawalla, L., Luckin, R., Seljeflot, S., & Woolard, A. (2006). "Making it real": Exploring the potential of augmented reality for teaching primary school science. *Virtual Reality (Waltham Cross)*, *10*(3-4), 163–174. DOI: 10.1007/s10055-006-0036-4

Kessler, G. (2018). Technology and the future of language teaching. *Foreign Language Annals*, *51*(1), 205–218. DOI: 10.1111/flan.12318

Khalid, A. M., & Owusu-Boateng, O. (2024). *Navigating through Technology in Modern Education*. Bentham Science Publishers. DOI: 10.2174/97898152387851240101

Khan, M. S. H., Hasan, M., & Clement, C. K. (2012). Barriers to the introduction of ICT into education in developing countries: The example of Bangladesh. *International Journal of Instruction*, *5*(2), 61–80.

Khan, N. M., & Kuddus, K. (2020). Integrating ICT in English language teaching in Bangladesh. *Rupkatha Journal on Interdisciplinary Studies in Humanities*, *12*(5), 1–10. DOI: 10.21659/rupkatha.v12n5.rioc1s23n1

Kim, H., & Kwon, Y. (2012). Exploring smartphone applications for effective mobile-assisted language learning. *Multimedia-Assisted Language Learning*, *15*(1), 31–57. DOI: 10.15702/mall.2012.15.1.31

Kinshuk, (2016). Designing Adaptive and Personalized Learning Environments (1st ed.). Routledge. DOI: 10.4324/9781315795492

Kissling, E. M. (2015). Phonetics instruction improves learners' perception of L2 sounds. *Language Teaching Research*, *19*(3), 254–275. DOI: 10.1177/1362168814541735

Klašnja-Milićević, A., & Ivanović, M. (2021). E-learning personalization systems and sustainable education. *Sustainability (Basel)*, *13*(12), 6713. DOI: 10.3390/su13126713

Klaus, D. (2022). *Implementing a trauma-informed compassionate classroom to increase student engagement and academic achievement*. ProQuest LLC.

Klingenberg, S., Fischer, R., Zettler, I., & Makransky, G. (2023). Facilitating learning in immersive virtual reality: Segmentation, summarizing, both or none? *Journal of Computer Assisted Learning*, *39*(1), 218–230. DOI: 10.1111/jcal.12741

Koehler, M. J., Mishra, P., & Cain, W. (2017). What is Technological Pedagogical Content Knowledge (TPACK)? Journal of Education, 193(3), 13-19. (Original work published 2013)DOI: 10.1177/002205741319300303

Kökver, Y., Pektaş, H. M., & Çelik, H. (2025). Artificial intelligence applications in education: Natural language processing in detecting misconceptions. *Education and Information Technologies*, *30*(3), 3035–3066. DOI: 10.1007/s10639-024-12919-1

Kot, S. O., & Nykyporets, S. S. (2024). *Utilization of artificial intelligence in enhancing English language proficiency in tertiary education. Science and Education in the Third Millennium: Information Technology*. Education, Law, Psychology, Social Sphere, Management.

Kovalchuk, O. I., Bondarenko, M. P., Okhrey, A. G., Prybytko, I. Y., & Reshetnyk, E. M. (2020). Features of using immersive technologies (virtual and augumented reality) in medical education and practice. *Morfolohiia*, *14*(3), 158–164. Advance online publication. DOI: 10.26641/1997-9665.2020.3.158-164

Kovalenko, I., & Baranivska, N. (2024). Integrating Artificial Intelligence in English Language Teaching: Exploring the potential and challenges of AI tools in enhancing language learning outcomes and personalized education. Європейські соціо-правові та гуманітарні студії, (1), 86-95.

Kramsch, C., & Vork Steffensen, S. (2008). Ecological Perspectives on Second Language Acquisition and Socialization. In Hornberger, N. H. (Ed.), *Encyclopedia of Language and Education*. Springer., DOI: 10.1007/978-0-387-30424-3_194

Krashen, S. D. (1995). *Principles and practice in second language acquisition*. Phoenix ELT incorporating Prentice Hall Macmillan.

Krengel, M. E. (2021). *Instructor Perspectives of Digital Natives in Face-to-Face Classrooms: A Qualitative Descriptive Study*. Northcentral University.

Kuddus, K. (2018). Emerging technologies and the evolving roles of language teachers. *Language in India*, *18*(6), 81–86.

Kudo, M. F., Lussier, C. M., & Swanson, H. L. (2015). Reading disabilities in children: A selective meta-analysis of the cognitive literature. *Research in Developmental Disabilities*, *40*, 51–62. DOI: 10.1016/j.ridd.2015.01.002 PMID: 25725389

Kukulska-Hulme, A., Norris, L., & Donohue, J. (2015). Mobile pedagogy for English language teaching: a guide for teachers. *British Council*. https://www.teachingenglish.org.uk/publications/case-studies-insights-and-research/mobile-pedagogy-english-language-teaching-guide

Kukulska-Hulme, A. (2005). Introduction. In Kukulska-Hulme, A., & Traxler, J. (Eds.), *Mobile Learning: A Handbook for Educators and Trainers* (pp. 1–7). Routledge.

Kukulska-Hulme, A. (2013). Mobile-assisted language learning. In Chapelle, C. (Ed.), *The encyclopedia of applied linguistics* (pp. 3701–3709). Wiley.

Kukulska-Hulme, A., & Shield, L. (2008). An overview of mobile assisted language learning: From content delivery to supported collaboration and interaction. *ReCALL*, *20*(3), 271–289. DOI: 10.1017/S0958344008000335

Kukulska-Hulme, A., & Viberg, O. (2018). Mobile collaborative language learning: State of the art. *British Journal of Educational Technology*, *49*(2), 207–218. DOI: 10.1111/bjet.12580

Kullman, S. M., & Chudyk, A. M. (2025). Participatory Member Checking: A Novel Approach for Engaging Participants in Co-Creating Qualitative Findings. *International Journal of Qualitative Methods*, *24*, 16094069251321211. Advance online publication. DOI: 10.1177/16094069251321211

Kumar, R. (2023). English language and digital literacy: Navigating the information age. Journal of International English Research Studies (JIERS), ISSN: 3048-5231, 1(2), 25-31.

Kumar, B. A., & Goundar, S. (2023). Developing mobile language learning applications: A systematic literature review. *Education and Information Technologies*, *28*(5), 5651–5671. DOI: 10.1007/s10639-022-11377-x

Kumar, P., Gupta, A., & Budarina, A. O. (Eds.). (2023). *Innovative Teaching Learning Practices: A Paradigm Shift*. GLOBAL ACADEMY YAYINCILIK VE DANIŞMANLIK HİZMETLERİ SANAYİ TİCARET LİMİTED ŞİRKETİ.

Kurbanazarova, N., Shavkidinova, D., Khaydarov, M., Mukhitdinova, N., Khudoymurodova, K., Toshniyozova, D., Karimov, N., & Alimova, R. (2024). Development of Speech Recognition in Wireless Mobile Networks for An Intelligent Learning System in Language Education. *Journal of Wireless Mobile Networks, Ubiquitous Computing and Dependable Applications*, *15*(3), 298–311. DOI: 10.58346/JOWUA.2024.I3.020

Kuru Gönen, S. İ., & Zeybek, G. (2022). Training on multimodal mobile-assisted language learning: A suggested model for pre-service EFL teachers*. *Computer Assisted Language Learning*, *37*(7), 2202–2223. DOI: 10.1080/09588221.2022.2157016

Kusuma, I. P. I., Pratiwi, N., Paramartha, A., & Utami, I. (2022). Can e-portfolios reduce foreign language anxiety in online speaking courses during the COVID-19 pandemic? *Computer-Assisted Language Learning Electronic Journal (CALL-EJ)*, *23*(2), 38–55.

Kusuma, I. P. I., Mahayanti, N. W. S., Adnyani, N. L. D., & Budiarta, L. G. R. (2021). Incorporating e-portfolios with flipped classrooms: An in-depth analysis of students' speaking performance and learning engagement. *The JALT CALL Journal*, *17*(2), 93–111. DOI: 10.29140/jaltcall.v17n2.378

Kusuma, I. P. I., Mahayanti, N. W. S., Gunawan, M. H., Rachman, D., & Pratiwi, N. P. A. (2021). How well do e-portfolios facilitate students' learning engagement in speaking courses during the COVID-19 pandemic? *Indonesian Journal of Applied Linguistics*, *11*(2), 351–363. DOI: 10.17509/ijal.v11i2.30583

Kusuma, I. P. I., & Waluyo, B. (2023). Enacting e-portfolios in online English-speaking courses: Speaking performance and self-efficacy. *Iranian Journal of Language Teaching Research*, *11*(1), 75–95. DOI: 10.30466/ijltr.2023.121273

Lablidi, A., Nachit, B., Abourriche, A., Namir, A., & Talbi, M. (2013). ICT Practice in Morocco's innovative teachers. *American Journal of Engineering Research (AJER)*. https://www.ajer.org/papers/v2(5)/T0251450150.pdf

Lai, A. F., Chen, C. H., & Lee, G. Y. (2019). An augmented reality-based learning approach to enhancing students' science reading performances from the perspective of the cognitive load theory. *British Journal of Educational Technology*, *50*(1), 232–247. DOI: 10.1111/bjet.12716

Lamb, M. (2017). The motivational dimension of language teaching. *Language Teaching*, *50*(3), 301–346. DOI: 10.1017/S0261444817000088

Langin, F. (2022). *Teaching french as a second language to primary students online: implementing a gamified routine system*. McGill University.

Lantolf, J. P. (2000). *Sociocultural theory and second language learning* (Vol. 78). Oxford university press.

Lantolf, J. P., & Pavlenko, A. (1995). Sociocultural theory and second language acquisition. *Annual Review of Applied Linguistics*, *15*, 108–124. DOI: 10.1017/S0267190500002646

Lantolf, J. P., & Thorne, S. L. (2006). *Sociocultural theory and the genesis of second language development*. Oxford University Press.

Lantolf, J., & Thorne, S. L. (2006). *Sociocultural Theory and the genesis of second language development*. Oxford University Press.

Lan, Y.-J. E. (2022). A comparative study of computer and mobile-assisted pronunciation training: The case of university students in Taiwan. *Education and Information Technologies*, *27*(2), 1559–1583. DOI: 10.1007/s10639-021-10647-4

Lăpădat, M. M. (2023). Harnessing technological transformation: An exploration of the impact of mobile assisted language learning on English vocabulary acquisition. Analele Universităţii din Craiova, Seria Ştiinţe Filologice. *Limbi Străine Aplicate*, (1), 266–279.

Lapka, O., & Shcherbak, A. F. (2025). *Innovative approaches to learning foreign languages: The impact of virtual reality on overcoming the language barrier*. Gamification and Augmented Reality., DOI: 10.56294/gr202590

Larsen-Freeman, D., & Anderson, M. (2011). *Techniques and principles in language teaching* (3rd ed.). Oxford University Press.

Lau, B. T., & Sim, C. H. (2008). Exploring the extent of ICT adoption among secondary school teachers in Malaysia. *International Journal of Computing and ICT Research*, *2*(2), 19–36.

Lave, J., & Wenger, E. (1991). *Situated Learning: Legitimate Peripheral Participation*. Cambridge University Press., DOI: 10.1017/CBO9780511815355

Law, L. (2024). Application of generative artificial intelligence (GenAI) in language teaching and learning: A scoping literature review. *Computers and Education Open*, *6*, 100174. DOI: 10.1016/j.caeo.2024.100174

Lebert, A., & Vilarroya, Ó. (2024). The links between experiential learning and 4E cognition. *Annals of the New York Academy of Sciences*, *1541*(1), 37–52. DOI: 10.1111/nyas.15238 PMID: 39383206

Lee, H., & Hwang, Y. (2022). Technology-Enhanced Education through VR-Making and Metaverse-Linking to Foster Teacher Readiness and Sustainable Learning. *Sustainability (Basel)*, *14*(8), 4786. DOI: 10.3390/su14084786

Lee, J., & Jeong, H. (2023). Keyword analysis of artificial intelligence education policy in South Korea. *IEEE Access : Practical Innovations, Open Solutions*, *11*, 102408–102417. DOI: 10.1109/ACCESS.2023.3317261

Lee, M. J., Spryszynski, A., & Nersesian, E. (2019). Personalizing VR educational tools for English language learners. *CEUR Workshop Proceedings*, •••, 2327.

Lehtinen, A., Nieminen, P., & Viiri, J. (2016). Preservice teachers' TPACK beliefs and attitudes toward simulations. *Contemporary Issues in Technology & Teacher Education*, *16*(2). https://citejournal.org/volume-16/issue-2-16/science/preservice-teachers-tpack-beliefs-and-attitudes-toward-simulations

Leinonen, T., Keune, A., Veermans, M., & Toikkanen, T. (2014). Mobile apps for reflection in learning: A design research in K-12 education. *British Journal of Educational Technology*, *47*(1), 184–202. DOI: 10.1111/bjet.12224

Levis, J. M. (2005). Changing contexts and shifting paradigms in pronunciation teaching. *TESOL Quarterly*, *39*(3), 369–377. DOI: 10.2307/3588485

Levis, J. M. (2018). *Intelligibility, oral communication, and the teaching of pronunciation.* Cambridge University Press., DOI: 10.1017/9781108241564

Liakin, D., Cardoso, W., & Liakina, N. (2014). Learning L2 pronunciation with a mobile speech recognizer: French /y. *CALICO Journal*, *32*(1), 1–25. DOI: 10.1558/cj.v32i1.25962

Liang, E., Liaw, R., Nishihara, R., Moritz, P., Fox, R., Goldberg, K., & Stoica, I. (2018, July). RLlib: Abstractions for distributed reinforcement learning. In *International conference on machine learning* (pp. 3053-3062). PMLR.

Li, F., Fan, S., & Wang, Y. (2022). Mobile-assisted language learning in Chinese higher education context: A systematic review from the perspective of the situated learning theory. *Education and Information Technologies*, *27*(7), 9665–9688. DOI: 10.1007/s10639-022-11025-4

Lim, F. V., Toh, W., & Nguyen, T. T. H. (2022). Multimodality in the English language classroom: A systematic review of literature. *Linguistics and Education*, *69*, 101048. DOI: 10.1016/j.linged.2022.101048

Lincoln, Y. S., & Guba, E. G. (1985). *Naturalistic inquiry*. Sage. Crossref. DOI: 10.1016/0147-1767(85)90062-8

Lin, T. H. (2021). Revelations of service-learning project: Multiple perspectives of college students' reflection. *PLoS One*, *16*(9), e0257754. DOI: 10.1371/journal.pone.0257754 PMID: 34555105

Lissovoy, N. D. (2022). Capitalism, pedagogy, and the politics of being.

LIU. Y. (2024). Towards a Contextualised Interpretation of Chinese University Student Engagement with Teacher Written Feedback (Doctoral dissertation, University of Hong Kong).

Liu, P. H. E., & Tsai, M. K. (2013). Using augmented-reality-based mobile learning material in EFL English composition: An exploratory case study. *British Journal of Educational Technology*, *44*(1), E1–E4. DOI: 10.1111/j.1467-8535.2012.01302.x

Liu, S.-H. (2012). A multivariate model of factors influencing technology use by preservice teachers. *Journal of Educational Technology & Society*, *15*(4), 137–149.

Liu, T. Y. (2009). A context-aware ubiquitous learning environment for language listening and speaking. *Journal of Computer Assisted Learning*, *25*(6), 515–527. DOI: 10.1111/j.1365-2729.2009.00329.x

Loewen, S., Crowther, D., Isbell, D. R., Kim, K. M., Maloney, J., Miller, Z. F., & Rawal, H. (2019). Mobile-assisted language learning: A Duolingo case study. *ReCALL*, *31*(3), 293–311. DOI: 10.1017/S0958344019000065

Lou, Y. (2025). The impact of virtual reality environments on English language acquisition: Innovative immersive learning technologies for communication skills development. *Journal of Computational Methods in Sciences and Engineering*. DOI: 10.1177/14727978251337950

Lo, Y. Y., Shah, S. S. A., Hasim, Z., Senom, F., Saad, M. R. M., & Thienpermpool, P. (2025). Thai EFL university students' views on the effectiveness of Duolingo. *Studies in English Language and Education*, *12*(1), 154–171. DOI: 10.24815/siele.v12i1.40517

Lui, T. W., & Goel, L. (2022). Learning effectiveness of 3D virtual reality in hospitality training: A situated cognitive perspective. *Journal of Hospitality and Tourism Technology*, *13*(3), 441–460. DOI: 10.1108/JHTT-03-2021-0091

Lu, J. (2025). Research on interactive teaching of college English based on deep learning and virtual reality. *International Journal of High Speed Electronics and Systems*, 2540240. Advance online publication. DOI: 10.1142/S0129156425402402

Lu, M. (2008). Effectiveness of vocabulary learning via mobile phone. *Journal of Computer Assisted Learning*, *24*(6), 515–525. DOI: 10.1111/j.1365-2729.2008.00289.x

Luo, Y., & Watts, M. (2022). Exploration of university students' lived experiences of using smartphones for English language learning. *Computer Assisted Language Learning*, *37*(4), 608–633. DOI: 10.1080/09588221.2022.2052904

Luo, Z. (2022). Gamification for educational purposes: What are the factors contributing to varied effectiveness? *Education and Information Technologies*, *27*(1), 891–915. DOI: 10.1007/s10639-021-10642-9

Lutfi, N., Sunardi, A. I., Utami, A. N. E. W., & Nafiah, R. (2025). Integrating virtual reality into English language instruction: Assessing efficacy and student motivation. *Journal of Education*, *7*(2), 10975–10989. Advance online publication. DOI: 10.31004/joe.v7i2.8154

Lyster, R., & Ranta, L. (1997). Corrective feedback and learner uptake: Negotiation of form in communicative classrooms. *Studies in Second Language Acquisition*, *19*(1), 37–66. DOI: 10.1017/S0272263197001034

Maghni, A., & El Khannouss, A. (2024). *Digitalisation de l'enseignement supérieur au Maroc: Approche théorique et conceptualisation*. International Journal of Accounting, Finance, Auditing, Management and Economics., DOI: 10.5281/zenodo.14253421

Maghnouj, S., Bélanger, J., Clarke, M., Fordham, E., Kitchen, H., & McGregor, I. (2018). *Examens de l'OCDE du cadre d'évaluation de l'éducation*. OCDE., DOI: 10.1787/9789264301832-fr

Mahdi, S. H. (2017). Effectiveness of mobile devices on vocabulary learning: A meta-analysis. *Journal of Educational Computing Research*, *56*(1), 134–154. DOI: 10.1177/0735633117698826

Mahmood, R. Q. (2024a). ESL learners' perceptions of using High-Variability Phonetic Training (HVPT) through YouGlish to improve pronunciation skills. *Australian Journal of Applied Linguistics*, *7*(3), 1818. DOI: 10.29140/ajal.v7n3.1818

Mahmood, R. Q. (2024b). Still matters: Kurdish EFL learners' perceptions towards non-native English teachers' pronunciation skills. *Círculo de Lingüística Aplicada a la Comunicación*, *100*, 245–252. DOI: 10.5209/clac.81012

Mahmood, R. Q. (2025). Exploring educational technology in language education: Empirical evidence from teachers' perspectives. In Stevkovska, M., Klemenchich, M., & Kavaklı Ulutaş, N. (Eds.), *Reimagining Intelligent Computer-Assisted Language Education* (pp. 129–162). IGI Global., DOI: 10.4018/979-8-3693-4310-4.ch004

Mahmudulhassan, M., Muthoifin, M., & Begum, S. (2024). Artificial Intelligence in Multicultural Islamic Education: Opportunities, Challenges, and Ethical Considerations. *Solo Universal Journal of Islamic Education and Multiculturalism*, *2*(01), 19–26. DOI: 10.61455/sujiem.v2i01.114

Maisha, K., & Shetu, S. N. (2023). Influencing factors of e-learning adoption. *Future Business Journal*, *9*(1), 37. DOI: 10.1186/s43093-023-00214-3

Makhal, A. D. (2023). Role of Digital Social Stories in enhancing Campus Inclusivity.

Ma, M., & Zheng, H. (2011). Virtual reality and serious games in healthcare. In *Advanced Computational Intelligence Paradigms in Healthcare 6* (pp. 169–192). Springer. DOI: 10.1007/978-3-642-17824-5_9

Manyeredzi, T., & Mpofu, V. (2022). Smartphones as digital instructional interface devices: The teacher's perspective. *Research in Learning Technology*, *30*, 30. DOI: 10.25304/rlt.v30.2639

Marley, D. (2004). Language attitudes in Morocco following recent changes in language policy. *Language Policy*, *3*(1), 25–46. DOI: 10.1023/B:LPOL.0000017724.16833.66

Marougkas, A., Troussas, C., Krouska, A., & Sgouropoulou, C. (2023). Virtual Reality in Education: A Review of Learning Theories, Approaches and Methodologies for the Last Decade. In *Electronics (Switzerland)* (Vol. 12, Issue 13). DOI: 10.3390/electronics12132832

Martin, D., Malpica, S., Gutierrez, D., Masia, B., & Serrano, A. (2022). Multimodality in VR: A Survey. In *ACM Computing Surveys* (Vol. 54, Issue 10 s). DOI: 10.1145/3508361

Martínez Durán, L. M. (2024). Enhancing Speaking Skills through Small-talk2Me AI Tool in the English Language Learning Process among High School Students at a Public Institution in Riobamba City, the school year 2023-2024 (Master's thesis, Riobamba: Universidad Nacional de Chimborazo).

Mast, E. (2017). *The gamification of an online English composition course: An action research exploration of student motivation and engagement*. ProQuest Dissertations & Theses.

Mayer, R. E. (2002). Multimedia learning. [). Academic Press.]. *Psychology of Learning and Motivation*, *41*, 85–139. DOI: 10.1016/S0079-7421(02)80005-6

Mayer, R. E. (2009). *Multimedia learning* (2nd ed.). Cambridge University Press. DOI: 10.1017/CBO9780511811678

McHaney, R. (2023). *The new digital shoreline: How Web 2.0 and millennials are revolutionizing higher education*. Taylor & Francis. DOI: 10.4324/9781003447979

Medina-Gómez, D. C., Cely-Betancourt, B. L., & Abreus-González, A. (2023). Octalysis framework: A favourable environment to increase successful practices in English as a foreign language (EFL) classrooms. *Revista Metropolitana de Ciencias Aplicadas*, *6*, 70–78. DOI: 10.61707/p9156r22

Mee, M., Shahdan, R., Ismail, T., Ghani, M., Pek, K., Von, L., & Rao, R. (2020). Role of gamification in classroom teaching: Pre-service teachers' view. *International Journal of Evaluation and Research in Education*, *9*(3), 1–7. DOI: 10.11591/ijere.v9i3.20622

Mei, H. H., & Sheng, L. S. (2011). Applying situated learning in a virtual reality system to enhance learning motivation. *International Journal of Information and Education Technology (IJIET)*, *1*(4), 298–302.

Meng, N., Li, Q., Knight, L., & Hayes, A. (2020). Panel - Improve foreign language speaking proficiency in immersive environment. In *Proceedings of the 2020 IEEE International Conference on e-Learning, e-Management, and e-Services* (pp. 411–412). IEEE. https://doi.org/DOI: 10.23919/iLRN47897.2020.9155192

Merriam, S. B., & Tisdell, E. J. (2016). *Qualitative Research: A Guide to Design and Implementation* (4th ed.). Jossey Bass.

Meşe, E., & Sevilen, S. (2021). Factors influencing EFL students' motivation in online learning: A qualitative case study. *Journal of Educational Technology and Online Learning*, *4*, 11–22. https://dergipark.org.tr/en/pub/jetol/issue/60134/817680

Metruk, R. (2024). Mobile-assisted language learning and pronunciation instruction: A systematic literature review. *Education and Information Technologies*, *29*(13), 16255–16282. DOI: 10.1007/s10639-024-12453-0

Meyer, R. H. (1997). Value-added indicators of school performance: A primer. *Economics of Education Review*, *16*(3), 283–301. DOI: 10.1016/S0272-7757(96)00081-7

Mihai, M., Albert, C. N., Mihai, V. C., & Dumitras, D. E. (2022). Emotional and social engagement in the English language classroom for higher education students in the COVID-19 online context. *Sustainability (Basel)*, *14*(8), 1–20. DOI: 10.3390/su14084527

Mills, A., Bali, M., & Eaton, L. (2023). How do we respond to generative AI in education? Open educational practices give us a framework for an ongoing process. *Journal of Applied Learning and Teaching*, *6*(1), 16–30.

Mirazchiyski, P. V. (2024). *Contemporary gaps in research on digital divide in education: a literature review*. Univ Access Inf Soc., DOI: 10.1007/s10209-024-01166-3

Mishra, P., & Koehler, M. J. (2006). Technological Pedagogical Content Knowledge: A Framework for Teacher Knowledge. Teachers College Record, 108(6), 1017-1054. (Original work published 2006)DOI: 10.1111/j.1467-9620.2006.00684.x

Mizumoto, M., Nguyen, D. T., Sytsma, J., Alfano, M., Izumi, Y., Fujita, K., & Minh, N. L. (2025). Cross-linguistic disagreement as a conflict of semantic alignment norms in multilingual AI~ Linguistic Diversity as a Problem for Philosophy, Cognitive Science, and AI~. arXiv preprint arXiv:2503.04792.

Moesl, B., Schaffernak, H., Vorraber, W., Braunstingl, R., & Koglbauer, I. V. (2023). Multimodal Augmented Reality Applications for Training of Traffic Procedures in Aviation. *Multimodal Technologies and Interaction*, *7*(1), 3. Advance online publication. DOI: 10.3390/mti7010003

Mohamed, M. S. P. (2024). Exploring ethical dimensions of AI-enhanced language education: A literature perspective. *Technology in Language Teaching & Learning*, *6*(3), 1813–1813. DOI: 10.29140/tltl.v6n3.1813

Morgan, J. D. (2024). Developing a capacity for lifelong learning: self-regulation and autonomous learning competencies within the European Framework (Doctoral dissertation).

Motteram, G., & Sharma, P. (2009). Blending learning in web 2.0 world. *International Journal of Emerging Technologies and Society*, *7*(2), 83–96.

Moumen, A. (2024). TICE and Their Role in Achieving the Objectives of Morocco's Strategic Vision 2015-2030. *International Journal of Scientific Research and Innovative Studies*, *3*(4), 82–85. DOI: 10.5281/zenodo.14178955

Mountasser, T. (2023). Smartphones Usefulness in Mobile Learning: The Moroccan University Case - Students of High School of Education and Training, Agadir. In: Ben Ahmed, M., Boudhir, A.A., Santos, D., Dionisio, R., Benaya, N. (eds) Innovations in Smart Cities Applications Volume 6. SCA 2022. Lecture Notes in Networks and Systems, vol 629. Springer, Cham. https://doi.org/DOI: 10.1007/978-3-031-26852-6_39

Moya, S., & Camacho, M. (2024). Leveraging AI-powered mobile learning: A pedagogically informed framework. *Computers and Education Artificial Intelligence*, *7*, 100276. DOI: 10.1016/j.caeai.2024.100276

Mudra, H. (2020). Digital literacy among young learners. *Teaching English with Technology*, *20*(3), 3–24.

Munro, M. J., & Derwing, T. M. (2015). Intelligibility in research and practice. In Reed, M., & Levis, J. M. (Eds.), *The handbook of English pronunciation* (pp. 375–396). Wiley-Blackwell., DOI: 10.1002/9781118346952.ch21

Mustafoyeva, N. (2025). TEACHING FOREIGN LANGUAGES USING MOBILE APPS. *Modern Science and Research*, *4*(2), 181–187.

Muthmainnah, M., Cardoso, L., Marzuki, A. G., & Al Yakin, A. (2025). A new innovative metaverse ecosystem: VR-based human interaction enhances EFL learners' transferable skills. *Discover Sustainability*, *6*(1), 156. Advance online publication. DOI: 10.1007/s43621-025-00913-7

Nadif, B., & Bidari, S. (2023). Advanced tools in the EFL classroom: Friend or foe for novice teachers. *The Journal of Quality in Education*, *13*(21), 12–20. DOI: 10.37870/joqie.v13i21.336

Namaziandost, E., Alekasir, S., Hassan Mohammed Sawalmeh, M., & Miftah, M. Z. (2020). Investigating the Iranian EFL learners' attitudes towards the implementation of e-portfolios in English learning and assessment. *Cogent Education*, *7*(1), 1856764. Advance online publication. DOI: 10.1080/2331186X.2020.1856764

Namaziandost, E., Razmi, M. H., Ahmad Tilwani, S., & Pourhosein Gilakjani, A. (2022). The impact of authentic materials on reading comprehension, motivation, and anxiety among Iranian male EFL learners. *Reading & Writing Quarterly*, *38*(1), 1–18. DOI: 10.1080/10573569.2021.1892001

Nasser, R. (2008). A formative assessment of ICT in Lebanese schools. *International Journal of Education and Development Using ICT*, *4*(3), 63–77.

National Telecommunications Regulatory Agency (ANRT). (2023). ICT *Equipment and uses for households and individuals: Survey*. Morocco. www.anrt.ma/sites/default/files/2024-06/enquete_tic_2023_-_synthese_ang.pdf

Nazari, Z., Vahidi, A. R., & Musilek, P. (2024). Blockchain and Artificial Intelligence Non-Formal Education System (BANFES). *Education Sciences*, *14*(8), 881. DOI: 10.3390/educsci14080881

Ndimbo, G. K., Gu, J., Haulle, E., & Yu, L. (2024). Why mobile phone matters: The role of ICT in promoting farmers' access to agricultural information and extension services in a tea outgrowing scheme in Tanzania. *Information Development*, •••, 02666669241284235. DOI: 10.1177/02666669241284235

Nechifor, A. (2024). Gamified Romanian for Specific Purposes: A Journey to Engaged Language Acquisition. Analele Universităţii Ovidius din Constanţa. *Seria Filologie*, *35*(2), 348–371.

Neri, A., Mich, O., Gerosa, M., & Giuliani, D. (2008). The effectiveness of computer assisted pronunciation training for foreign language learning by children. *Computer Assisted Language Learning*, *21*(5), 393–408. DOI: 10.1080/09588220802447651

Neureiter, D., Klieser, E., Neumayer, B., Winkelmann, P., Urbas, R., & Kiesslich, T. (2020). Feasibility of Kahoot! as a Real-Time Assessment Tool in (Histo-)pathology Classroom Teaching. *Advances in Medical Education and Practice, 11*, 695–705. DOI: 10.2147/AMEP.S264821 PMID: 33117039

Nguyen, L. T., Hung, B. P., Duong, U. T. T., & Le, T. T. (2021). Teachers' and learners' beliefs about pronunciation instruction in tertiary English as a foreign language education. *Frontiers in Psychology, 12*, 739842. DOI: 10.3389/fpsyg.2021.739842 PMID: 34489841

Ngwenya, S. C. (2022). Affordances of mobile learning technologies in addressing language issues in Grade 7 Natural Sciences Teaching and Learning (Doctoral dissertation, University of Johannesburg).

Nikimaleki, M., & Rahimi, M. (2022). Effects of a collaborative AR-enhanced learning environment on learning gains and technology implementation beliefs: Evidence from a graduate teacher training course. *Journal of Computer Assisted Learning, 38*(3), 758–769. Advance online publication. DOI: 10.1111/jcal.12646

Nikoletta, V. (2023). The effects of EFL variables on motivation and the learning outcome.

Ningrum, B., & Rokhuma, C. (2023). Positive and negative emotions of high school students in learning English through google classroom. *Journal of English Teaching and Learning Issues, 6*(1), 19–28. DOI: 10.21043/jetli.v6i1.19113

Novawan, A., Aisyiyah, S., Miqawati, A. H., Wijayanti, F., & Rinda, R. K. (2019). Implementing an Ecological Teaching through Mobile-Assisted Language Learning in EFL Education. *Advances in Social Science, Education and Humanities Research, 254*. Advance online publication. DOI: 10.2991/conaplin-18.2019.247

Novawan, A., Walker, S. A., & Ikeda, O. (2024). The new face of technology-enhanced language learning (TELL) with artificial intelligence (AI): Teacher perspectives, practices, and challenges. *Journal of English in Academic and Professional Communication, 10*(1), 1–18. DOI: 10.25047/jeapco.v9i1.3754

Nowell, L. S., Norris, J. M., White, D. E., & Moules, N. J. (2017). Thematic analysis: Striving to meet the trustworthiness criteria. *International Journal of Qualitative Methods, 16*(1), 1609406917733847. DOI: 10.1177/1609406917733847

Noy, S., Capetola, T., & Patrick, R. (2021). The wheel of fortune as a novel support for constructive alignment and transformative sustainability learning in higher education. *International Journal of Sustainability in Higher Education, 22*(4), 854–869. DOI: 10.1108/IJSHE-08-2020-0289

Nushi, M., & Eqbali, M. H. (2018). 50 languages: a mobile language learning application (App Review). *Teaching English with Technology*, *18*(1), 93–104.

Nushi, M., & Jenabzadeh, H. (2016). Busuu: The Mobile App. *The TESL Reporter*, *49*(2), 30–38.

O'Malley, J. M., & Pierce, L. V. (1996). *Authentic assessment for English language learners: Practical approaches for teachers*. Addison-Wesley Pub. Co.

Oberoi, Z. (2024). Using gamification to enhance development.

Ogata, H., Akamatsu, R., & Yano, Y. (2005). Computer Supported Ubiquitous Learning Environment for Vocabulary Learning Using RFID Tags. In Courtiat, J. P., Davarakis, C., & Villemur, T. (Eds.), *Technology Enhanced Learning. IFIP WCC TC3 2004. IFIP International Federation for Information Processing* (Vol. 171). Springer., DOI: 10.1007/0-387-24047-0_10

Ogata, H., & Yano, Y. (2004). Context-aware support for computer-supported ubiquitous learning. In *Proceedings of the 2nd IEEE International Workshop on Wireless and Mobile Technologies in Education* (pp. 27–34). IEEE. DOI: 10.1109/WMTE.2004.1281330

Oh, C. M., Krish, P., & Hamat, A. (2022). Reading on smartphones. *CALL-EJ*, *23*(1), 259–277.

Omidian, H. (2024). Synergizing blockchain and artificial intelligence to enhance healthcare. *Drug Discovery Today*, *29*(9), 104111. DOI: 10.1016/j.drudis.2024.104111 PMID: 39034026

Otto, B. W. (2017). From past to present: A hundred years of technology for L2 learning. In Chapelle, C. A., & Sauro, S. (Eds.), *The handbook of technology in second language teaching and learning* (pp. 10–25). Wiley-Blackwell. DOI: 10.1002/9781118914069.ch2

Oved, O., & Alt, D. (2025). Teachers' technological pedagogical content knowledge (TPACK) as a precursor to their perceived adopting of educational AI tools for teaching purposes. *Education and Information Technologies*, *30*(10), 14095–14121. Advance online publication. DOI: 10.1007/s10639-025-13371-5

Pachler, N., Cook, J., & Bachmair, B. (2010). Appropriation of mobile cultural resources for learning. *International Journal of Mobile and Blended Learning*, *2*(1), 1–21. DOI: 10.4018/jmbl.2010010101

Pack, A., & Maloney, J. (2024). Using artificial intelligence in TESOL: Some ethical and pedagogical considerations. *TESOL Quarterly*, *58*(2), 1007–1018. DOI: 10.1002/tesq.3320

Palamar, S., Brovko, K., & Semerikov, S. (2023). Enhancing foreign language learning in Ukraine: Immersive technologies as catalysts for cognitive interest and achievement. *CEUR Workshop Proceedings*, *3624*, 69–81.

Panagiotidis, P., Krystalli, P., & Arvanitis, P. (2023). Technology as a motivational factor in foreign language learning. *European Journal of Education*, *6*(1), 69–84. DOI: 10.2478/ejed-2023-0007

Pardede, P. (2018). Improving EFL students' English pronunciation by using the explicit teaching approach. *Journal of English Teaching*, *4*(3), 143–155. DOI: 10.33541/jet.v4i3.852

Park, J. J., Kim, B. H., Wong, N., Zheng, J., Breen, S., Lo, P., & Poon, O. A. (2023). Inequality beyond standardized tests: Trends in extracurricular activity reporting in college applications across race and class. *American Educational Research Journal*, •••, 00028312241292309.

Parry, S., & Metzger, E. (2023). Barriers to learning for sustainability: A teacher perspective. *Sustainable Earth Reviews*, *6*(1), 2. DOI: 10.1186/s42055-022-00050-3

Parvin, S. (2013). Integration of ICT in education sector. *International Journal of Computer Science and Information Technologies*, *5*(4), 81–92. DOI: 10.5121/ijcsit.2013.5406

Pataquiva, F. D. P. F., & Klimova, B. (2022). A systematic review of virtual reality in the acquisition of second language. [iJET]. *International Journal of Emerging Technologies in Learning*, *17*(15), 43–53. DOI: 10.3991/ijet.v17i15.31781

Patty, J. (2024). The use of AI in language learning: What you need to know. Jurnal Review Pendidikan dan Pengajaran (JRPP), 7(1), 642-654.

Pekrun, R. (2006). The control-value theory of achievement emotions: Assumptions, corollaries, and implications for educational research and practice. *Educational Psychology Review*, *18*(4), 315–341. DOI: 10.1007/s10648-006-9029-9

Peng, W., & Yang, Y. (2024). Framework Design of Blockchain Intelligent Technology for Chinese Language Teaching Management System. *Procedia Computer Science*, *243*, 306–312. DOI: 10.1016/j.procs.2024.09.038

Pennington, M. C. (1999). Computer-aided pronunciation pedagogy: Promise, limitations, directions. *Computer Assisted Language Learning*, *12*(5), 427–440. DOI: 10.1076/call.12.5.427.5693

Pennington, M. C. (2021). Teaching pronunciation: The state of the art 2021. *RELC Journal*, *52*(1), 3–21. DOI: 10.1177/00336882211002283

Pennington, M. C., & Rogerson-Revell, P. (2019). *English pronunciation teaching and research* (Vol. 10). Palgrave Macmillan London. DOI: 10.1057/978-1-137-47677-7

Pennycook, A. (2017). The cultural politics of English as an international language. In *Routledge eBooks*. https://doi.org/DOI: 10.4324/9781315225593

Pentaris, P., & Christodoulou, P. (2021). Qualities of culturally and religiously sensitive practice: A cross-sectional study. *Journal of Palliative Care*, •••, 08258597211050742. PMID: 34841969

Permatasari, N. P. I., & Wulandari, C. I. S. A. (2025). English Speaking Fluency and Confidence for Global Competitive Work: The Highlight of Authentic Role-play Activity for University Students. *Jurnal Multidisiplin Sahombu*, *5*(02), 400–411.

Peterson, M., & Jabbari, N. (2023). *Digital games in language learning: Case studies and applications* (New directions in computer assisted language learning). Routledge.

Petkevičius, M. (2024). The impact of gamification elements on the intention to use health related applications (Doctoral dissertation, Vilniaus universitetas.).

Pham, A. T. V., & Pham, H. T. (2022, September). Developing the habit of learning English using Duolingo application. In *Proceedings of the 6th International Conference on Digital Technology in Education* (pp. 131-136). DOI: 10.1145/3568739.3568763

Pham, A. T., & Nguyen, T. B. (2024). English as a foreign language students' acceptance of Google Classroom in writing classes: A case study in Vietnam. *Heliyon*, *10*(8), e29832. DOI: 10.1016/j.heliyon.2024.e29832 PMID: 38699708

Phanwiriyarat, K., Anggoro, K. J., & Chaowanakritsanakul, T. (2025). Exploring AI-powered gamified flipped classroom in an English-speaking course: A case of Duolingo. *Cogent Education*, *12*(1), 2488545. Advance online publication. DOI: 10.1080/2331186X.2025.2488545

Phirangee, K., & Foster, L. (2024). Decolonizing digital learning: Equity through intentional course design. *Distance Education*, *45*(3), 357–364. DOI: 10.1080/01587919.2024.2362413

Pim, C. (2013). Emerging technologies, emerging minds: Digital innovations within the primary sector. In Motteram, G. (Ed.), *Innovations in learning technologies for English language learning* (pp. 17–42). British Council.

Pineda, V. S. (2024). Radical Inclusion: The Key to Urban Transformation. In *Inclusion and Belonging in Cities of Tomorrow: Governance and Access by Design* (pp. 1–21). Springer Nature Singapore. DOI: 10.1007/978-981-99-3856-8_1

Pirker, J., & Dengel, A. (2021). The Potential of 360° Virtual Reality Videos and Real VR for Education - A Literature Review. In *IEEE Computer Graphics and Applications* (Vol. 41, Issue 4). DOI: 10.1109/MCG.2021.3067999

Poehner, M. E., & Lantolf, J. P. (2024). *Sociocultural theory and second language developmental education.* Cambridge University Press., https://doi.org/DOI, DOI: 10.1017/9781009189422

Prasojo, L. D.. (2018). Learning to teach in a digital age. *Teaching English with Technology*, *18*(3), 18–32.

Pratolo, B., Habibie, A., & Setiawan, A. (2019). Speaking Skill Challenges Encountered by Indonesian EFL Learners'. *Proceedings of the 2019 Ahmad Dahlan International Conference Series on Education & Learning, Social Science & Humanities (ADICS-ELSSH 2019).* DOI: 10.2991/adics-elssh-19.2019.7

Procel, G. J. O., Medina, M. L. F., Sotomayor, D. J., & Sanchez, M. A. (2024). Using Technology in English Teaching. *Journal of Environmental Research and Public Health*, *17*(9), 9.

Qarooni, N. (2023). Process: The Journey is the Learning. In Nourishing Caregiver Collaborations (pp. 43-69). Routledge.

Qasserras, L., Qasserras, M., Anasse, K., & Slimani, N. (2023). Exploring Pre-Service Teachers' Experience With The Use Of Smartphones During The Internship Period - CASE STUDY: CRMEF OF RABAT, MOROCCO. *European Journal of Open Education and E-learning Studies*, *8*(2). Advance online publication. DOI: 10.46827/ejoe.v8i2.4765

Qutieshat, A. (2025). Towards a Liberated Higher Education Landscape. In *Artificial Intelligence in Higher Education: A Contemporary Examination of Illich's Theories* (pp. 23–48). Springer Nature Singapore. DOI: 10.1007/978-981-97-9989-3_3

Rabbani Yekta, R., & Kana'ni, M. A. (2020). Using Google Drive as the e-portfolio for the self-assessment of speaking fluency. *International Journal of Research in English Education*, *5*(2), 49–60. DOI: 10.29252/ijree.5.2.49

Raman, K., Hashim, H., & Ismail, H. H. (2024). *Exploring the impact of VR integration on ESL learners' English verbal communication skills: A case study in a Malaysian high school.* Arab World English Journal.

Ramya, R., & Ramamoorthy, S. (2025). AI in Communication Technology and Innovation of AR/VR in Communication. In Impact of Digitalization on Communication Dynamics (pp. 1-28). IGI Global Scientific Publishing.

Rane, N. (2024). Enhancing the quality of teaching and learning through Gemini, ChatGPT, and similar generative Artificial Intelligence: Challenges, future prospects, and ethical considerations in education. *TESOL and Technology Studies*, *5*(1), 1–6. DOI: 10.48185/tts.v5i1.1000

Rashid, A., Zafar, N., Mir, B., Asim, S., Sheikh, M. G., & Jahan, M. (2025). Artificial Intelligence and Smart Pedagogy: Machine Learning in Digital Media for Language Education. *Metallurgical and Materials Engineering*, *31*(2), 7–22. DOI: 10.63278/1324

Rasyid, F., Ibna Seraj, P. M., Ghofur, A., & Asrifan, A. (2023). Students' Perception toward Teaching Strategies of Native and Nonnative English-Speaking Teachers: A Case Study in Indonesia. *Education Research International*, *2023*(1), 7827917. DOI: 10.1155/2023/7827917

Rauschnabel, P. A., Babin, B. J., tom Dieck, M. C., Krey, N., & Jung, T. (2022). What is augmented reality marketing? Its definition, complexity, and future. In *Journal of Business Research* (Vol. 142). DOI: 10.1016/j.jbusres.2021.12.084

Redecker, C. (2017). Digital Competence of Educators. https://doi.org/DOI: 10.2760/159770

Redlener, I., Dudashvili, K., Viontsek, O., Grant, R., & Roca, E. (2025). Intervention for Psychological Trauma in Children Impacted by War in Ukraine. *JAMA Network Open*, *8*(3), e253057. DOI: 10.1001/jamanetworkopen.2025.3057 PMID: 40085087

Redondo, B., Cózar-Gutiérrez, R., González-Calero, J. A., & Ruiz, R. S. (2020). Integration of augmented reality in the teaching of English as a foreign language in early childhood education. *Early Childhood Education Journal*, *48*(2), 147–155. DOI: 10.1007/s10643-019-00999-5

Reinders, H., & White, C. (2010). The theory and practice of technology in materials development and task design. In Harwood, N. (Ed.), *Materials in ELT: Theory and practice* (pp. 58–80). Cambridge University Press.

Remolar, I., Rebollo, C., & Fernández-Moyano, J. A. (2021). Learning history using virtual and augmented reality. *Computers*, *10*(11), 146. DOI: 10.3390/computers10110146

Resta, P., & Laferrière, T. (2015). Digital equity and intercultural education. *Education and Information Technologies*, *20*(4), 743–756. DOI: 10.1007/s10639-015-9419-z

Rhaya, F., & El Ouazzani, Y. (2025). ICT and the Digital Divide: Analyzing ICT Indicators in Morocco Using PCA. In Bach Tobji, M. A., Jallouli, R., Sadok, H., Lajfari, K., Mafamane, D., & Mahboub, H. (Eds.), *Digital Economy. Emerging Technologies and Business Innovation. ICDEc 2024. Lecture Notes in Business Information Processing* (Vol. 531). Springer., DOI: 10.1007/978-3-031-76368-7_11

Richards, J. C. (2020). Exploring emotions in language teaching. *RELC Journal*, *53*(1), 1–15. DOI: 10.1177/0033688220927531

Richter, J., Scheiter, K., & Eitel, A. (2016). Signaling text-picture relations in multimedia learning: A comprehensive meta-analysis. *Educational Research Review*, *17*, 19–36. DOI: 10.1016/j.edurev.2015.12.003

Rinda, R. (2019). Implementing an Ecological Teaching through Mobile-Assisted Language Learning in EFL Education. *Proceedings of the Eleventh Conference on Applied Linguistics (CONAPLIN 2018)*.

Rinehart, T. D. (2023). *Developing the Personal and the Professional Addressing Teacher Attrition Through Self-Care & Leisure Programming*. University of Dayton.

Robinson-Jones, C. (2024). Tension in the linguistic landscape: The implications of language choices for diversity and inclusion in multilingual museums representing minorities. *International Journal of Multilingualism*, *21*(2), 881–905. DOI: 10.1080/14790718.2022.2135710

Roblyer, M. D., & Doering, A. H. (2010). *Integrating technology into teaching* (6th ed.). Pearson Education Limited.

Rocchi, E., Ferrarotti, A., & Carli, M. (2025). A Comparison of the Meta Quest Pro and HTC Vive Focus 3 Eye-Tracking Systems: Analysis of Data Accuracy and Spatial Precision. *IEEE Access : Practical Innovations, Open Solutions*, *13*, 71995–72010. DOI: 10.1109/ACCESS.2025.3562672

Rocco, T. S., & Plakhotnik, M. S. (2009). Literature Reviews, Conceptual Frameworks, and Theoretical Frameworks: Terms, Functions, and Distinctions. *Human Resource Development Review*, *8*(1), 120–130. DOI: 10.1177/1534484309332617

Rodríguez Moreno, J., Agreda Montoro, M., & Ortiz Colón, A. M. (2019). Changes in Teacher Training within the TPACK Model Framework: A Systematic Review. *Sustainability (Basel)*, *11*(7), 1870. DOI: 10.3390/su11071870

Roediger, H. L.III, & Butler, A. C. (2011). The critical role of retrieval practice in long-term retention. *Trends in Cognitive Sciences*, *15*(1), 20–27. DOI: 10.1016/j.tics.2010.09.003 PMID: 20951630

Rojas-Sánchez, M. A., Palos-Sánchez, P. R., & Folgado-Fernández, J. A. (2023). Systematic literature review and bibliometric analysis on virtual reality and education. *Education and Information Technologies*, *28*(1), 155–192. Advance online publication. DOI: 10.1007/s10639-022-11167-5 PMID: 35789766

Rosell-Aguilar, F. (2007). Top of the pods—In search of a podcasting "podagogy" for language learning. *Computer Assisted Language Learning*, *20*(5), 471–492. DOI: 10.1080/09588220701746047

Rosendahl, P., & Wagner, I. (2024). 360° videos in education – A systematic literature review on application areas and future potentials. *Education and Information Technologies*, *29*(2), 1319–1355. Advance online publication. DOI: 10.1007/s10639-022-11549-9

Rudnik, Y. (2023). The use of augmented reality and virtual reality technologies in teaching foreign languages. *Educological Discourse*, *1*(40), 165–183. DOI: 10.28925/2312-5829.2023.110

Rustemi, A., Dalipi, F., Atanasovski, V., & Risteski, A. (2024, June). Enhancing Academic Credentials: The Synergy of Blockchain and Artificial Intelligence. In 2024 7th International Balkan Conference on Communications and Networking (BalkanCom) (pp. 206-211). IEEE.

Sadaf, M. (2023). Influences of Duolingo English test. *Open Journal of Modern Linguistics*, *13*(3), 392–405. DOI: 10.4236/ojml.2023.133024

Safar, A. H., Al-Jafar, A. A., & Al-Yousefi, Z. H. (2016). The effectiveness of using augmented reality apps in teaching the English alphabet to kindergarten children: A case study in the State of Kuwait. *Eurasia Journal of Mathematics, Science and Technology Education*, *13*(2), 417–440. DOI: 10.12973/eurasia.2017.00624a

Saito, K. (2012). Effects of instruction on L2 pronunciation development: A synthesis of 15 quasi-experimental intervention studies. *TESOL Quarterly*, *46*(4), 842–854. DOI: 10.1002/tesq.67

Sajja, R., Sermet, Y., Cikmaz, M., Cwiertny, D., & Demir, I. (2024). Artificial intelligence-enabled intelligent assistant for personalized and adaptive learning in higher education. *Information (Basel)*, *15*(10), 596. DOI: 10.3390/info15100596

Saleem, A. N., Noori, N. M., & Ozdamli, F. (2022). Gamification applications in E-learning: A literature review. Technology. *Knowledge and Learning*, *27*(1), 139–159. DOI: 10.1007/s10758-020-09487-x

Salehi, H., & Salehi, Z. (2012). Integration of ICT in language teaching. In *Proceedings of the 3rd International Conference on e-Education, e-Business, e-Management and e-Learning* (pp. 215–219).

Salimi, H. (2023). Beyond Universal Gamification: Exploring Persian Gamification Strategies (Doctoral dissertation, Concordia University).

Salloum, S. A., Alomari, K. M., Alfaisal, A. M., Aljanada, R. A., & Basiouni, A. (2025). Emotion recognition for enhanced learning: Using AI to detect students' emotions and adjust teaching methods. *Smart Learning Environments*, *12*(1), 21. DOI: 10.1186/s40561-025-00374-5

Samson, S., & Karthiga, D. S. V. (2020). Gamification as a tool in English language teaching. *International Journal of Early Childhood Special Education*, *12*(2), 99–102. DOI: 10.9756/INT-JECSE/V12I2.201061

Sánchez, E., Burgos, M., & Saborido, A. (2023). Virtual Reality in Linguistic Teaching to Immigrants. *Journal of Education For Sustainable Innovation*, *1*(2), 95–102. DOI: 10.56916/jesi.v1i2.561

Santos, M. E. C., Chen, A., Taketomi, T., Yamamoto, G., Miyazaki, J., & Kato, H. (2013). Augmented reality learning experiences: Survey of prototype design and evaluation. *IEEE Transactions on Learning Technologies*, *7*(1), 38–56. DOI: 10.1109/TLT.2013.37

Sari, S. N., Aminatun, D., Sari, S. N., Aminatun, D., Sari, S. N., & Aminatun, D. (2021). Students' perception on the use of English movies to improve vocabulary mastery. *J. English Lang. Teach. Learn*, *2*(1), 16–22. DOI: 10.33365/jeltl.v2i1.757

Sattarova, M. (2024). A COMPARATIVE ANALYSIS OF TRADITIONAL AND MODERN APPROACHES IN LANGUAGE TEACHING. *Mental Enlightenment Scientific-Methodological Journal*, *5*(09), 214–223.

SaveSchools.in.ua. (2025). Education under attack: Damage to Ukrainian educational institutions. https://saveschools.in.ua/en/

Scheeler, M. C. (2008). Generalising effective teaching skills. *Journal of Behavioral Education, 17*(2), 145–159. DOI: 10.1007/s10864-007-9051-0

Schweitzer, L., Lyons, S., & Smith, C. J. (2023). Career sustainability: Framing the past to adapt in the present for a sustainable future. *Sustainability (Basel), 15*(15), 11800. DOI: 10.3390/su151511800

Sciberras, M., & Dingli, A. (2023). Research Analysis—Triangulation Approach. In *Investigating AI Readiness in the Maltese Public Administration. Lecture Notes in Networks and Systems* (Vol. 568). Springer., DOI: 10.1007/978-3-031-19900-4_9

Seidlhofer, B. (2017). Standard English and the dynamics of ELF variation. In J. Jenkins, W. Baker, & M. Dewey (Eds.), *The Routledge handbook of English as a lingua franca* (pp. 85-100). Routledge. DOI: 10.4324/9781315717173-8

Selwyn, N. (2011). *Education and Technology: Key Issues and Debates.* Continuum International Publishing Group.

Şendurur, P., & Sendurur, E. (2022). Students as gamers: Design, code, and play. In Lane, C. (Ed.), *Handbook of research on acquiring 21st century literacy skills through game-based learning* (pp. 868–887). IGI Global., DOI: 10.4018/978-1-7998-7271-9.ch044

Shadiev, R., Hwang, W. Y., Huang, Y. M., & Liu, T. Y. (2018). Facilitating application of language skills in authentic environments with a mobile learning system. *Journal of Computer Assisted Learning, 34*(1), 42–52. Advance online publication. DOI: 10.1111/jcal.12212

Shadiev, R., & Yang, M. (2020). Review of studies on technology-enhanced language learning and teaching. *Sustainability (Basel), 12*(2), 524. DOI: 10.3390/su12020524

Shah, K. B. (2023). Traversing the City: Examining the Impacts of Extended Commute on Student Wellbeing & High School Experience in Chicago.

Shanmughan, P., & Joseph, J. S, B. N., SM, A., CS, A., & Jose, J. (2024, November). The AI revolution in micro-credentialing: personalized learning paths. In Frontiers in Education (Vol. 9, p. 1445654). Frontiers Media SA.

Sharples, M., Taylor, J., & Vavoula, G. (2010). A Theory of Learning for the Mobile Age. In Bachmair, B. (Ed.), *Medienbildung in neuen Kulturräumen.* VS Verlag für Sozialwissenschaften., DOI: 10.1007/978-3-531-92133-4_6

Sheen, Y. (2010). Differential effects of oral and written corrective feedback in the ESL classroom. *Studies in Second Language Acquisition, 32*(2), 203–234. DOI: 10.1017/S0272263109990507

Shew, A. (2023). *Against technoableism: rethinking who needs improvement.* WW Norton & Company.

Shi, W. (2024). Analysis of Local Governance and Legal Protection of Students' Information Security in the Context of Sustainable Development of Digital Education. Lex localis-Journal of Local Self-Government, 22(4), 435-458.

Shirinova, S. (2025). Leveraging Artificial Intelligence in Linguistics: Innovations in Language Acquisition and Analysis. *EuroGlobal Journal of Linguistics and Language Education*, *2*(1), 50–57. DOI: 10.69760/egjlle.250006

Shi, Z., Luo, G., & He, L. (2017). Mobile-assisted language learning using WeChat. *International Journal of Emerging Technologies in Learning*, *12*(2), 16. DOI: 10.3991/ijet.v12i02.6681

Shrestha, S.. (2022). Preparations for online education during Covid-19. *Education and Information Technologies*, *27*(1), 243–265. DOI: 10.1007/s10639-021-10659-0 PMID: 34341654

Shukla, T., Duthade, A., & Deshbhratar, N. (2023). Improving Kids English Vocabulary Through Immersive Mobile Games. *INTERANTIONAL JOURNAL OF SCIENTIFIC RESEARCH IN ENGINEERING AND MANAGEMENT*, *07*(10), 1–11. Advance online publication. DOI: 10.55041/IJSREM26473

Silva, C. V. C. (2024). Communicative language skills: a comparative analysis of duolingo and memrise applications for language learning.

Şimşek, E. E. (2024). The effect of augmented reality storybooks on the story comprehension and retelling of preschool children. *Frontiers in Psychology*, *15*, 1459264. DOI: 10.3389/fpsyg.2024.1459264 PMID: 39439758

Sinthiya, B. (2023). English language learning in the metaverse: Exploring the potential of AR and VR. *Shanlax International Journal of English*, *12*(S1-Dec, S1), 52. DOI: 10.34293/rtdh.v12iS1-Dec.52

Şi ianu, A., & Pu ca u, A. (2024). Methods of assessing students' knowledge in english language lessons. *Journal of Social Sciences*, *7*(3), 234–242. DOI: 10.52326/jss.utm.2024.7(3).16

Slaouti, D., & Barton, A. (2007). Opportunities for practice and development. *Journal of In-service Education*, *33*(4), 405–424. DOI: 10.1080/13674580701687807

Slimane, H. K. (2024). Digital linguistic landscapes: Unveiling youth code-switching in computermediated communication. *Akofena Journal*, *12*(3), 297–308.

Song, T., Li, X., Wang, B., & Han, L. (2024). Research on intelligent application design based on artificial intelligence and adaptive interface.

Song, J., Feng, Y., & Cheng, J. (2024). Dongji he renzhicelue dui qingshaonian shuziyuedusuyang yingxiang yanjiu [investigating motivational and cognitive influences on adolescent digital Reading literacy: a fuzzy-set qualitative comparative analysis of 510 survey responses]. *Tushuguanxue Yanjiu*, *45*, 96–107. DOI: 10.15941/j.cnki.issn1001-0424.2024.05.013

Song, Y. (2021). A review of how class orchestration with technology has been conducted for pedagogical practices. *Educational Technology Research and Development*, *69*(3), 1477–1503. DOI: 10.1007/s11423-021-10001-y

Soye, E. (2024). Funny Language? Curiosity, Contact, and Humour. In Peer Relationships at School (pp. 71-88). Bristol University Press. DOI: 10.2307/jj.12348141.9

Spradley, J. P. (1980). *Participant observation*. Holt, Rinehart and Winston.

Sriprakash, A., Williamson, B., Facer, K., Pykett, J., & Valladares Celis, C. (2024). Sociodigital futures of education: Reparations, sovereignty, care, and democratisation. *Oxford Review of Education*, •••, 1–18.

Stake, R. E. (1995). *The art of case study research*. SAGE Publications.

Stanley, G. (2013). *Language learning with technology: Ideas for integrating technology in the classroom*. Cambridge University Press.

Stark, B. C. (2025). Speech and language rehabilitation: The potential role of immersive VR interventions. *[Preprint]*. https://doi.org/DOI: 10.31219/osf.io/rhz3c

Stockwell, G. (2007). Vocabulary on the move: Investigating an intelligent mobile phone-based vocabulary tutor. *Computer Assisted Language Learning*, *20*(4), 365–383. DOI: 10.1080/09588220701745817

Stockwell, G. (2010). Using mobile phones for vocabulary activities: Examining the effect of the platform. *Language Learning & Technology*, *14*(2), 95–110. http://llt.msu.edu/vol14num2/stockwell.pdf. DOI: 10.64152/10125/44216

Stockwell, G. (2013). Tracking learner usage of mobile phones for language learning outside of the classroom. *CALICO Journal*, *30*, 118–136. https://www.jstor.org/stable/calicojournal.30.118. DOI: 10.1558/cj.v30i0.118-136

Stockwell, G. (2022). *Mobile assisted language learning: Concepts, contexts and challenges*. Cambridge University Press. DOI: 10.1017/9781108652087

Stockwell, G. (Ed.). (2012). *Computer-assisted language learning*. Cambridge University Press. DOI: 10.1017/CBO9781139060981

Stockwell, G., & Hubbard, P. (2013). *Some Emerging Principles for Mobile-Assisted Language Learning*. The International Research Foundation for English Language Education., https://www.tirfonline.org/wp-content/uploads/2013/11/TIRF_MALL_Papers_StockwellHubbard.pdf

Stoughton, A. M., & Kang, O. (2024). A systematic review of empirical mobile-assisted pronunciation studies through a perception–production lens. *Languages (Basel, Switzerland)*, *9*(7), 251. https://www.mdpi.com/2226-471X/9/7/251. DOI: 10.3390/languages9070251

Sulispera, T., & Recard, M. (2020). Octalysis gamification framework for enhancing students' engagement in language learning. *Jurnal Dialektika Program Studi Pendidikan Bahasa Inggris, 8*, 103–128. https://scispace.com/pdf/octalysis-gamification-framework-for-enhancing-students-1wmk7scims.pdf

Sung, Y.-T., Chang, K.-E., & Liu, T.-C. (2016). The effects of integrating mobile devices with teaching and learning on students' learning performance: A meta-analysis and research synthesis. *Computers & Education*, *94*, 252–275. DOI: 10.1016/j.compedu.2015.11.008

Sun, T. (2023). Jiyu tubianjishufa de shehuihua yuedu xingwei fenxi [analysis of social Reading user behavior based on CPM: a case study of we read]. *Qingbao Kexue*, *41*, 148–154. DOI: 10.13833/j.issn.1007-7634.2023.10.017

Sun, Y. C., & Yang, F. Y. (2015). I help, therefore, I learn: Service learning on Web 2.0 in an EFL speaking class. *Computer Assisted Language Learning*, *28*(3), 202–219. DOI: 10.1080/09588221.2013.818555

Supyan, H. (2017, March 4). Mobile Apps, Mobile Learning Apps, Mobile-Assisted Learning Apps, MobileAssisted Language Learning Apps. Retrieved January 22, 2023 from https://supyanhussin.wordpress. com/2017/03/04/mobile-apps-mobile-learning-apps-mobile-assisted-learning-apps-mobile-assisted-language learning-apps/

Sweller, J. (1988). Cognitive load during problem solving: Effects on learning. *Cognitive Science*, *12*(2), 257–285. DOI: 10.1207/s15516709cog1202_4

Sweller, J., Ayres, P., & Kalyuga, S. (2011). *Cognitive load theory*. Springer. DOI: 10.1007/978-1-4419-8126-4

Syafri, N. A., & Ramli, S. (2025). Design development of Arabic-Rumi transliteration module with virtual reality concept to improve primary school students' Arabic reading skills: Design and development. *Multidisciplinary Applied Research and Innovation, 6* (2), 53–61. Retrieved from https://publisher.uthm.edu.my/periodicals/index.php/mari

Symonenko, S., Zaitseva, N., Osadchyi, V., Osadcha, K., & Shmeltser, E. (2020). Virtual reality in foreign language training at higher educational institutions.

Szabó, F., & Kopinska, M. (2023). Gamification in foreign language teaching: A conceptual introduction. *Hungarian Educational Research Journal, 13*(3), 1–11. DOI: 10.1556/063.2023.00202

Taber, K. S. (2018). The use of Cronbach's alpha. *Research in Science Education, 48*(6), 1273–1296. DOI: 10.1007/s11165-016-9602-2

Tabuenca, B., Uche-Soria, M., Greller, W., Hernández-Leo, D., Balcells-Falgueras, P., Gloor, P., & Garbajosa, J. (2024). Greening smart learning environments with Artificial Intelligence of Things. *Internet of Things : Engineering Cyber Physical Human Systems, 25*, 101051. DOI: 10.1016/j.iot.2023.101051

Tan, Y., Xu, W., Li, S., & Chen, K. (2022). Augmented and Virtual Reality (AR/VR) for Education and Training in the AEC Industry: A Systematic Review of Research and Applications. In *Buildings* (Vol. 12, Issue 10). DOI: 10.3390/buildings12101529

Tarng, W., & Ou, K. L. (2012). A study of campus butterfly ecology learning system based on augmented reality and mobile learning. In *2012 IEEE Seventh International Conference on Wireless, Mobile and Ubiquitous Technology in Education* (pp. 62–66). IEEE. DOI: 10.1109/WMUTE.2012.17

Tarteer, S., Badah, A., & Khlaif, Z. N. (2021). Employing Google Classroom to Teach Female Students during the COVID-19 Pandemic. *Computers in the Schools, 38*(4), 300–321. DOI: 10.1080/07380569.2021.1988318

Teng, M. F., Wang, C., & Wu, J. G. (2021). Metacognitive Strategies, Language Learning Motivation, Self-Efficacy Belief, and English Achievement During Remote Learning: A Structural Equation Modelling Approach. *RELC Journal*, 54(3), 648-666. (Original work published 2023).DOI: 10.1177/00336882211040268

Teräs, M., Suoranta, J., Teräs, H., & Curcher, M. (2020). Post-Covid-19 education and education technology 'solutionism': A seller's market. *Postdigital Science and Education, 2*(3), 863–878. DOI: 10.1007/s42438-020-00164-x PMID: 40477107

Tergujeff, E. (2021). Second language comprehensibility and accentedness across oral proficiency levels: A comparison of two L1s. *System*, *100*, 102567. https://doi.org/https://doi.org/10.1016/j.system.2021.102567. DOI: 10.1016/j.system.2021.102567

Thomson, R. I., & Derwing, T. M. (2014). The effectiveness of L2 pronunciation instruction: A narrative review. *Applied Linguistics*, *36*(3), 326–344. DOI: 10.1093/applin/amu076

Thornton, P., & Houser, C. (2005). Using mobile phones in English education in Japan. *Journal of Computer Assisted Learning*, *21*(3), 217–228. http://onlinelibrary.wiley.com. DOI: 10.1111/j.1365-2729.2005.00129.x

Thurairasu, V. (2022). Gamification-based learning as the future of language learning: An overview. *European Journal of Humanities and Social Sciences*, *2*(6), 62–69. DOI: 10.24018/ejsocial.2022.2.6.353

Tilden, F. (2007). *Interpreting Our Heritage*. Caroline: University of North Carolina Press.

Tiwari, H. P. (2024). Artificial intelligence in the classroom: Revolutionizing English language teaching. [JET Li]. *Journal of English Teaching and Linguistics Studies*, *6*(1), 42–59. DOI: 10.55215/jetli.v6i1.9757

Tobias, S., Fletcher, J. D., & Wind, A. P. (2014). Game-Based Learning. In Spector, J., Merrill, M., Elen, J., & Bishop, M. (Eds.), *Handbook of Research on Educational Communications and Technology*. Springer., DOI: 10.1007/978-1-4614-3185-5_38

Tobin, E., & Hieker, C. (2021). What the EdTech experience in refugee camps can teach us in times of school closure. Blended learning, modular and mobile programs are key to keeping disadvantaged learners in education. *Challenges*, *12*(2), 19. DOI: 10.3390/challe12020019

Toledo Palomino, P., Nacke, L., & Isotani, S. (2023, October). Gamification of virtual learning environments: a narrative and user experience approach. In *Proceedings of the XXII Brazilian Symposium on Human Factors in Computing Systems* (pp. 1-10). DOI: 10.1145/3638067.3638103

Tomé Klock, A. C., Palomino, P. T., Rodrigues, L. A. L., Toda, A. M., Simanke, S., Spors, V., ... & Hamari, J. (2024). Gamification towards and alongside equity, diversity and inclusion: Looking back to move forward. new media & society, 14614448241254028.

Tomeschek, S., Schaberreiter, T., Kandlhofer, M., & Lieckfeld, A. S. (2024). Assessing data literacy and AI education in digital education policies—Insights from the TrainDL Erasmus+ project. *Informatik Spektrum*, *47*(3-4), 113–119. DOI: 10.1007/s00287-024-01576-w

Tong, G., Li, Y., Zhang, H., & Xiong, N. (2023). A fine-grained channel state information-based deep learning system for dynamic gesture recognition. *Information Sciences*, *636*, 118912. DOI: 10.1016/j.ins.2023.03.137

Tong, Q., & Liu, S. (2025). Intelligent Classroom Environment: Application of Internet of Things and Artificial Intelligence in English Teaching. *International Journal of High Speed Electronics and Systems*, 2540364. DOI: 10.1142/S012915642540364X

Torresan, S., & Hinterhuber, A. (2023). Continuous learning at work: The power of gamification. *Management Decision*, *61*(13), 386–412. DOI: 10.1108/MD-12-2020-1669

Traxler, J. (2018). Learning with mobiles: The Global South. *Research in Comparative and International Education*, *13*(1), 152–175. DOI: 10.1177/1745499918761509

Traxler, J. M., & Crompton, H. (2015). Mobile learning. In *Encyclopedia of Mobile Phone Behavior* (pp. 506–518). IGI Global. DOI: 10.4018/978-1-4666-8239-9.ch042

Traxler, J., & Kukulska-Hulme, A. (Eds.). (2015). *Mobile Learning: The Next Generation* (1st ed.). Routledge., DOI: 10.4324/9780203076095

Traxler, J., & Vosloo, S. (2014). Introduction: The prospects for mobile learning. *Prospects*, *44*(1), 13–28. DOI: 10.1007/s11125-014-9296-z

Trifanova, A., Knapp, J., Ronchetti, M., & Gamper, J. (2004). *Mobile ELDIT: Challenges in the transitions from an e-learning to an m-learning system*. http://eprints.biblio.unitn.it/archive/00000532/01/paper4911.pdf

Tromp, C., & Baer, J. (2022). Creativity from constraints: Theory and applications to education. *Thinking Skills and Creativity*, *46*, 101184. DOI: 10.1016/j.tsc.2022.101184

Tsichouridis, C., Batsila, M., Vavougios, D., & Ioannidis, G. (2020). Virtual and augmented reality in science teaching and learning. In *The Impact of the 4th Industrial Revolution on Engineering Education* (pp. 193–205). Springer. DOI: 10.1007/978-3-030-40274-7_20

Tyagi, P., & Shrivastava, N. Sakshi, & Jain, V. (2024). Synergizing Artificial Intelligence and Blockchain. In Next-Generation Cybersecurity: AI, ML, and Blockchain (pp. 83-97). Singapore: Springer Nature Singapore.

Tyagi, A. K. (Ed.). (2023). *Automated secure computing for next-generation systems*. John Wiley & Sons.

Ulutaş, N. K., Durmaz, M. E., Elbuğa, N., Sezgin, S., Kaya, K., & İsmailoğulları, B. (2025). The Use of Augmented Reality (AR), Virtual Reality (VR), and Artificial Intelligence (AI) in Language Education: A Systematic Literature Review. Reimagining Intelligent Computer-Assisted Language Education, 401-430.

Umarova, Z. I. (2024). THE IMPORTANCE OF THE CONTINUITY PRINCIPLE IN TEACHING FOREIGN LANGUAGES. *PEDAGOG*, *7*(11), 242–246.

UNESCO. (2018). *ICT competency framework for teachers*. https://unesdoc.unesco.org/ark:/48223/pf0000265721

UNESCO. (2022). *Reimagining our futures together: A new social contract for education*. https://unesdoc.unesco.org/ark:/48223/pf0000379707

Unwin, T. (Ed.). (2009). *ICT4D: Information and Communication Technology for Development*. Cambridge University Press.

Uppoor, N., Banerjee, D., Shah, D., Mishra, P., & Saha, I. (2022). Interactive language learning with VR and NLP assistance. In *2022 IEEE 7th International Conference for Convergence in Technology (I2CT)* (pp. 1–6). IEEE. https://doi.org/ DOI: 10.1109/I2CT54291.2022.9824754

Valentina, V. (2024). Teaching grammar to young learners.

Valtonen, T., Sointu, E., Kukkonen, J., Kontkanen, S., Lambert, M. C., & Mäkitalo-Siegl, K. (2017). TPACK updated to measure pre-service teachers' twenty-first century skills. *Australasian Journal of Educational Technology*, *33*(3). Advance online publication. DOI: 10.14742/ajet.3518

van Meerkerk, E. (2017). Teacher Logbooks and Professional Development: A Tool for Assessing Transformative Learning Processes. *International Journal of Qualitative Methods*, *16*(1), 1609406917735255. Advance online publication. DOI: 10.1177/1609406917735255

Vargas Florez, J. L. (2024). Analysis of mobile applications for language learning: unpacking discourses about second language learning trends and marketing strategies.

Vasarainen, M., Paavola, S., & Vetoshkina, L. (2021). A Systematic Literature Review on Extended Reality: Virtual, Augmented and Mixed Reality in Working Life. *Internationa Journal of Virtual Reality*, *21*(2), 1–28. DOI: 10.20870/IJVR.2021.21.2.4620

Vasselinov, R., & Groge, J. (2016). The Babbel Efficacy Study. *Final Report of Reseacrh Team*. pp. 3

Venet, S. (2021). *Equity-centered trauma-informed education* (Equity and Social Justice in Education). WW Norton & Company.

Viberg, O., & Grönlund, Å. (2012). Mobile assisted language learning: A literature review. In Specht, M., Sharples, M., & Multisilta, J. (Eds.), *Proceedings of the 11th International Conference on Mobile and Contextual Learning* 2012, vol. 955, (pp. 9-16), Helsinki, Finland.

Viberg, O., & Grönlund, Å. (2013). Cross-cultural analysis of users' attitudes toward the use of mobile devices in second and foreign language learning in higher education: A case from Sweden and China. *Computers & Education*, *69*, 169–180. DOI: 10.1016/j.compedu.2013.07.014

Vicente González, M. E. (2021). Learning English as a Second Language: A Comparison between Spanish and Adopted Chinese Teenagers.

Vidergor, H. E. (2022). Coping with teaching in innovative learning spaces: Challenges, insights and practices. *Learning Environments Research*, *25*(3), 707–724. DOI: 10.1007/s10984-021-09396-5

Villegas, E., Fonseca, D., Peña, E., Bonet, P., & Fernández-Guinea, S. (2021). Qualitative assessment of effective gamification design processes using motivators to identify game mechanics. *Sensors (Basel)*, *21*(7), 2556. DOI: 10.3390/s21072556 PMID: 33917409

Vintaloro, G. (2021). Teaching English in Italy: A Tale of Two Systems.

Vygotsky, L. S. (1978). *Mind in Society: Development of Higher Psychological Processes* (Cole, M., Jolm-Steiner, V., Scribner, S., & Souberman, E., Eds.). Harvard University Press., DOI: 10.2307/j.ctvjf9vz4

Vygotsky, L. S. (1978). *Mind in society: The development of higher psychological processes*. Harvard University Press.

Walsh, M. (2010). Multimodal Literacy: What does it mean for classroom practice? *Australian Journal of Language and Literacy*, *33*(3), 211–239. DOI: 10.1007/BF03651836

Walsh, R. (2022). Teaching communicatively with mobile-sourced materials. *Advances in Language and Literary Studies*, *12*(5), 23–29. DOI: 10.7575/aiac.alls.v.12n.5.p.23

Waluyo, B., & Bucol, J. L. (2021). The impact of gamified vocabulary learning using Quizlet on low-proficiency students. *Computer Assisted Language Learning*, *22*(1), 158–179.

Wang, L. (2024). A study on the application of virtual reality technology in English language education relying on immersive learning environments and interactive experience design. *Journal of Electrical Systems*. https://doi.org/DOI: 10.52783/jes.2626

Wang, P., Ganushchak, L., Welie, C., & van Steensel, R. (2024). The dynamic nature of emotions in language learning context: Theory, method, and analysis. *Educational Psychology Review*, *36*(4), 105. DOI: 10.1007/s10648-024-09946-2

Wanner, T., & Palmer, E. (2015). Personalizing learning: Exploring student and teacher perceptions about flexible learning and assessment in a flipped university course. *Computers & Education*, *88*, 354–369. DOI: 10.1016/j.compedu.2015.07.008

Wardoyo, J., Iyati, W., & Citraningrum, A. (2025). *UAV-Based Thermal Mapping for Urban Heat Island Intensity Assessment in Malang*.

Warschauer, M. (2004). Technological change and the future of CALL. In *New perspectives on CALL for second language classrooms* (pp. 15–26).

Warschauer, M., & Liaw, M.-L. (2011). Emerging Technologies for Autonomous Language Learning. *Studies in Self-Access Learning Journal*, 107–118. DOI: 10.37237/020302

Warschauer, M., & Healey, D. (1998). Computers and language learning. *Language Teaching*, *31*(2), 57–71. DOI: 10.1017/S0261444800012970

Wasie, T. (2024). *Disconnect to reconnect: Finding balance to reclaim well-being*. Teshome Wasie.

Weinberg, L. (2024). *Smart University: Student Surveillance in the Digital Age*. JHU Press. DOI: 10.56021/9781421450018

Welch, B. F. (2024). In Conclusion, I Don't Know: Humility as the Beginnin g and End of Every Class. The Art of Teaching Philosophy, 173.

Weng, Y., Schmidt, M., Huang, W., & Hao, Y. (2024). The effectiveness of immersive learning technologies in K-12 English as second language learning: A systematic review. *ReCALL*, *36*(2), 210–229. Advance online publication. DOI: 10.1017/S0958344024000041

White, R. (1987). Managing innovation. *ELT Journal*, *41*(3), 211–218. DOI: 10.1093/elt/41.3.211

Whyte, S. (2011). Learning to teach with videoconferencing in primary foreign language classrooms. *ReCALL*, *23*(3), 271–293. DOI: 10.1017/S0958344011000188

Wijnen-Meijer, M., Brandhuber, T., Schneider, A., & Berberat, P. O. (2022). Implementing Kolb's Experiential Learning Cycle by Linking Real Experience, Case-Based Discussion and Simulation. *Journal of Medical Education and Curricular Development*, *9*, 23821205221091511. Advance online publication. DOI: 10.1177/23821205221091511 PMID: 35592131

Williamson, B., Eynon, R., & Potter, J. (2020). Pandemic politics, pedagogies and practices: Digital technologies and distance education during the coronavirus emergency. *Learning, Media and Technology*, *45*(2), 107–114. DOI: 10.1080/17439884.2020.1761641

Wirentake, M. P., & Arfani, S. (2024). *Exploring English through tourist attractions: Motivating students to speak*. Jakad Media Publishing.

Wong, L.-H., & Looi, C.-K. (2011). What seams do we remove in mobile-assisted seamless learning? A critical review of the literature. *Computers & Education*, *57*(4), 2364–2381. DOI: 10.1016/j.compedu.2011.06.007

Wu, D., Yang, Z., Zhang, P., Wang, R., Yang, B., & Ma, X. (2023). Virtual-Reality Interpromotion Technology for Metaverse: A Survey. *IEEE Internet of Things Journal*, *10*(18), 15788–15809. DOI: 10.1109/JIOT.2023.3265848

Wu, J. G., Zhang, D., Lee, S. M., & Xian, J. (2025). Game on! Digital gaming and augmented reality/virtual reality in language learning. *International Journal of Computer-Assisted Language Learning and Teaching*, *15*(1), 1–15. Advance online publication. DOI: 10.4018/IJCALLT.367601

Wulandari, I., Purwati, O., Setiawan, S., & Anam, S. U. (2021). AN INQUIRY INTO EFL LEARNERS'PERCEPTIONS OF WHATSAPP FOR ORAL PEER ASSESSMENT. Academic Journal Perspective: Education. *Language and Literature*, *9*(1), 37–49.

Wulantari, N. P., Rachman, A., Sari, M. N., Uktolseja, L. J., & Rofi'i, A. (2023). The role of gamification in English language teaching: A literature review. *Journal of Education*, *6*, 2847–2856. DOI: 10.31004/joe.v6i1.3328

Wu, P. H., & Marek, M. (2016). Incorporating LINE smartphone affordances: Cross-cultural collaboration, willingness to communicate, and language learning. [IJCALLT]. *International Journal of Computer-Assisted Language Learning and Teaching*, *6*(2), 56–73. DOI: 10.4018/IJCALLT.2016040104

Wu, R. (2024). Teaching Strategies for Vocabulary Acquisition in Junior High School English Language Learning. *Frontiers in Educational Research*, *7*(9).

Xiangming, L., & Song, S. (2017). Mobile technology affordance and its social implications: A case of "Rain Classroom.". *British Journal of Educational Technology*, *49*(2), 276–291. DOI: 10.1111/bjet.12586

Xiao, H. (2022). Innovation of digital multimedia VR technology in music education curriculum in colleges and universities. *Scientific Programming*, *2022*(1), 6566144. DOI: 10.1155/2022/6566144

Xie, Y., Ryder, L., & Chen, Y. (2019). Using interactive virtual reality tools in an advanced Chinese language class: A case study. *TechTrends*, *63*(3), 251–259. DOI: 10.1007/s11528-019-00389-z

Xu, P., Zhu, X., & Clifton, D. A. (2023). Multimodal Learning With Transformers: A Survey. *IEEE Transactions on Pattern Analysis and Machine Intelligence*, *45*(10), 12113–12132. DOI: 10.1109/TPAMI.2023.3275156 PMID: 37167049

Xu, X., Daniel, Z., & Li, Y. (2024). 2023 niandu zhongguo shuziyuedu baogao fabu—shuziyuedu shichang guimo maishang xintaijie [the "2023 China digital Reading report" has been released: The scale of the digital Reading market has reached a new level]. *Guang Ming Ri Bao*. Advance online publication. DOI: 10.28273/n.cnki.ngmrb.2024.001843

Yadav, R., Nghiem, X. H., & Yadav, M. (2025). Adaptive AI in Metaverse Language Education. In Harnessing Chatbots for Effective Language Teaching (pp. 49-74). IGI Global Scientific Publishing. DOI: 10.4018/979-8-3693-9974-3.ch003

Yadav, S. (2024). Designing Effective E-Portfolios for Authentic Assessment: Encouraging Reflection and Promoting Student Ownership. In Cases on Authentic Assessment in Higher Education (pp. 235-258). IGI Global.

Yangambi, M. W. (2024). Culturally Relevant Teaching in the Digital Age: Student-Centered and Personalized Learning in Developing Countries. *Creative Education*, *16*(1), 71–89. DOI: 10.4236/ce.2025.161005

Yang, C., Luo, L., Vadillo, M. A., Yu, R., & Shanks, D. R. (2021). Testing (quizzing) boosts classroom learning: A systematic and meta-analytic review. *Psychological Bulletin*, *147*(4), 399–435. DOI: 10.1037/bul0000309 PMID: 33683913

Yang, L., & Gottlieb, M. (2023). Gamification Mobile Applications: A Literature Review of Empirical Studies. In Auer, M. E., Pachatz, W., & Rüütmann, T. (Eds.), *Learning in the Age of Digital and Green Transition. ICL 2022. Lecture Notes in Networks and Systems* (Vol. 633). Springer., DOI: 10.1007/978-3-031-26876-2_88

Yang, S., & Mei, B. (2018). Understanding learners' use of augmented reality in language learning: Insights from a case study. *Journal of Education for Teaching*, *4*(4), 511–513. DOI: 10.1080/02607476.2018.1450937

Yan, L., Sha, L., Zhao, L., Li, Y., Martinez-Maldonado, R., Chen, G., Li, X., Jin, Y., & Gašević, D. (2024). Practical and ethical challenges of large language models in education: A systematic scoping review. *British Journal of Educational Technology*, *55*(1), 90–112. DOI: 10.1111/bjet.13370

Yastibas, A. E., & Cepik, S. (2015). Teachers' attitudes toward the use of e-portfolios in speaking classes in English language teaching and learning. *Procedia: Social and Behavioral Sciences*, *176*, 514–525. DOI: 10.1016/j.sbspro.2015.01.505

Yates, L., & Zielinski, B. (2009). *Give it a go: Teaching pronunciation to adults*. Adult Migrant English Program (AMEP) Research Centre, Macquarie University.

Yefymenko, T., Bilous, T., Zhukovska, A., Sieriakova, I., & Moyseyenko, I. (2024). Technologies for using interactive artificial intelligence tools in the teaching of foreign languages and translation. *Revista Amazonia Investiga*, *13*(74), 299–307. DOI: 10.34069/AI/2024.74.02.25

Yılmaz, A., & Söğüt, S. (2022). Language education for social justice: Reproductions or disruptions through technology. *Computers & Education*, *187*, 104535. DOI: 10.1016/j.compedu.2022.104535

Yin, R. K. (2018). *Case Study Research and Applications: Design and Methods* (6th ed.). Sage.

Yuan, Y., Li, Z., & Zhao, B. (2025). A Survey of Multimodal Learning: Methods, Applications, and Future. *ACM Computing Surveys*, *57*(7), 1–34. Advance online publication. DOI: 10.1145/3713070

Yudintseva, A. (2023). Virtual reality affordances for oral communication in English as a second language classroom: A literature review. *Computers & Education: X Reality, 2*, 100012. https://doi.org/DOI: 10.1016/j.cexr.2023.100012

Yuen, A. H. K., Law, N., & Wong, K. C. (2003). ICT implementation and school leadership. *Journal of Educational Administration*, *41*(2), 158–170. DOI: 10.1108/09578230310464666

Yu, Z. (2019). A meta-analysis of use of serious games in education over a decade. *International Journal of Computer Games Technology, 2019. International Journal of Computer Games Technology*, *4797032*, 1–8. DOI: 10.1155/2019/4797032

Zain, D. S. M., & Bowles, F. A. (2021). Mobile-assisted language learning (MALL) for higher education instructional practices in EFL/ESL contexts: A recent review of literature. *Computer-Assisted Language Learning Electronic Journal*, *22*(1), 283–307.

Zainuddin, A. A., Handayani, D., Ridza, I. H. M., Rahman, S. H. A., Kamarudin, S. I., Ahmad, K. Z., . . . Dhuzuki, N. H. M. (2024, May). Converging for Security: Blockchain, Internet of Things, Artificial Intelligence-Why Not Together? In 2024 IEEE 14th Symposium on Computer Applications & Industrial Electronics (ISCAIE) (pp. 181-186). IEEE.

Zakian, M., Xodabande, I., Valizadeh, M., & Yousefvand, M. (2022). Out-of-the-classroom learning of English vocabulary. *Asian-Pacific Journal of Second and Foreign Language Education*, *7*(1), 16. DOI: 10.1186/s40862-022-00143-8

Zhang, C., Liu, Z., Aravind, B. R., & Hariharasudan, A. (2024). Synergizing language learning: SmallTalk AI In industry 4.0 and Education 4.0. *PeerJ. Computer Science*, *10*, e1843. DOI: 10.7717/peerj-cs.1843 PMID: 38435575

Zhang, W. (2022). Language, Culture, and Ecology: An Exploration of Language Ecology in Pragmatics. *English Language Teaching*, *15*(6), 80–87. DOI: 10.5539/elt.v15n6p80

Zhang, X., & Hu, J. (2024). A study on the learning behaviors and needs of design-maker communities of practice in the era of mobile learning. *Library Hi Tech*, *42*(2), 580–606. DOI: 10.1108/LHT-12-2021-0486

Zhao, X., & McClure, C. D. (2022). Gather.Town: A Gamification Tool to Promote Engagement and Establish Online Learning Communities for Language Learners. *RELC Journal, 55*(1), 240-245. https://doi.org/ (Original work published 2024)DOI: 10.1177/00336882221097216

Zhao, X., Ren, Y., & Cheah, K. S. L. (2023). Leading Virtual Reality (VR) and Augmented Reality (AR) in Education: Bibliometric and Content Analysis From the Web of Science (2018–2022). *SAGE Open*, *13*(3), 21582440231190821. Advance online publication. DOI: 10.1177/21582440231190821

Zheng, Y., Ogata, H., & Yano, Y. (2005). A conceptual framework of computer-supported ubiquitous learning environment L. Li.

Zheng, L., & Yang, Y. (2024). Research perspectives and trends in Artificial Intelligence-enhanced language education: A review. *Heliyon*, *10*(19), e38617. DOI: 10.1016/j.heliyon.2024.e38617 PMID: 39416824

Zheng, Y., & Barrot, J. S. (2022). Social media as an e-portfolio platform: Effects on L2 learners' speaking performance. *Language Learning & Technology*, *26*(1), 1–19. https://hdl.handle.net/10125/73487. DOI: 10.64152/10125/73487

Zhou, J., & Ao, L. (2020). Digital gamification in language teaching. *Dialog on Language Instruction*, *30*, 25–40.

Zhu, J., & Fan, Y. (2018). Daily travel behavior and emotional well-being: Effects of trip mode, duration, purpose, and companionship. *Transportation Research Part A, Policy and Practice*, *118*, 360–373. DOI: 10.1016/j.tra.2018.09.019

Zimmerman, B. J. (2002). Becoming a self-regulated learner: An overview. *Theory into Practice*, *41*(2), 64–70. DOI: 10.1207/s15430421tip4102_2

Zimmerman, B. J., & Risemberg, R. (1997). Becoming a self-regulated writer: A social cognitive perspective. *Contemporary Educational Psychology*, *22*(1), 73–101. DOI: 10.1006/ceps.1997.0919

Zou, D., Luo, S., Xie, H., & Hwang, G. J. (2020). A systematic review of research on flipped language classrooms: Theoretical foundations, learning activities, tools, research topics and findings. *Computer Assisted Language Learning*, *35*(8), 1811–1837. DOI: 10.1080/09588221.2020.1839502

Zouhir, A. (2013). Language situation and conflict in Morocco. In Orie, O. O., & Sanders, K. W. (Eds.), *Selected proceedings of the 43rd annual conference on African linguistics* (pp. 271–277). Cascadilla Proceedings Project., http://www.lingref.com/cpp/acal/43/paper2975.pdf

Zuo, Y. (2024). Exploring the Synergy: AI Enhancing Blockchain, Blockchain Empowering AI, and their Convergence across IoT Applications and Beyond. *IEEE Internet of Things Journal*.

About the Contributors

Ehsan Namaziandost, Ph.D. is a lecturer at the Department of General Courses, Ahvaz Jundishapur University of Medical Sciences, and the Department of English, Islamic Azad University of Ahvaz, Iran. His research primarily centers on Computer-Assisted Language Learning (CALL), language learning and technology, and educational psychology. Dr. Namaziandost has published extensively in prestigious international journals such as System, Computer Assisted Language Learning, Journal of Research on Technology in Education, The Asia-Pacific Education Researcher, British Educational Research Journal, Teaching and Teacher Education, Thinking Skills and Creativity, Instructional Science, Current Psychology, European Journal of Education, The International Review of Research in Open and Distributed Learning, Acta Psychologica, Frontiers in Psychology, and BMC Psychology, among others. Dr. Namaziandost has been included in the list of the world's top 2% of scientists in 2024.

***

Munir Ahmad is a seasoned professional in the realm of Spatial Data Infrastructure (SDI), Geo-Information Productions, Information Systems, and Information Governance, boasting over 25 years of dedicated experience in the field. With a PhD in Computer Science, Dr. Ahmad's expertise spans Spatial Data Production, Management, Processing, Analysis, Visualization, and Quality Control. Throughout his career, Dr. Ahmad has been deeply involved in the development and deployment of SDI systems specially in the context of Pakistan, leveraging his proficiency in Spatial Database Design, Web, Mobile & Desktop GIS, and Geo Web Services Architecture. His contributions to Volunteered Geographic Information (VGI) and Open Source Geoportal & Metadata Portal have significantly enriched the geospatial community. As a trainer and researcher, Dr. Ahmad has authored over 50 publications, advancing the industry's knowledge base and fostering innovation in Geo-Tech, Data Governance, and Information Infrastructure, and Emerging Technologies. His commitment to Research and Development (R&D) is evident in his role as a dedicated educator and mentor in the field.

Ferdi Çelik is a researcher specializing in English as a foreign language (EFL) teaching. With a keen focus on language acquisition, pedagogy, and innovative teaching methodologies, he explores ways to enhance language learning effectiveness. His work delves into curriculum development, assessment strategies, and the integration of technology in language education. Passionate about advancing the field, Ferdi contributes through research, publications, and active engagement in the academic community.

Sholihatul Hamidah Daulay, was born in Surabaya on June 22, 1975. She completed her bachelor's degree in English Language Education Department of the North Sumatra State Islamic Institute (IAIN SU) in 1999, then continued her Master's degree in the Linguistics Study Program at the University of North Sumatra (USU) in 2002, and then completed a Ph.D. in the Linguistics Study Program at the University of North Sumatra (USU) in 2014. The author is a permanent lecturer at the Tadris English Study Program at the State Islamic University of North Sumatra (UINSU) from 2003 till now. The author actively conducts research and writes books including Let's Study English (Cita Pustaka, 2007), Introduction to General Linguistics (La-Tansa, 2011), Language and Society (LPPP Medan, 2019), Reading Comprehension (Kencana, 2020), Basic Morphology (UMSU Press, 2021), Developing Vocabulary by Educational Game (Cakrawala Satria Mandiri, 2021), Introduction to Linguistics (UMSU Press, 2022), Student Difficulties in Learning Speaking (Muslim Intellectual Education Foundation, 2022), Fun Writing for Children: Games and Activities (Muslim Scholars Education Foundation, 2023), Designing an ICT-Based Practical Interactive Guide Interactive Quiz for Educators (Akademia Pustaka, 2023), Quick on the Draw for Writing Recount Text (Muslim Scholars Education Foundation, 2023). E-mail: sholihatulhamidah@uinsu.ac.id.

Tetiana Holovatenko, PhD in Education, TESOL, is an associate professor at V. N. Karazin Kharkiv National University (Ukraine). Her research interests include technology-enhanced teaching of English, trauma-informed language teaching, and comparative education studies.

S. Ida Evangeline, presently working as an Assistant Professor in the Department of Electrical and Electronics Engineering at the Government College of Engineering, Tirunelveli, Tamilnadu, India. She obtained her Bachelor's in Electrical and Electronics Engineering from Manonmaniam Sundaranar University in 2003, and Master's in Power Systems Engineering from Anna University, Chennai,

Tamilnadu, India in 2005. She completed her PhD degree in 2023 at Anna University, Chennai, Tamilnadu, India. She has published six research papers in international journals. Her current research interests include optimal power flow, renewable energy, soft computing, machine learning, and deep learning.

Luh Indrayani earned her Bachelor and Master degrees in English Education from Universitas Pendidikan Ganesha in 2015 and 2017 respectively. Now she is teaching in the English Education Department at Universitas Pendidikan Ganesha. Her research interests include Technology Integration in English Language Teaching (ELT), Artificial Intelligence (AI) in Education and Innovative Pedagogical Practices. She teaches several disciplines in education and literature. She also actively participates in service as action for the community.

Gres Desika Br Kaban is a student in the English Language Education Department, Universitas Pendidikan Ganesha, Indonesia. Her academic interests include academic and creative writing, language learning strategies, and English language pedagogy. During her studies, she was actively involved in various academic projects and organization. She aspires to contribute to the field of education and continue growing as a lifelong learner.

Nargiza Kadirova is a professional teacher and linguist with nearly 20 years of experience in teaching English and foreign languages at colleges and universities. She currently serves as an Associate Professor at TIIAME National Research University, where she teaches ESL to undergraduate students. Previously, she worked as a Senior Lecturer and Lecturer at Bukhara State University, where she taught English, developed syllabuses and educational materials, supported students in their research, and published academic articles and study guides. She also served as Head of the Foreign Languages Department at Bukhara State Pedagogical Institute, managing a team of over 30 educators and overseeing the quality of instruction in English, French, and German. Dr. Kadirova holds a PhD in Literary Criticism from Bukhara State University and is widely respected for her leadership, analytical thinking, and commitment to excellence in education.

Mustafa Kayyali is an ardent advocate for excellence in higher education, driven by a relentless pursuit of quality, recognition, and innovation. Their journey in academia started with a Master's in Quality Management and Evaluation in Higher Education from Universitat Oberta de Catalunya, followed by a Ph.D. in Quality Management from Azteca University. These academic pursuits have laid a strong foundation for their expertise in Accreditation, Quality Assurance, and Higher Education Rankings. As an entrepreneur, researcher, translator, and publisher, they are deeply involved in various facets of the academic world. Their diverse interests encompass Management, Translation, Interpretation, and Academic consulting, contributing to a well-rounded understanding of the industry. Throughout their career, they have made significant contributions to academic literature, with more than 40 published papers and 30 book chapters to my name. Additionally, they take pride in having translated 6 books, bridging language gaps, and promoting knowledge exchange on a global scale. They are committed to fostering positive change in higher education and contributing to its continuous improvement.

Nafis Mahmud Khan currently serves as an Associate Professor in Teaching Theory and Methodology at TIIAME National Research University, located in Tashkent, Uzbekistan. Before this, he was a faculty member at Daffodil International University in Dhaka, Bangladesh. He earned his Ph.D. in English from the School of Humanities at KIIT Deemed to be University in India. With more than seven years of experience in teaching various educational levels and conducting research, Dr. Khan is actively involved in both educational and scholarly activities. His research focuses primarily on Information and Communication Technology (ICT) and its application in English language learning and teaching. Dr. Khan's interests span a wide range of areas, including English Language Teaching (ELT), Computer Assisted Language Learning (CALL), Educational Technology, Applied Linguistics, and Professional and Business Communication Skills. Dr. Khan is affiliated with renowned academic bodies such as IATEFL Kuwait, Eudoxia Research University USA, and Eudoxia Research Centre, India. He has a strong publication record, with research papers and book chapters featured in international journals indexed in Web of Science (WoS) and SCOPUS. Having presented papers at both national and international conferences, Dr. Khan is experienced in conducting workshops and seminars on various topics, both in physical and hybrid modes, nationally and internationally. He has extensive experience in delivering keynote speeches in international conferences. Currently, he supervises PhD students working on foreign language teaching and educational technology.

I Putu Indra Kusuma is an associate professor in English Language Education, Universitas Pendidikan Ganesha, Indonesia. He received his doctoral degree from Western Michigan University, USA. His research interests are in English education, technology-enhanced language learning, technology-enhanced assessment, and teacher education program using technology. He has several publications, including book chapters and articles in those areas mentioned above.

Rizgar Mahmood pursued his academic endeavours at California State University, Northridge in the USA, where he earned a master's degree in Linguistics/ TESL. After graduating in 2019, he worked as an assistant lecturer at Salahaddin

University, Erbil in Iraqi Kurdistan until 2022. Currently, he is pursuing his PhD in Applied Linguistics/Second Language Acquisition, focusing on the effectiveness of pronunciation instruction, corrective feedback, and technology-assisted language learning on the development of L2 learners' pronunciation.

Maslathif Dwi Purnomo is a Magister lecturer in the English Education Department at the State Islamic University of North Sumatra (UINSU). He can be contacted by email: maslathifdwipurnomo@uinsu.ac.id

Ali Rahmanipur holding a Master's degree in Educational Technology from Bu-Ali Sina University, Hamadan. Ali is active in learning technologies, researching serious games, digital games, artificial intelligence in education, entrepreneurial mindset, and educational technologies.

Belia Damayanti br Sembiring is an undergraduate student at the English Language Education Department, Ganesha University of Education, specializing in English Language Teaching. Her research interests include digital assessment, students' emotional well-being, and speaking performance. She has explored the use of E-portfolios to enhance student engagement and reflective learning in speaking classes. This chapter is part of her final thesis project.

Shaumiwaty Shaumiwaty is an English lecturer in the English Education Department at the IAIN Takengon, Aceh. She can be contacted by email: shaumiwaty26@gmail.com

Moein Shokri with a Master's degree in Translation Studies from the SAzerbaijan Shahid Madani University in a double-degree program, Moein's research interests include intercultural communication, linguistics and literature, the application of artificial intelligence in language learning, and Translation.

Jasur Umirzokov was born on 26 January 1993 in Samarkand, Uzbekistan. He is currently the Head of the International Relations Department and an Associate Professor (DSc) in the Department of World Economy and International Economic Relations at Tashkent State University of Economics.

Index

A

Adaptive Learning 39, 40, 72, 106, 124, 126, 239, 255, 379, 381, 391, 398
AI Tutors 264, 381
Artificial Intelligence 26, 38, 72, 75, 91, 102, 105, 109, 121, 173, 255, 260, 286, 296, 374, 379, 380, 385, 393, 395, 396, 397, 398, 399, 400
Augmented Reality (AR) 7, 38, 102, 106, 113, 133, 173, 219, 220, 224, 225, 228, 229, 231, 238, 240, 241, 242, 243, 244, 245, 247, 248, 251, 252, 256, 263, 264, 265, 266, 268, 269, 270, 271, 273, 275, 277, 280, 281, 283, 285, 286, 287, 288, 289, 290, 291, 292, 293, 294, 296, 297, 298, 299, 301, 302, 305, 306, 308, 309, 310, 311, 312, 383, 389, 399

B

Babbel 111, 112, 113, 115, 116, 117, 118, 119, 120, 121, 122, 123, 124, 125, 126, 127, 128, 129, 130, 131, 132, 135, 160
Blockchain 379, 380, 382, 383, 385, 386, 387, 388, 389, 390, 391, 392, 393, 394, 395, 396, 397, 398, 399, 400

C

CALL 32, 46, 68, 69, 70, 73, 74, 122, 136, 141, 142, 160, 162, 243, 283, 340, 376, 377
Communicative Language Teaching 229, 254, 259, 279, 345, 346, 348, 350, 351, 353, 366, 369, 374

D

digital divide 11, 56, 61, 68, 74, 76, 87, 387, 388, 393
Digital Gamification 189, 191, 199, 210, 214, 217, 218
Digital Inclusion 34, 42, 64, 68, 72
Digital Literacy 2, 22, 26, 31, 33, 42, 43, 47, 50, 52, 53, 55, 56, 66, 69, 81, 100, 157, 158, 331, 344, 349, 369, 376
Digital Pedagogy 35, 37, 53, 56, 66, 218, 368
Digital Portfolios 18, 63, 338, 339

E

Educational Technology 27, 33, 44, 45, 62, 64, 65, 68, 69, 71, 72, 73, 78, 79, 103, 105, 124, 125, 128, 142, 160, 163, 216, 233, 234, 271, 276, 278, 281, 309, 310, 339, 374, 376, 399
Eportfolios 315, 316, 317, 319, 320, 321, 322, 323, 324, 325, 326, 331, 334, 335, 336, 338, 339
Experiential Learning 46, 219, 220, 223, 224, 226, 238, 242, 243
Extended Reality (XR) 243, 274, 276, 286, 287, 296, 312

G

Game-Based Learning 184, 203, 204, 216, 217, 268
Gamification 27, 33, 79, 97, 106, 108, 169, 170, 171, 172, 173, 174, 175, 176, 179, 180, 181, 182, 183, 184, 185, 186, 187, 189, 190, 191, 192, 193, 194, 195, 196, 197, 198, 199, 200, 201, 202, 203, 204, 205, 206, 207, 208, 210, 211, 212, 213, 214, 215, 216, 217, 218, 241, 253, 255, 257, 260, 261, 262, 263, 266, 267, 268, 270, 271, 272, 273, 274, 275, 281, 306

I

ICT in Education 344, 349, 350, 370, 376
Immersive Learning 41, 220, 230, 233, 234, 239, 242, 243, 275, 279, 282, 379, 383
Information and Communication Technology 69, 78, 344, 348, 373
Interactive Learning 113, 155, 157, 240,

246, 270, 307, 308
Internet of Things 244, 379, 380, 383, 385, 393, 399, 400

K

Kurdish EFL learners 137, 139, 140, 141, 145, 151, 156, 157, 163

L

language acquisition 8, 21, 32, 38, 41, 72, 83, 84, 87, 88, 89, 91, 93, 95, 99, 103, 105, 106, 115, 117, 123, 124, 125, 128, 133, 146, 160, 161, 162, 163, 164, 170, 184, 186, 205, 215, 222, 224, 227, 229, 238, 239, 242, 245, 247, 250, 254, 256, 257, 258, 267, 269, 271, 275, 276, 279, 285, 292, 294, 295, 305, 313, 373, 384, 385, 386, 399
Language Education 2, 3, 6, 7, 10, 14, 19, 32, 35, 38, 39, 40, 46, 77, 83, 84, 87, 88, 103, 109, 114, 116, 125, 126, 128, 129, 160, 163, 164, 166, 169, 170, 183, 219, 220, 236, 237, 245, 247, 249, 252, 255, 256, 257, 259, 262, 263, 267, 268, 269, 270, 271, 275, 276, 282, 285, 287, 292, 295, 296, 304, 305, 306, 307, 312, 313, 348, 377, 379, 380, 382, 383, 384, 385, 387, 388, 389, 390, 391, 392, 393, 394, 395, 396, 397, 398, 399, 400
Language Learning 1, 2, 3, 8, 9, 10, 11, 12, 14, 16, 20, 22, 23, 24, 25, 28, 29, 31, 32, 34, 36, 39, 41, 44, 45, 46, 48, 49, 51, 59, 62, 68, 69, 70, 71, 72, 73, 74, 75, 76, 77, 78, 79, 83, 84, 85, 86, 87, 88, 90, 91, 92, 94, 95, 96, 97, 98, 99, 100, 101, 102, 103, 104, 105, 106, 107, 109, 110, 111, 112, 113, 114, 115, 116, 119, 122, 123, 124, 125, 126, 127, 128, 130, 131, 132, 133, 134, 135, 136, 138, 140, 141, 142, 143, 146, 148, 156, 158, 160, 161, 162, 163, 164, 165, 166, 169, 170, 171, 172, 173, 174, 176, 177, 178, 179, 180, 182, 183, 184, 185, 186, 187, 191, 192, 193, 199, 204, 205, 213, 214, 216, 218, 219, 220, 221, 222, 223, 224, 225, 227, 229, 230, 231, 232, 233, 235, 237, 238, 239, 241, 243, 246, 247, 248, 249, 250, 251, 252, 253, 254, 255, 256, 257, 258, 259, 260, 261, 262, 263, 264, 265, 266, 267, 268, 269, 270, 271, 273, 274, 275, 276, 277, 278, 279, 280, 281, 282, 283, 285, 286, 287, 289, 290, 292, 295, 296, 297, 299, 301, 302, 305, 306, 307, 309, 311, 312, 316, 317, 318, 334, 335, 336, 339, 340, 341, 344, 345, 346, 347, 353, 369, 371, 373, 375, 377, 379, 380, 381, 383, 385, 386, 387, 388, 389, 390, 391, 392, 393, 394, 396, 397, 398, 400
Language Teacher Education 31, 42, 49, 212
Learner Autonomy 22, 36, 39, 41, 47, 48, 95, 99, 103, 110, 116, 122, 123, 124, 138, 265, 272
learner engagement 46, 55, 80, 113, 125, 144, 169, 174, 218, 230, 267, 269, 270, 271, 273, 312, 335
Learner Motivation 33, 47, 83, 132, 183, 255, 257, 260, 262, 288

M

Mobile-Assisted Language Learning (MALL) 1, 22, 29, 31, 32, 34, 36, 39, 45, 46, 48, 49, 51, 59, 62, 72, 73, 74, 75, 76, 77, 84, 99, 101, 102, 103, 106, 109, 110, 113, 114, 115, 122, 123, 125, 126, 127, 138, 141, 142, 162, 163, 166, 346, 369, 373, 377
mobile-assisted pronunciation learning (MAPL) 137, 139, 140, 145, 146, 147, 148, 149, 150, 153, 154, 155, 156, 158, 161, 166, 167
Mobile Language Learning 102, 134, 162, 169, 170, 171, 172, 179, 182, 305, 307, 381, 383, 387, 389, 391
Mobile Learning 4, 6, 8, 9, 13, 16, 18, 21, 26, 33, 40, 47, 50, 56, 58, 59, 60, 64,

65, 66, 67, 68, 69, 72, 75, 77, 85, 87, 89, 90, 92, 93, 94, 95, 96, 97, 98, 99, 100, 101, 102, 103, 106, 109, 135, 143, 146, 151, 155, 157, 162, 163, 165, 170, 173, 176, 177, 182, 217, 243, 261, 293, 297, 305, 310, 311, 349, 374, 379, 380, 383, 385, 386, 387, 392
Mobile Technology 1, 2, 3, 4, 6, 7, 10, 11, 13, 14, 15, 16, 19, 22, 23, 31, 32, 35, 36, 39, 48, 51, 52, 54, 56, 57, 58, 59, 60, 61, 63, 65, 67, 79, 80, 81, 84, 85, 87, 88, 91, 95, 100, 103, 139, 141, 143, 166, 169, 171, 175, 371, 379, 380, 386
Multimedia Learning 290, 291, 306, 308, 310, 312

P

pedagogical integration 47, 248, 261, 276
Personalized Learning 10, 21, 39, 72, 83, 84, 85, 89, 90, 104, 106, 107, 109, 113, 122, 230, 286, 296, 381, 390, 395, 398
Pre-service Teacher Education 33, 34, 42, 47, 59, 62, 72, 81
Professional Development 9, 11, 12, 13, 21, 23, 25, 35, 42, 45, 57, 62, 63, 78, 101, 202, 211, 347, 370, 395

R

Reading Comprehension 115, 278, 340

S

self-regulated learning 33, 140, 156, 158, 248, 251, 252, 259, 265, 272, 273, 277, 278
Situated Cognition 219, 224, 225, 226, 238, 240
Situated learning 39, 40, 73, 113, 221, 226, 285, 287, 305, 309, 310, 313, 379
Smart Classrooms 345, 379, 380, 383, 384, 392
sociocultural theory 36, 37, 38, 39, 59, 62, 70, 73, 140, 156, 158, 162, 164
Speaking Performance 318, 319, 324, 328, 332, 335, 339, 340, 341, 396

T

Teachers' Awareness of ICT Devices 354, 366
teacher training 31, 36, 43, 45, 48, 50, 62, 63, 65, 67, 77, 163, 189, 192, 195, 196, 210, 211, 215, 217, 218, 233, 236, 242, 245, 279, 285, 304, 307
technology-enhanced language learning (TELL) 128, 140, 141, 156, 158, 311, 375, 397
traditional instruction 8, 85, 144
Trauma-Informed Language Teaching 191, 192

U

User Feedback 119, 122, 125, 127, 128, 129, 130, 262

V

Virtual Reality (VR) 38, 79, 102, 219, 220, 224, 225, 229, 240, 241, 242, 243, 244, 245, 247, 249, 251, 256, 259, 277, 280, 281, 282, 283, 285, 286, 287, 288, 289, 290, 291, 292, 293, 294, 296, 299, 300, 303, 304, 305, 306, 308, 309, 310, 311, 312, 313, 389, 392, 398, 399

www.ingramcontent.com/pod-product-compliance
Lightning Source LLC
LaVergne TN
LVHW062332300926
849161LV00007B/1660